IDOLS OF
PERVERSITY

Jean Delville (1867–1953), ''The Idol of Perversity,'' drawing (1891)

IDOLS OF PERVERSITY

Fantasies of Feminine Evil in Fin-de-Siècle Culture

BRAM DIJKSTRA

OXFORD UNIVERSITY PRESS
New York   Oxford

Oxford University Press

Oxford   New York   Toronto
Delhi   Bombay   Calcutta   Madras   Karachi
Petaling Jaya   Singapore   Hong Kong   Tokyo
Nairobi   Dar es Salaam   Cape Town
Melbourne   Auckland

and associated companies in
Berlin   Ibadan

First published in 1986 by Oxford University Press, Inc.,
200 Madison Avenue, New York, New York 10016

First issued as an Oxford University Press paperback, 1988

Oxford is a registered trademark of Oxford University Press

Library of Congress Cataloging-in-Publication Data
Dijkstra, Bram.
   Idols of perversity.
   Bibliography: p.
   Includes index.
   1. Women in art.   2. Sexism in art.   3. Arts.
Victorian.   I. Title
NX652.W6D55   1986   700'.9'034   85-31076
ISBN 0-19-503779-0
ISBN 0-19-505652-3 (pbk.)

6 8 10 9 7 5

Printed in the United States of America

This book is dedicated to the many brilliant women who have taught me most of what I know about the world. Among them Inge Marcuse must always occupy a very special place. She showed me by example that true intellectual strength expresses itself best in quiet determination, and that to love life is to want to make its living better for all.

Without the help of the companion of my sweetest years, my best friend and most demanding critic, this book would probably have remained no more than just a lot of talk. Therefore, as ever: to Sandra, of course.

Preface

This is a book filled with the dangerous fantasies of the Beautiful People of a century ago. It contains a few scenes of exemplary virtue and many more of lurid sin. Much of it deals with magnificent dreams of intellectual achievement doomed to wither before the tempting presence of woman, who, throughout these pages, is to be found dragging man into a grim trough of perversion. In short, this book is about what and how, during the second half of the nineteenth century, men came to think about women—and why.

As I hope to be able to show, scientific advances, economic developments, and the cultural environment conjoined to create a unique set of intellectual conditions which were to have a fundamental influence on twentieth-century modes of thinking about sex, race, and class. Flushed with the sense of possibility which inevitably accompanies poorly digested and partly understood new knowledge, the artists and intellectuals of the second half of the nineteenth century saw themselves as standing in the vanguard of a new era of evolutionary progress. Science had proved to them that inequality between men and women, like that among races, was a simple, inexorable law of nature.

When women became increasingly resistant to men's efforts to teach them, in the name of progress and evolution, how to behave within their appointed station in civilization, men's cultural campaign to educate their mates, frustrated by women's "inherently perverse" unwillingness to conform, escalated into what can truthfully be called a war on woman—for to say "women" would contradict a major premise of the period's antifeminine thought. If this was a war largely fought on the battlefield of words and images, where the dead and wounded fell without notice into the mass grave of lost human creativity, it was no less destructive than many real wars. Indeed, I intend to show that the intellectual assumptions which underlay the turn of the century's cultural war on woman also permitted the implementation of the genocidal race theories of Nazi Germany.

This book has its origins in my interest in the still neglected academic schools of painting of the late nineteenth century. The wholesale dismissal, during the past sixty years, of the work of thousands of highly accomplished artists who chose not to paint impressionist confections for a clientele whose main claim to discernment in matters of art was that they had grown tired of looking at the slick concoctions Bouguereau once used to produce for them, has created in our own time a peculiarly skewed conception of the parameters of cultural production at the turn of the century.

Over the past fifteen years I have done a great deal of archeological spadework among the exhibition catalogues and art magazines of this first period in the history of art for which there exists an extensive photographic documentation. Whenever possible, I also sought out the (usually moldering and now terminally obscure) shrines which after their death were established by a devoted public in the homes or studios of artists who had once lived like kings upon the fruits of an extraordinary adoration bestowed upon them by their contemporaries—an adoration which was not unlike that currently bestowed on the great gods of the minimal modernist gesture.

During these explorations I was struck, time and again, by the endless recurrence of what I soon came to recognize as a veritable iconography of misogyny. Over the years I found that the images of the painters also echoed in the words of the writers working during this period, and from the latter I let myself be guided into the morass of the nineteenth-century's assault upon women. Books such as Richard Hofstadter's *Social Darwinism in American Thought*, Viola Klein's *The Feminine Character*, John and Robin Haller's *The Physician and Sexuality in Victorian America*, and a number of similar pioneering studies by G. J. Barker-Benfield, Barbara Ehrenreich, Deirdre English, and others helped illuminate my steps.

While I was finding my own path into this realm, others were inevitably doing the same. But the field of late nineteenth-century cultural opinion on women, while not as variegated as one might expect, is so voluminous—and yet so little known—that extensive dredging still needs to be done. The materials I present here overlap, to some extent, with such important recent studies as Sandra Gilbert and Susan Gubar's *The Madwoman in the Attic*, Stephen Jay Gould's *The Mismeasure of Man*, Lillian Faderman's *Surpassing the Love of Men*, Charles C. Eldredge's *American Imagination and Symbolist Painting*, and Nina Auerbach's *Woman and the Demon*. But all of these works, and numerous other recent studies not mentioned, focus on precisely circumscribed facets of the nineteenth-century's social paranoia concerning women. None, with the exception of Eldredge, deal more than marginally with the visual art of this period.

Idols of Perversity tries to pull many strands together in order to show that virulent misogyny infected all the arts to an extent understood by very few specialists in the cultural history of the turn of the century—perhaps precisely because it was so endemic and therefore so completely taken for granted by everyone. The fundamental shift in the art public's interest from strict representational styles to more modern tendencies has served to hide the work dealt with in this book from historical scrutiny. As a result, the often brilliant but, due to its current obscurity, less accessible work of these turn-of-the-century artists still remains largely ignored by the general public. The misogynist aspect of part of their production, therefore, also continues to be hidden. But while we may have forgotten the specific content of these images, they nevertheless had a fundamental influence on the development of our preconceptions regarding the nature of women; as anyone familiar with the imagery of the popular arts of our own time can attest, their legacy still festers among us, kept alive by Hollywood and the world of advertising.

In order to disclose the pervasive nature of turn-of-the-century misogyny and its international ramifications, I have chosen to roam far and wide in this book, without regard to national boundaries. Of course, there were local differences of focus in individual countries, but the conformity of international opinion is frankly much more striking than the local differences. And the art of the period, which by this time had become an international language familiar to intellectuals everywhere, speaks volumes in favor of a comprehensive approach.

Many of the texts I use for documentation have not yet found their way into the work of other students of the period. Most of this material is not widely available, and I have had to translate a good number of the French and German sources myself. This is already a long book, and the original sources I have used need to be brought to the attention of contemporary readers. That is why I have chosen not to refer to, or quote from, the many admirable studies previously mentioned or listed in the bibliography of secondary sources. I do not wish to minimize my indebtedness to these works in any way, but inevitably some of my readings of certain texts are markedly at variance with those of other students of the period. To avoid lengthy polemics on points of relatively minor importance, and to preserve space for the citation of original sources, I have, with very few exceptions, limited my in-text references to works written during the period under discussion. To improve the readability of my book I have also dispensed with footnotes. However, in the text I have identified as much of the source of each quotation I use as might be necessary to locate it in the bibliography.

I would also like to stress that if in this book some of the most revered names of recent art history are placed unceremoniously beside the host of now virtually forgotten painters whose work makes up the bulk of my illustrative material, such a juxtaposition should not be construed as an attempt to cast aspersion on the artistic excellence or historical prominence of these important painters. My point is merely that exceptional talent in art does not necessarily imply an exceptional analytic capacity on the part of the artist in question. Most often the great artists have been intellectually in tune with their own time rather than in advance of it in any significant fashion. Astonished and inspired by the creative fervor and imaginative sweep of an exceptional talent, we tend to assume that innovative thought accompanies innovative form. We are consequently inclined to exempt the content of the great works of modernist art from critical scrutiny and allow their implicit messages to stand unchallenged.

Unfortunately, a work of art whose content is left unexamined (presumably because that content is seen as secondary to its form) continues to have a dangerous power of persuasion, something which is generally much less of a concern in the case of works whose ideological import is clearly understood. We can dialogue with—and therefore learn a great deal from—paintings whose messages we understand within their historical framework. To do so does not necessarily diminish their stature as works of art in any way. But those works whose narrative content we privilege or try to ignore because they have become cultural icons are allowed to continue to disseminate, without serious challenge, their often quite inhumane proposals concerning the nature of human interaction. This book amply demonstrates that there is cause to assume that works of art are by no means beyond involvement in the dominant ideological movements of the time in which they were created. My presentation is historical, not—at least not primarily—art-historical, and I have therefore made it a point to include famous and obscure artists side by side, without permitting myself to become involved in questions of aesthetic preference or judgment without relevance to the issues at hand.

In my choice of illustrations, furthermore, I have let myself be guided, in principle, only by their usefulness as historical documents and their popularity among a broad spectrum of turn-of-the-century viewers.

Between 1875 and 1900 a wide range of new processes of photographic reproduction was making the work of contemporary painters accessible to millions of eager viewers all over the world. In earlier days these paintings could, with few exceptions, have been known only to a relatively small group of cognoscenti: those with enough wealth, motivation, and leisure to visit the sites of the yearly grand exhibitions. Certainly, there had been limited use of engravings of works of art in periodicals and books before 1880, but the last two decades of the nineteenth century saw a veritable craze for photogravures of the works of contemporary painters. Art magazines and popular monthlies vied for the right to reproduce the favorites of the yearly exhibitions, and often these photographic reproductions were presented with meticulous care and with far greater attention to detail and shading than is customary among publishers today.

The visual creations of the painters therefore came to occupy a position of international importance not unlike that occupied by television in our own time. To emphasize the range and significance of the images presented to the turn-of-the-century public, I have made it a point to select my own illustrations, in all but a handful of instances, directly from widely disseminated books and magazines of the period—even when the current location of the painting in question was known to me. Therefore, unless otherwise indicated, all reproductions in this book have been made directly from turn-of-the-century sources. Especially where the source was a gravure-assisted photographic reproduction of the sort so popular among publishers of the time, this use of period reproductions has the added benefit of allowing the reader to examine narrative details which, given current practices of photographic reproduction, might otherwise simply have disappeared into the murk of halftone.

An important additional feature of many of the paintings reproduced here—which might escape the attention of an uninitiated viewer but could not but have impressed the many thousands of visitors to the yearly exhibitions of a century ago—is that these pictures were often formidable in size. For instance an installation photograph of the Fifth Berlin Secession Exhibition of 1902 [Preface, 1] conveys a sense of the overpowering presence of such a painting as Max Liebermann's ''Samson and Delilah'' (see XI, 12), its popular importance further emphasized by the lone chair facing it—placed there, no doubt, to receive the weight of a weary admirer eager to study the philosophical message of this monumental production.

The reason for the imposing size of many of these works is not difficult to determine. By the turn of the century the number of paintings exhibited at the major annual shows habitually ran into the thousands. Paintings were often displayed virtually from floor to ceiling in tightly packed rows. Hanging committees became as important as the selection juries themselves, and nasty factional rivalries in the art world made it possible for even works of the most famous contemporary artists to end up ''skied,'' that is, hung above the normal line of sight. The work of lesser known, though not necessarily less skillful, artists was habitually banished to those upper regions. The artists retaliated by making certain

Preface, I. Installation photograph, Fifth Berlin Secession Exhibition (1902)

that, at least in terms of sheer size, their contributions to the major shows would be hard to overlook, even if they were hung high "above the line." The anonymous writer of a note in *The Art Journal* of September 1903 pointed out with amusement that "Painters of pictures skied in the Academy exhibitions [had] the consolation that many of the visitors to the crowded galleries—for instance on a Bank Holiday—[could] see completely only those works hung high" (288).

Whatever their original size, I have made it a point to reproduce only those paintings that represent a true cross-section of turn-of-the-century works dealing with the representation of women such as one was likely to encounter in the houses of the wealthy and on the walls of the yearly exhibitions in London, Paris, Munich, New York, Vienna, and (biennially) Venice. I have not reproduced—although in my text I have occasionally referred to—images which either were censored or might have been censored by the generally accessible print media of the period. The range of images reproduced in this book can therefore be regarded as typical of the materials made available to the readers of the major art magazines between 1880 and 1920.

As for the images themselves, some clearly represent very accomplished works of art, while others must always remain rather maudlin productions. There are, in virtually every case, literally thousands of comparable images to be found in the exhibition catalogues, popular monthlies, and art magazines of this period. The originals of many of the works reproduced here have not been located, and not a few, undoubtedly, have been destroyed over time. Despite the nature of their subject matter, a good many did not deserve such a fate. A few, one might hope, were destroyed by embarrassed inheritors. It is to be feared, however, that on balance these paintings suffered more from changing tastes in style than from righteous indignation concerning their content.

University of California, San Diego   B.D.
April 1986

Contents

Acknowledgments

Elizabeth Sisco rephotographed nearly all the illustrations in this book from turn-of-the-century sources. Her superb talents as a photographer, and her concern for quality, precision, and detail have been an essential factor in the successful transfer of these images. Financial support for the production of the photographs was provided by the Research Committee of the Academic Senate at UC San Diego. Without the patient cooperation of many librarians I would not have been able to obtain these reproductions. Joyce Ludmer and Olga Morales of the UCLA Art Library and William Treese and Felipe Cervera of the Art Library at UC Santa Barbara were especially helpful in this respect. Among the graduate students at UC San Diego who, over the last decade, participated in my seminars on fin-de-siècle culture, there are many who helped focus my ideas. Those friends and coworkers, in particular, who during the seventies formed the core of the Art and Literature Study Group will find in these pages many echoes of our deliberations. A special thanks, therefore, to Andrea Hattersley, Suzanne Lewald, Karin Dillman, Patrick Condon, Donald Johns, Cecilia Ubilla-Arenas, Susan Self, Frank Langer, Steve Dopkin, Alfred Weber, and Kevin O'Byrne. Many others had an important role in the shaping of this book. Without the determined efforts of Sandra Dijkstra and the early support of Curtis Church at Oxford University Press, the book might never have been published. Michel de Certeau—friend, colleague, and celebrant of the imagination—encouraged my explorations beyond the artificial and restrictive boundaries of narrow intellectual specialization. Michel lived for the spirit, and his spirit will live with us as long as humane concerns continue to find a voice in this world.

IDOLS OF
PERVERSITY

Raptures of Submission: The Shopkeeper's Soul Keeper and the Cult of the Household Nun

H er eyes are glazed with the terror of understanding. The pallor of sudden knowledge has settled on her face. A paralyzing consciousness of her entrapment has turned her body into a wedge of fear. Wracked by dark foreboding, she pits the force of newborn moral responsibility against the soul-destructive lure of the senses. The eternal battle between God and the devil finds agonized expression in the struggle of her tensing limbs against the importunate arms of her illicit lover, arms which form a playful chain around her trembling loins. The clean, golden light of virtue, which plays through the fresh and tender leaves of a springtime tree beyond the open window of her lavishly gilded cage, reaches out to her with the promise of wholesome regeneration, of cleansing immersion in the unpolluted vegetable world of God's garden. She knows that this is the moment of her last chance. She must break the chains of material temptation now or be doomed to die a long, slow death of perpetual dishonor. She must fly from her young man's thoughtlessly deceptive caresses or become the abject companion to the little sparrow at her feet, which, clawed and battered, a helpless victim of the heartless sense of play of an evil-eyed, grossly materialistic cat, flutters its broken wings in a last, desperate, useless effort at escape.

William Holman Hunt undoubtedly meant to shock his female viewers into virtuous conformity by means of this painted melodrama of sin and sudden recognition titled "The Awakening Conscience" [I, 1]. Clearly, this cautionary shapshot, dated 1853, was meant to teach woman vital lessons about her appointed role in civilized society. Still, the painting must have stirred salutary feelings of revolt and angry recognition, the spirit of Seneca Falls, rather than the moral terror Hunt had wished to evoke, in at least some of the women who might have seen the painting in the year in which it was executed. By this date the ever-increasing enclosure of women within the ornate walls of the middle-class household, and their ever-greater disenfranchisement from virtually all forms of intellectual and social choice—a pattern which had been developing for more than a century—had been virtually

3

I, 1. William Holman Hunt
(1827–1910), "The Awakening
Conscience" (1853)

completed. The expulsion of the middle-class woman from participation in practical life
had become fact; woman had never been placed on a more lofty pedestal. An apparently
insuperable plateau had been reached in her canonization as a priestess of virtuous inanity.
Yet, underlying Hunt's still basically cheerful assessment of erring femininity's ability to
become appalled at sin, was a festering suspicion that woman was too weak a creature to
be able to sustain man's lofty dreams of her material sainthood. Not just Hunt's chubby,
domesticated prostitute but woman in general was heading for another cultural "fall"
which, toppling her from her place among the household gods of bourgeois society, would
first drive her out of the window of domesticity into the trees and the dubious freedom of
"nature" and, finally, by the end of the nineteenth century, straight into the primordial
lair of the devil.

The relentlessly dualistic sensibility of most nineteenth-century intellectuals was to
make that second fall a foregone conclusion. Still, those who hitch their fortunes to the
cultural pendulum of absolute opposites never realize that they are riding the devil's tail,
not the gods' own chariot to the sun. Around the middle of the nineteenth century, therefore, the promise of material progress and the cultural success of the functional marginalization of women, had made males heady with confidence that they might actually succeed

in changing "earthly woman, with a woman's weakness, a woman's faults" back into "the unforgotten Eve of Paradise." Owen Meredith would insist in *After Paradise* that "Adam bequeath'd" a vision of this perfect primal Eve

> . . . to his posterity,
> Who call'd it THE IDEAL. And Mankind
> Still cherish it, and still it cheats them all.
> For, with the Ideal Woman in his mind,
> Fair as she was in Eden ere the Fall,
> Still each doth discontentedly compare
> The sad associate of his earthly lot;
> And still the Earthly Woman seems less fair
> Than her ideal image unforgot.

Not-so-ideal women, of course, had been around ever since that first Fall, and they had been very much in evidence in the annals of culture long before the industrialization of Europe had begun to gather momentum. Nor is there any reason to suggest that examples of sharp distrust between men and women had been absent before the eighteenth century. The world of industrialism by no means invented what the late nineteenth century was fond of calling the "Battle of the Sexes." Some of the most vicious expressions of male distrust of, and enmity toward, women can be found in the writings of the medieval church fathers which late nineteenth-century writers liked to quote. These tireless purveyors of culture were also forever delving into the large fund of antifeminine lore to be found in classical mythology and the Bible.

But the economic rise to power of the middle classes, which was an integral feature of the development of the mercantile-industrial society of the eighteenth and nineteenth centuries, while creating new wealth, also created a new pattern of social relationships. The middle classes needed to establish new conditions for self-identification and self-justification in a world which, just when they confidently thought it was theirs, and would remain so forever, was already slipping out of their grasp. For the world of commerce, toward the second half of the nineteenth century, was becoming increasingly the stomping ground of disembodied corporations rather than of individual entrepreneurs—and this just at a time when the ideology of capitalism had established the primacy of "individual" worth as the justification for human existence.

The tremendous changes in social relationships which were taking place throughout Europe, especially during the first half of the nineteenth century, gave many intellectuals of the time a feeling that the world was on a breathless ride to glory. "Man," said Jules Michelet in 1859 in his book *Woman,* is "on a train of ideas, inventions, and discoveries, so rapid that the sparks dart from the burning rail" (13). However, this rapid train ride to the would-be promised land of industrialism and of that fundamental "pecuniary decency" which—as Adam Smith had promised as early as 1776—would, in the process, trickle down to all, posed certain moral problems to the individual passengers on the train, who were, on the whole, well aware that the ties with which the rails were held together were constructed of the bodies of their fellow human beings. This awareness led to crucial adjustments to the middle class's sense of cultural and moral motivation in what had been perceived, from at least the time of Hobbes and Locke, as a world of necessary mutual depredation. These adjustments, in turn, led to the establishment of a fundamentally new, massively institutionalized, ritual-symbolic perception of the role of woman in society which was, as we shall see, a principal source of the pervasive antifeminine mood of the late nineteenth century—and, by logical extension, the source of a number of elements of sexist mythology which still survive.

There is no more accurate documentation of the manner in which a culture perceives its moral and social mission than in the works of visual art it produces. If, for instance, we compare Frans Hals' early seventeenth-century portrait of a Dutch couple [I, 2] with Holman Hunt's mid-nineteenth-century painting of a similar couple, the fundamental differences in the psychology of visual composition expressed in these two paintings tell us more about the actual changes in the relationship between men and women which were instituted during the intervening two centuries than the statistical analysis of a hefty armful of historical data ever could. Those changes, clearly, are fundamental and tragic.

Frans Hals' man and woman—married or not—are friends. It is evident that they tease each other, argue, have opinions, are companions. They are equals—and that fact does not bother them a bit. One can be very sure that this young woman has her say in the couple's business decisions too. Nor could one possibly imagine her to be terribly prudish about sex. She is not afraid of this man's body (and the casual position of her hand on his shoulder clearly shows that he is hers). The man, in turn, is equally unafraid of this woman's presence. There is a naturalness in their relationship which late twentieth-century advertising executives—in their attempts to dream up images of "liberated" women in order to make consumers buy their hair sprays—have yet to match. But just as Hals' painting loudly proclaims the functional cultural equality of certain middle-class women during this stage of Holland's economic development, so Holman Hunt's painting expresses the abject economic dependence of most women under mid-nineteenth-century British cultural and economic conditions.

But in England, too, a little more than a century earlier things had been different. The rowdy court-centered culture of late seventeenth-century London still reflected a far more complex, variegated range of male-female relationships than was considered acceptable among the middle classes of the second half of the nineteenth century. Indeed, it is almost possible to pinpoint the decade in which British attitudes began to change in earnest and the cultural conception of woman as relatively equal companion—and, in some cases, as economic competitor—faded. In the early part of the eighteenth century a writer such as Daniel Defoe, for instance, showed virtually no indication of having ever underestimated the intellectual and commercial prowess of women, nor of having attributed to them any particular capacity for extraordinary feats of virtue. The women in his novels, such as Moll Flanders, and especially the heroine of his *Roxana* (1724) are levelheaded, intelligent, and extraordinarily tenacious businesswomen, perfectly capable of surviving—and thriving—in a predatory world. His twentieth-century critics have tried to make Defoe's heroines into neurotics and criminals, but in Defoe's eyes they were admirable participants in the world's great rush to financial gain. Indeed, the economic activities of such a personage as Roxana demonstrate that Defoe still fully accepted the statement which, when Tennie C. (Tennessee) Claflin was to make it again in 1871 in her book *Constitutional Equality,* had come to seem like the absurdly exaggerated ravings of an extremist, namely, that "were women trained to business pursuits from childhood, as men are, the largest successes in business would be obtainable by them" (26).

At the same time, it is clear that the development of the concept of a market society, whose motives were of necessity predatory, was to be a major factor in altering the cultural status of the middle-class woman from that of a participant in active life to that of a

I, 2. Frans Hals (1580/6–1666), "The Painter and His Wife" (or "Dutch Couple") (ca. 1622)

I, 3. William Hogarth (1697–1764), "Garrick and His Wife" (1757)

plangent, helpless, consumptive invalid. For, as the pace of economic development quickened, and businessmen everywhere became involved in marketing schemes and industrial projects for which they needed credit, their credit worthiness needed to be amply visible to prospective lenders. There were no automated credit checks, and the Bank of England had only just been founded. One's worth was established by word of mouth, and the words in everyone's mouth were determined by the evidence of the eye. Wives who earlier, in plain dress, would have helped run the family business now became fashion plates: The more yards of silk, lace, and brocade they could wear, the fancier their newly acquired carriages could look, the greater their capacity for what Veblen later was to call "conspicuous leisure," the more likely it was that their husbands would secure the credit they needed for their next business venture. Indeed, any chronological study of British female portraiture from the late seventeenth through the late eighteenth centuries would amply demonstrate the rapid proliferation among middle-class women of, again in Veblen's words, "divers contrivances going to show that the wearer does not and, as far as it may conveniently be shown, cannot engage in productive labor" (*Theory of the Leisure Class,* 122).

The businessmen who in this fashion built their credit ratings, in large measure, upon the shoulders of their women by imprisoning them in unwieldy clothing also took a giant step toward entrapping them permanently in virtue. And, indeed, for many of these businessmen virtue became an ever more desirable commodity in their women. Most British middle-class males of the earlier eighteenth century had their moral roots in Puritan theology and were firmly convinced, with John Wesley, that it was a direct tribute to God for "all Christians to gain all they can, and to save all they can; that is, in effect, to grow rich" (Southey, *Life of Wesley,* II, 308). They were, however, also well aware that this spiritual responsibility to enrich themselves entailed considerable moral dangers to their immortal souls. For, as Defoe had pointed out in 1710, the world in which one could become wealthy was a predatory, nasty place, and upright moral behavior was not likely to get you very far in it: "Trade knows no friends, in commerce there is correspondence of nations, but no confederacy; he is my friend in trade, who I can trade with, that is, can get by; but he that would get *from me,* is my mortal enemy in trade, tho' he were my father, brother, friend, or confederate" (*Review,* 7, no 54 [1710], 210).*

In a world of this sort, in which it was a virtual everyday necessity for the ambitious middle-class male to risk his soul, the notion that the family was, as it were, a "soul unit," that man and wife shared one soul, rapidly gained appeal. A man's wife, it was thought, could, by staying at home—a place unblemished by sin and unsullied by labor—protect her husband's soul from permanent damage; the very intensity of her purity and devotion would regenerate, as it were, its war-scarred tissue and thus keep his personal virtue protected from the moral pitfalls inherent in the world of commerce.

By 1740, with the publication of Samuel Richardson's *Pamela,* the new middle classes were given their first fully developed symbolic narrative of woman as housekeeper of the male soul. Given the fragility of this undernourished spiritual appendage in most men, it was, of course, of the utmost importance that a gentleman test his prospective wife's capacity to withstand the onslaught of temptation—pecuniary as well as physical—from the predatory world around her. Squire B., Pamela's boss, does just that. He moves heaven and earth to get at what Pamela's parents delicately term "that jewel, your virtue," while Pamela, who is extremely well versed in matters of moral economy for a fifteen-year-old housemaid, continues to misunderstand her employer's intentions: "I am sure my master would not demean himself, so as to think upon such a poor girl as I, for my harm. For such a thing would ruin his credit, as well as mine, you know" (10).

Knowing what's what in the world of commerce, Pamela succeeds, albeit with considerable difficulty, in withstanding Squire B.'s dastardly attempts to carry off her "jewel" without benefit of clergy. Secure, therefore, in the knowledge that Pamela's jewel is as sturdy a carrying case for his soul as he is ever likely to encounter, our B. marries her. And, indeed, no sooner is this nasty specimen of predatory manhood united in wedded bliss with our virtuous little lady than he becomes a paragon of virtue. "He pities my weakness of mind, allows for all my little foibles, endeavors to dissipate all my fears; his words are so pure, his ideas so chaste, and his whole behavior so sweetly decent, that never, surely, was so happy a creature as your Pamela!" (372), she exclaims contentedly, satisfied that her passive virtue and his predatory energies will create a virtually unbeatable team in the arena of economic survival. That this is indeed so Richardson demonstrates over and over again in the second half of Pamela's story, which delineates the practical details of our heroine's duties as the keeper of her husband's soul. In fact, the second half of the novel will only be seen as the pointless continuation of a narrative of pointed struggle if one misunderstands the symbolic significance of the first half.

Richardson's contemporaries eagerly absorbed his lessons, and it is clear that his writings responded neatly to the moral dilemmas facing the eighteenth-century middle classes. Gradually a woman's physical inaccessibility came to be seen as the primary guarantee of her moral purity. Any public—or even private—display of levity or physical energy on the part of women was a clear indication of the spiritual frivolity of such women and their

---

*NB: For a detailed discussion of the psychological contexts of the business environment in early eighteenth-century Britain, see my book *Defoe and Economics* (London: Macmillan, in press).

concomitant inability to serve as efficient vessels for the care and feeding of their hus-bands' souls.

During much of the second half of the eighteenth century older, freer habits of inter-action between men and women still coexisted with the new morality, but it was a confus-ing uneasy coexistence, and the older attitudes were doomed to be superseded by the new cult of plangent feminine domesticity. In that respect Hogarth's playful 1757 portrait of David Garrick and his wife [I, 3], which is in many ways an eighteenth-century counter-part to Frans Hals' double portrait of 1622, can be seen as the swan song of an outdated notion that men and women could be friends, that they did not have to form dualistic poles in a world of relentless depredation. Painted late in Hogarth's life and clearly influenced by French "frivolity" and Italian "brio"—both terms of extreme opprobrium in England by this time— the portrait affirms that Hogarth knew too much about the everyday world of average people to bother with the allegorical representations of vestal virtue such younger contemporaries as Reynolds were already making an integral part of feminine portraiture.

In Hogarth's portrait Garrick and his wife interact and enjoy each other quite as much as the couple in Hals' painting. As the wife's hand reaches stealthily for her husband's pen, and her wrist rests lightly and irreverently upon his head, we become vividly aware of that intimate pleasure of shared experience which validates the marriage of equals. Hogarth uses the cheerful sense of mischief and play which lights up the faces of this couple to inform us that there is little false expectation between them. Nor are they merely playacting: This is not just another role in the theater; their pose may be deliberately theatrical, but their practical equality speaks forcefully in the image. We can hunt far and wide through the annals of portrait painting after Hogarth created this image of joyful companionship in marriage, but we are not likely to find another painting in which per-spective is used, as it is by Hogarth, to express the equal and nonfetishized conjunction of man and woman in a world of practical experience. Least of all, from this point on, is one likely to find a woman who forms the top of a visual triangle as she does in Hogarth's composition.

Instead, we now find the mythologized ladies of Reynolds, women surrounded by children, women looking up admiringly at their husbands, women who have become pale creatures with curved necks and weak knees. At this point all we have left is Mary Woll-stonecraft, angrily yet helplessly arguing against the destructive, manipulative, and all too attractive power of the image of woman as the sole keeper of man's virtue. She had now become the prisoner of male symbolism, forced to buy herself into her imprisonment with the only object of value she was thought to have, that overromanticized jewel, her virtue. There was no greater evil, Wollstonecraft pointed out, than to impose on woman "the impossibility of regaining respectability by a return to virtue, though men preserve theirs during the indulgence of vice. It was natural for women then to endeavour to preserve what once lost—was lost forever, till this care swallowing up every other care, reputation for chastity, became the only thing needful to the sex" (*A Vindication of the Rights of Woman*, 133).

The oppressive male sentimentality about the soul-healing power of female virtue had already become much too firmly entrenched by this point—had, indeed, become much too useful a means with which to sidestep any responsibility for one's predatory actions in the world of affairs—for most males to want to change matters as the nineteenth century began to unfold. Nor were things much different on the continent. Jules Michelet, writing in the 1850s, was urging the French to be more assiduous in the imitation of British marital arrangements. To emphasize that other nations had also heard the call, he pointed out that since 1830, when he had first visited Germany, he had seen things change there "in the higher classes." Now, he exulted, "there is everywhere the humble obedient wife, anxious to obey—in a word, the loving woman" (209). And in 1851 Auguste Comte, in his *System of Positive Polity,* had already declared that "female life, instead of becoming independent of the family, is being more and more concentrated in it" [I, 199].

Handbooks began to appear to instruct young women how to be properly submissive, how to give their prospective or actual husbands full value in return for their investment in spiritual solvency. Sarah Stickney Ellis' book *The Women of England: Their Social Duties and Domestic Habits,* published in 1839 and an immediate hit in both England and the United States, was a characteristic example of these guidebooks to the new morality. Ellis spelled out matters very precisely. She was, she emphasized, speaking primarily to the middle class, "to those who belong to that great mass of the population of England which is connected with trade and manufactures." She pointed proudly to the fact that, in England at least, "the acquisition of wealth, with the advantages it procures, is all that is necessary for advancement to aristocratic dignity." Thus, well before Veblen, but without any of the latter's critical distance, she recognized the extraordinary importance of "pecuniary decency" in establishing the reputability of a family. Ellis stressed the fact that in Britain access to the ranks of the socially reputable was a factor of a person's financial potency.

This well-defined road to success, however, also created extraordinary pressures and pitfalls: "It is no uncommon thing to see individuals who lately ranked amongst the aristocracy, suddenly driven, by the failure of some bank or some merchantile speculation, into the lowest walks of life, and compelled to mingle with the laborious poor" (21). On the other hand, Ellis admitted, it was a reconized fact "that gentlemen may employ their hours of business in almost any degrading occupation, and, if they have but the means of supporting a respectable establishment at home, may be gentlemen still" (266). Consequently in England, "long before the boy has learned to exult in the dignity of man, his mind has become familiarized to the habit of investing with supreme importance, all considerations relating to the acquisition of wealth" (45).

The middle-class male learned rapidly that the world of affairs was a Hobbesian jungle of predatory encounters in which the only viable strategy for survival was "every man for himself," and by whatever means necessary. In a passage worthy of Daniel Defoe, Ellis described the war zone:

> There is no union in the great field of action in which he is engaged, but envy, and hatred, and opposition, to the close of day—every man's hand against his brother, and each struggling to exalt himself, not merely by trampling upon his fallen foe, but by usurping the place of his weaker brother, who faints by his side, from not having brought an equal portion of strength unto the conflict, and who is consequently borne down by numbers, hurried over, and forgotten (45–46).

Women, Ellis stressed, should realize these facts of the marketplace, and accept that they constitute an unavoidable hazard of male existence, for "to men belongs the potent— (I had almost said the *omnipotent*)—consideration of worldly aggrandisement; and it is constantly misleading their steps, closing their ears against the voice of conscience, and beguiling them with the promise of peace, where peace was never found." However, this very necessity made it the wife's duty to replenish her mate. She should use her "moral power'' to counteract the destructive influence of the business world, that arena in which man ''sees before him, every day and every hour, a strife, which is nothing less than deadly to the highest impulses of the soul'' (45). Thus, woman, by becoming the guardian of the businessman's conscience, by becoming, as it were, the safekeeper of his soul while she stayed at home, ''cherishing and protecting the minor morals of life,'' could become to him ''a kind of second conscience, for mental reference, and spiritual counsel, in moments of trial'' (46).

At the same time it was incumbent upon the women of England to recognize that ''such is the nature of commerce and trade, as at present carried on in this country, that to slacken in exertion, is altogether to fail'' (48). Hence it was equally important that, instead of trying to reform her husband's public energies, a wife should be quiet and huddle at home, there to guard for him an environment to which, after he had pursued ''the necessary avocations of the day,'' he might return, ''and keep as it were a separate soul for his family, his social duty, and his God'' (50).

Indeed, argued Ellis, woman should rejoice that she had been assigned this role of being the keeper of the ''omnipotent'' male's soul, for without that task she would basically have remained no more than a meaningless speck of dust on the great commercial battlefield of time:

> Woman, with all her accumulation of minute disquietudes, her weakness, and her sensibility, is but a meagre item in the catalogue of humanity; but, roused by a sufficient motive to forget all these, or rather, continually forgetting them, because she has other and nobler thoughts to occupy her mind, woman is truly and majestically great.
>
> Never yet, however, was woman great, because she had great requirements; nor can she ever be great in herself—personally, and without instrumentality—as an object, not an agent (54–55).

Given the eloquence of women such as Sarah Ellis and her companions in arguing the men's case to their sisters, it is not surprising to discover that the period immediately following Mrs. Ellis' stirring call to the women of England to become the holding agents for their men's souls represented the heyday of the cult of woman as resident household nun. Ellis' words were echoed everywhere by other purveyors of quick-fix solutions to male overindulgence in the pleasures of predatory capital accumulation. For the edification of young women these moralists continued, throughout the nineteenth century, to paint lurid pictures of the fate of men on the economic battlefield. Still, as early as 1869, when Horace Bushnell, addressing himself to an American audience, published his *Women's Suffrage: The Reform Against Nature,* there were clear signs that not everything was ideal in the hushed corridors of man's nunnery. Sighed Bushnell,

> Why, if our women could but see what they are doing now, what superior grades of beauty and power they fill, and how far above equality with men they rise, when they keep their own pure atmosphere of silence and their field of peace, how they make a realm into which the poor bruised fighters, with their passions galled and their minds scarred with wrong—their hates, disappointments, grudges, and hard-worn ambitions—may come in, to be quieted, and civilized, and get some touch of the angelic, I think they would be very little apt to disrespect their womanly subordination (62–63).

French intellectuals such as Jules Michelet and Auguste Comte had by mid-century taken the lead in chanting the praises of woman as resident nun in the bourgeois family. ''The man,'' intoned Michelet in 1859, ''passes from drama to drama, not one of which resembles another, from experience to experience, from battle to battle. History goes forth, ever far-reaching, and continually crying to him: 'Forward!'

''The woman, on the contrary, follows the noble and serene epic that nature chants in her harmonious cycles, repeating herself with a touching grace of constancy and fidelity'' (104).

Sitting there in her cell outside of history, buoyed by her ''relative changelessness,'' it was, Michelet insisted, woman's role to energize the male as he forged ahead in history:

> She it is who, at sixteen, may with a word of proud enthusiasm, exalt a man far above himself, and make him cry, 'I *will* be great!' She it is who, at twenty, and at thirty, and all her life long, will renew her husband, every night, as he returns deadened by his labor, and make his wilderness of interests and cares blossom like the rose. She again, who, in the wretched days, when the heavens are dark, and everything is disenchanted, will bring God back to him, making him find and feel Him on her bosom (80).

Michelet's description of woman's role corresponds, in its implications, to the drooping

I, 4. Charles Alston Collins (1828–1873), ''Convent Thoughts'' (1851)

image of a nun in a timeless garden surrounded by lilies, portrayed by Charles Alston Collins in his 1851 painting "Convent Thoughts" [I, 4]. This work accurately depicts the fantasy of woman as the essence of unearthly purity and sacrifice Michelet had in mind when he wrote, "She is the altar." The French historian insisted that it is woman's capacity for self-sacrifice "which places her higher than man, and makes her a religion" (80).

Critics generally see Collins' painting as a slightly esoteric religious statement, but the work has much more to do with the secular manipulation of the concept of what woman should be than with a genuine expression of religious feeling. Collins has portrayed woman in the ideal, as the lily of purity, whose natural realm was the flower garden, and to whose denizens, in her fragile constitution and petal-like sensitivity, she corresponded. Even the painting's frame has an elaborately carved pattern of long-stemmed lilies. Although the painting seeks its inspiration in the medieval notion of the nun as the bride of God, Collins' nun is in fact an expression of the mid-nineteenth-century male's desire to find in his wife that paragon of self-sacrifice endlessly praised by writers from Sarah Ellis to Michelet, that perfect jewel box for the safekeeping of the male soul, nestled in the walled garden of the family home.

Ruskin, too, in *Sesame and Lilies* (1865) urged woman (the lily of his title, of course) to regard herself as a household nun: "The man's power," he said, echoing favorite mid-century clichés, "is active, progressive, defensive. He is eminently the doer, the creator, the discoverer, the defender. His intellect is for speculation and invention." Woman's talent, however, was for "modesty of service." Her capacities were not suited "for invention or creation"; instead she should "be enduringly, incorruptibly good; intinctively, infallibly wise—wise, not for self-development, but for self-renunciation." It was her job to turn the family home into "a sacred place, a vestal temple, a temple of the hearth watched over by households gods." Ruskin saw it as woman's role to undo the inevitable ravages wrought upon the male soul by his involvement in the necessary processes of predatory acquisition: "The man, in his rough work in open world, must encounter all peril and trial: —to him therefore, the failure, the offence, the inevitable error; often he must be wounded, or subdued, often misled, and *always* hardened" (59–60).

In the literature of England, the cult of woman as household nun was taken up far and wide, with Dickens doing much of the most effective flag waving. His endlessly tolerant, endlessly supportive feminine burden bearers, like the women in *Hard Times* (1854), had a grand total of three "positive" options in life. If a woman were of the working classes, like Rachael, she could prove her true value in the order of things by "working, ever working, but content and preferring to do it, as her natural lot, until she should be too old to labour anymore." A woman of the middle or upwardly mobile classes could look forward to such a life as we are promised for Sissy Jupe, who will pass her days "cultivating the flowers of existence [and being] a wife—a mother—lovingly watchful of her children." Finally, if, as in Louisa's case, through tragic parental mismanagement, the woman's redemptive jewel of virtue had been misspent, she could redeem herself—if, inevitably, only partly—by becoming a nun in deed, "trying hard to know her humbler fellow-creatures, and to beautify their lives of machinery and reality with those imaginative graces and delights, without which the heart of infancy will wither up, the sturdiest physical manhood will be morally stark dead" (226).

The self-sacrificial woman was, of course, most deliciously desirable if she—as in Collins' painting—clearly had turned quiet suffering and self-denial into a consummate art. Consequently the theme of woman as nun found its true apotheosis in *Lucile* (1860), a verse novel by Owen Meredith (the pen name of Robert, first earl of Lytton, son of Edward Bulwer, the novelist) which is now deservedly forgotten, but which upon its publication became a huge success in England and America and remained a popular favorite throughout the second half of the nineteenth century. Lucile, the novel's sacrificial heroine, is caught in a crossfire of misunderstanding which makes her true love, Alfred Vargrave, marry another, "an innocent child." But rather than waste time thinking about her own broken heart, Lucile sets about turning her confused would-be former lover's less-than-perfect marital arrangement into a true Victorian bower of wedded bliss. For, as we soon learn, Alfred's wife, Mathilda, feeling slighted by her husband's continuing interest in Lucile and being a not-yet-fully-domesticated specimen of femininity, begins to experience remarkably Eve-like forebodings of sin: "To the heart of Mathilda the trees seem'd to hiss / Wild instructions" (101). The devil, in the guise of a snakelike duke, who vilely tickles her dormant ego and makes her blush, is about to lead her out of domestic paradise (her "folly fast growing into crime") when Lucile quite literally jumps out of the bushes and chases the tempter away, thus saving for a still-confused husband "the love of an innocent wife." An enthusiastic convert to the virtues of feminine self-sacrifice, Lucile becomes what every woman should aspire to be. For to make certain that Alfred, her former lover-who-never-was, will understand that she will remain forever unattainable in her devotion to him, Lucile now becomes a bona fide nun. This enables her, toward the end of Meredith's paean to sacrificial love, to save Alfred and Mathilda's son from death on the battlefield by nursing him back to health in true Florence Nightingale fashion. After committing various other angelic deeds of devotion, she fades modestly into the mists, giving Meredith his opening for a sententious finale about "woman's mission":

> . . . to watch, and to wait,
> To renew, to redeem, and to regenerate.
> The mission of woman on Earth! to give birth
> To the mercy of Heaven descending on earth.
> The mission of woman: permitted to bruise
> The head of the serpent, and sweetly infuse,
> Through the sorrow and sin of earth's register'd curse,
> The blessing which mitigates all: born to nurse,
> And to soothe, and to solace, to help and to heal
> The sick world that leans on her. This was Lucile (149).

Women, then, both inside and outside of marriage, were to aspire to the vestal purity of the nun. The mid-century's acknowledged master of dualistic thought, Auguste Comte, godfather of the "science" of sociology and seeker after feminine purity, delineated the polarity. In the second volume of his *System of Positive Polity* (1852) he spelled out the "complementary" functions of "the active and the affective sex": Clearly, the essential duty and "mission of woman is to save man from the corruption, to which he is exposed in his life of action and of thought." In return, it was the man's responsibility to relieve "the loving sex from every anxiety which can interrupt the force of those affections" [II, 171–72].

What better place to make certain that none of these "anxieties" of a worldly nature could impinge on "the loving sex" than the walled garden of domesticity? After all, wasn't woman fragile, and didn't she need the same sort of care as gardeners were wont to lavish on flowers of domestic cultivation? The husband should regard himself as the

I, 5. Robert Reid (1862–1929), "Fleur de Lys" (ca. 1897)

I, 6. Edgard Maxence (1871–1954), "Rosa Mystica," watercolor (1903)

gardener and his wife as his flower. As a matter of fact, in her very essence, her fragility, her physical beauty, her passivity and lack of aptitude for practical matters, woman was virtually a flower herself. "The path of a good woman is indeed strewn with flowers; but they rise behind her steps, not before them," said Ruskin (76).

In the paintings of the second half of the nineteenth century, it was virtually impossible for a "pure" woman to stir without finding herself up to her neck in flowers. Woman became the personification of Flora in numerous paintings, and as the artists never tired of indicating, she lived and died like the petals of the flowers she was forever toting about. She was, in the eyes of painters, the bluebell, the rose, the fleur de lys made flesh. In fact, Robert Reid, in his "Fleur de Lys" [I, 5] dramatically suggested the unity of woman and flower by making her, as it were, "grow" among the flowers she is contemplating, the elegant curves of her body and dress serving as the stem and leaves of this delicate representative of the static existence of woman as flower. A host of other painters used far less imaginative means to make the same point. Images with descriptive titles such as "Spring Flowers," "Blossoms," and so on, would show elegant young ladies frolicking, spring blossoms themselves, among nature's petaled host.

Michelet, indeed, was of the opinion that all a female child needed to know in life could be conveyed to her by having her study a flower: "The care and assiduous contemplation of this flower, the relations which shall be pointed out to her between her plant and such or such an influence of the atmosphere or season—with these alone her entire education may be carried on." Michelet was convinced that for woman "a flower is a whole world, pure, innocent, peace-making; the little human flower harmonizes with it so much the better for not being like it in its essential point. Woman, especially the female child, is all nervous life; and so the plant, which has no nerves, is a sweet companion to it, calming and refreshing it, in a relative innocence'' (86).

Woman, generally speaking, could be counted upon to learn more about her proper position in society by dwelling among the flowers. ''Her thoughts,'' explained Michelet, ''grow calmer in such discreet society, for they are not inquisitive—they smile, but they are silent. At least they speak so low, these flowers, that we can hardly hear them. They are the earth's silent children'' (264).

It is clear that most nineteenth-century males would have liked their women to take a few pointers from the flowers and learn to be silent too. Thus, for them the true woman was a Lucile who let herself be planted in Collins' convent garden, carrying a lily to calm her nerves. Ruskin was stirred with a fervor that had a decidedly hothouse flavor in describing the contrast between the world of the flower-woman and the male vale of tears beyond the domestic garden:

> This is wonderful—oh, wonderful!—to see her with every innocent feeling fresh within her, go out in the morning into her garden to play with the fringes of its guarded flowers, and lift their heads when they are drooping, with her happy smile upon her face, and no cloud upon her brow, because there is a little wall around her place of peace: and yet she knows, in her heart, if she would look for its knowledge, that outside of that little rose-covered wall, the wild grass, to the horizon, is torn up by the agony of men, and beat level by the drift of their life-blood (76).

It is true that Ruskin, in speaking in this fashion to the women in his audience, was trying to stir a sense of sympathy in them for women less fortunate than they, for ''flowers that could bless you for having blessed them, and will love you for having loved them; flowers that have eyes like yours, and thoughts like yours, and lives like yours'' (77), but, in the end, the overwhelming perfume of the flowers in his mind drugged him and wafted his message away from its call to social responsibility and into the manipulative realm of the repressive concept of woman as an integral part of the domestic flora. The image of woman as flower which so addled Ruskin's brain obviously also stirred the gray matter of Edgard Maxence into a fine commercial sense of the sort of ''mystic rose'' his viewers coveted most [I, 6]. His very popular image of 1903, depicting woman as a white rose in a sea of white roses, was a true turn-of-the-century compendium of visual platitudes concerning the moonlit purity and virtuous, nunlike passivity of the Eternal Feminine on its best behavior.

The ''pure'' woman, the woman who, with her passive, submissive, imitative, tractable qualities, seemed to share with the flowers all the features characteristic of the plant life of the domestic garden, thus came very generally to be seen as a flower herself, to be cultivated in very much the same way a flower ought to be cultivated in order to thrive.

It is this idea, as a matter of fact, which underlies the symbolism of Verena Tarrant, the true heroine of Henry James' *The Bostonians* (1886). Present-day readers generally resist seeing her as a heroine, because in a book so full of powerful characterizations, she seems weak, unindividuated, and dull. But those were not negative characteristics, in James' eyes, for a woman destined by nature to be an exemplary representative of the ''household nun.'' James wanted to praise Verena when he portrayed her as the quintessential tractable woman, a late-nineteenth-century flower child. He saw her as America's hope in turbulent days: She is passive, she habitually ''gives herself up'' to others and she is pale (''anemic'' in the eyes of the practical Dr. Prance). ''It was in her nature to be easily submissive, to like being overborne'' (337), James tells us. Indeed, Olive Chancellor sees her as ''perfectly uncontaminated'' and is convinced that ''she would never be touched by evil'' (85). She is, in fact, ''divinely docile.'' James himself saw in her the as yet unfulfilled potential of the ideal domestic woman, the woman who could fulfill the male dream of the wife as household nun, for in her passivity she represented the soft, gentle, nonviolent, ''humane'' values he longed for and could not find in the male world. Thus, Verena became for him a characteristic soul-surrogate, a guardian angel of tenderness, of the ''spirit,'' always ''honest and natural'' (253) and intuitively right about things of the heart. Indeed, of all the characters in *The Bostonians*, she is the only one yearning to do what a good woman ought to do, namely, ''to take men as they are, and not to think about their badness. It would be nice not to have so many questions, but to think they were all comfortably answered, so that one could sit there on an old Spanish leather chair, with the curtains drawn and keeping out the cold, the darkness, all the big, terrible, cruel world— sit there and listen forever to Schubert and Mendelssohn'' (158–59).

It is because of these aspects of her personality that James saw her as ''the sweetest flower of character . . . that ever bloomed on earth'' (108). Verena was that ideal child-woman who, in her simplicity, inherent purity, and tractability, might have become the perfect domestic antidote to what James considered ''the detestable tendency toward the complete effacement of privacy in life and thought everywhere so rampant with us nowadays'' (*Literary Reviews and Essays*, 266). The tragedy which unfolds in *The Bostonians* is the destruction of Verena's potential as a household nun, as that flower of domesticity whose role in life was to bloom in the garden of man's privacy and tranquility. In James' eyes it was the tragedy of American womanhood and, by extension, the tragedy of America itself which was here symbolized by the sad undoing of Verena—she *was* the innocence of America as she was being warred over by ideologues, torn from the garden of quiet domestic contemplation and left to shed her tears forever in ''the cold, the darkness; all the big, terrible, cruel world'' which, in her flowerlike simplicity, she had dreamt of shutting out.

With all this emphasis on the saving function of the nunlike purity of the woman in marriage, it was inevitable that she should be given the task of aspiring to the position of the one figure in history who had clearly managed to be a complete success at being simultaneously virgin, mother, and wife: Mary, Mother of God. Woman as the virginal bride of the spirit, the mother of purity, became the preferred subject of numerous painters toward the end of the nineteenth century, a time when, as we shall see, the public image of woman as vestal had already been tarnished quite badly. Abbott Handerson Thayer's ''Virgin Enthroned'' of 1891 [I, 7] is characteristic of the numerous images of pale, listless, childlike feminine purity which came to be fancied as exquisite depictions of what Margaret Sangster, as late as 1900, called ''Winsome Womanhood.''

At every yearly exhibition of the Royal Academy, the Glas-Palast in Munich, the Paris salons, or the Pennsylvania Academy, painters exhibited these inspiring works, with titles such as ''A Modern Madonna,'' ''A Cottage Madonna,'' ''The Spirit of Christianity,'' and so on, showing not paintings with a genuinely religious sentiment but works representing Michelet's wish to make modern woman into a religion. T. C. Gotch made this sort of painting a specialty. His monumental Royal Academy entry of 1902, ''Holy Motherhood'' [I, 8], is a typical example of the genre, as is George Frederick Watts' relentlessly maudlin work entitled ''The Spirit of Christianity'' [I, 9]. Paintings such as these had the same ideological function as the writings of the poets and social critics who,

I, 7. Abbott Handerson Thayer (1849–1921), ''Virgin Enthroned'' (1891)

I, 8. Thomas C. Gotch (1854–1931), ''Holy Motherhood'' (ca. 1902)

like Horace Bushnell, were trying to keep woman in line at a time when the excesses of the earlier generation of idolators had already driven many women to organize in opposition to the joys of glorious subordination. For heaven's sake, exclaimed Bushnell, expressing the imagery of the painters in words, ''if she could only consent to be true gospel and woman together, to be gentle, and patient, and right, and fearless, how certainly would she come out superior and put him at her feet. There seems to me, in this view, I confess, to be a something sacred, or angelic, in such womanhood. The morally grandest sight we see in this world is a real and ideally true woman'' (67).

The madonna image as representative of the married woman's role in life was deemed especially appropriate because women and children formed, as it were, an inevitable continuity: The truly virtuous wife was, after all, as innocent as a child. Moreover, as Michelet had pointed out, ''from the cradle woman is a mother, and longs for maternity. To her everything in nature, animate or inanimate, is transformed into little children'' (82). Thus, to be surrounded by children was woman's fondest desire, a single indication of her madonnalike purity and docility. Indeed, when kept properly out of the fray of life, woman had a child's mind as well, and the more childlike her actions, the more obvious her childlike purity, her angelic, madonnalike transition from child to wife. In 1854 Coventry Patmore intoned in his *The Angel in the House:* ''She grows / More infantine, auroral, mild, / And still the more she lives and knows / The lovelier she's expressed as child'' (89). Indeed, argued Patmore, upon betrothal ''Back to the babe the woman dies'' (180).

With all the attention nineteenth-century men were paying to the virginal purity of their women as a guarantee of the continuance of their spiritual credit before the Almighty in a world of crude economic necessity, the practical impossibility of having their household nuns and modern madonnas duplicate the original Mary's immaculate conception must certainly have become rather an embarrassment. But even in this matter their capacity for the invention of ideal solutions was undaunted. In this case it was Comte who came to the rescue. Announcing what, in writing the fourth volume of his *System of Positive Polity* (1854), he certainly had a right to call "a daring hypothesis," and expressing the wish that it might in due time be "destined to become a reality in the course of our advance," Comte in effect suggested the exploration of artificial insemination as a means of keeping women as close to the madonna ideal as possible, while still allowing them to fulfill their function as mothers. In language which, given the strictures against direct statement during this period, was nonetheless relatively straightforward, he pointed out that since, "in human reproduction, the man contributes merely a stimulus, one that is but an incidental accompaniment of the real office of his generative system," it should be perfectly well possible to "substitute for this stimulus one or more which should be at woman's free disposal." For Comte this idea represented an exciting "presentiment, as it were, of the degree in which woman, even in her physical functions, may become independent of men." Comte was enthralled by the thought that "the highest species of production would no longer be at the mercy of a capricious and unruly instinct, the proper restraint of which has hitherto been the chief stumbling-block in the way of human discipline."

Even though Comte's remarks represent an audacious mid-nineteenth-century anticipation of late twentieth-century scientific reality, it should not be assumed that he was a genuine precursor of radical feminism, nor that his scheme was designed to liberate women from all forms of male dominance. Lest his reader misunderstand his radicalism as anything but a concern for feminine purity, he made haste to emphasize that

> the just independence of the sex may be regarded as resting upon two conditions in close connection with one another: the exemption of all women from work away from home, and their voluntary and complete renunciation of wealth. For women suffer more from the aspirations of ambition than they do from the pressure of poverty. Priestesses of Humanity in the family circle, born to mitigate by affection the rule, the necessary rule, of strength, women should shrink from any participation in power as in its very nature degrading (60–61).

Comte's scheme, then, was designed to make it possible for woman to become more effectively the "Angel in the House," that paragon of renunciation about whom Coventry Patmore was writing volubly and ploddingly while Comte was busy developing his schemes for the final removal of women from the work force.

Patmore's angel, about whose vicious long-range influence Virginia Woolf was to write so movingly, was "Marr'd less than man by mortal fall, / Her disposition is devout, / Her countenance angelical." But for the male the best thing about the angel in the house was that "Her will's indomitably bent / On mere submissiveness to him" (149). Indeed, "A rapture of submission lifts / Her life into celestial rest" (180). After all, "Man must be pleased; but him to please / Is woman's pleasure; down the gulf / Of his condoled necessities / She casts her best, she flings herself" (111). That, ultimately, was what the mid-nineteenth-century hoisting of woman onto a monumental pedestal of virtue was all about: a male fantasy of ultimate power, ultimate control—of having the world crawl at his feet. Michelet, for one, just couldn't get enough of that image. Finding French women still somewhat reluctant to become household slaves as well as household nuns, he allowed himself to fantasize about the greater pliability of African women:

> She is essentially young in blood, in heart, and in body—of gentle, childlike humility, never sure of pleasing, ready to do anything in order to displease less. No tyranny wearies her obedience; annoyed by her face, she is in no wise comforted by her perfect form, so full of touching languor, and elastic freshness. She throws at your feet what you were about to adore; she trembles and begs your pardon—she is so grateful for the pleasure she bestows! She loves, and her whole heart flows into her warm embrace (133–34).

What the mid-century male wanted most of all, then, was a woman who would not only be the safekeeper of his soul but who would, in fact, offer up her own being, her own soul completely to that task, a woman who would become a mere extension of himself, who would let herself be absorbed completely by him. What he wanted was the self-negation of Elsa, the pliant maiden in Wagner's *Lohengrin* (1850), who, upon being vindicated by the opera's titular hero from sundry ignominious accusations, warbles excitedly that her only desire is to let herself be absorbed into the being of her heroic defender: ''I will let myself fade into you—before you I become as nothing: Only then, when you absorb me totally, shall I be able to consider myself blessed!'' Whereupon Lohengrin modestly replies, ''I was only able to be victorious because you were so pure''—indicating once again that in mid-nineteenth-century mythology the worldly success of the male was deemed to be inextricably intertwined with the self-denial of woman. Elsa, indeed, is the personification of what Michelet insisted was an accurate image of the German middle-class woman, a woman who ''loves, and loves always. She is humble, obedient, and would like to obey still more. She is fitted for only one thing, love—but that is boundless'' (208). One wonders how many of the countless millions of women who have stepped toward the altar to the stately wedding march which Wagner composed to symbolize these sentiments were ever aware that in Wagner's eyes they were blithely stepping toward self-obliteration! As Bushnell explained in true Wagnerian spirit, woman in marriage became a factual nonentity, ''passing out of sight legally, to be a covert nature included henceforth in her husband'' (52). Indeed, Bushnell stressed contentedly, these were matters which English common law had institutionalized: ''Her personality is so far merged in his, that she cannot bring a suit any more in her own name, for it is a name no longer known to the law'' (88).

Everything, from simple common sense to the ample documentation provided by contemporary commentaries and literature, indicates clearly how insanely exaggerated the middle-class desire for visible evidence of domestic harmony had become between 1840 and 1860. Largely in response to the wholesale and all-too-convenient acceptance by middle-class males of the concept of the unavoidable necessity of predatory market conditions as part of the ''struggle for life,'' this desire had created an entirely new set of psychopathological responses in the women, who found themselves being forced into the position of having to prove their worthiness to be wives by means of impossible feats of virtue, and who, once they had become ''modern madonnas,'' could only retain their coveted position by playing the role of cringing household pets.

With willing female co-conspirators such as Sarah Ellis hammering into them the idea that there was no man in existence ''who would not rather his wife should be free from selfishness than to be able to read Virgil without the use of a dictionary,'' and telling them

I, 9. George Frederick Watts (1817–1904), ''The Spirit of Christianity'' (or ''To All Churches: A Symbolical Design'') (1875)

I, 10. Arthur Hacker (1858–1919), ''The Cloister or the World'' (1896)

to sacrifice their identity in deference to everyone in sight, and to seek their happiness ''only in the happiness of others,'' it was a foregone conclusion that many would fail in their attempts to conform, and that in their failure their mental equilibrium and their physical health would be the first casualties.

In fact, women everywhere were being set up very much in the way Wagner set up Elsa in *Lohengrin*. For Elsa was not only required to be a paragon of virtue and self-negation; Wagner also made her ability to hold on to her husband dependent on her willingness to trust him blindly, silently, abjectly, unconditionally (all very useful attributes, incidentally, for a wife whose husband does not want her to know much about his activities in the world of affairs). Wagner ultimately dumped Elsa on the trash heap of perfidious wives simply because she dared to ask her husband the simple and perfectly reasonable question, ''Who are you? Where are you from?'' That mild sign of insubordination was enough to send Lohengrin packing in disgust over the inability of women to keep their mouths shut. Wagner, siding with his hero, spitefully made Elsa, in despair over her own perfidy, die in her brother's arms. Step aside, Eurydice! The nineteenth century had found a better mousetrap.

Our assumptions always become a prison to our possibilities, and the mid-nineteenth-century's assumptions about the role of women in society turned the world of intelligent, thoughtful women everywhere into a grim, dank dungeon from which they yearned most desperately to escape. Emily Brontë's poem "The Prisoner"—published in 1846 as a somewhat abbreviated version of the narrative fragment "Julian M. and A. G. Rochelle"—stunningly expresses the plight of an intelligent woman trapped in a life lived according to man's self-serving assumptions about feminine virtue. It soon becomes clear that the poem, although ostensibly about an actual prisoner in a real dungeon, is in fact a commentary on woman's imprisonment in life.

> The captive raised her face; it was as soft and mild
> As sculptured marble saint or slumbering, unweaned child;
> It was so soft and mild, it was so sweet and fair,
> Pain could not trace a line nor grief a shadow there!

Outwardly the captive is a perfect representation of the mid-century ideal of the virtuous household nun: saint and child rolled into one, apparently hard as steel in her polished perfection. But it is a perfection gained in exchange for imprisonment within the "granite stones" of a stifling domesticity. The captive's jailor (mother? housekeeper?) describes the ruling male, her "master": "His aspect bland and kind, / But hard as hardest flint the soul that lurks behind" (237). It is a striking image of a manipulative, calculating, sadistic, aggressive middle-class male of the Coventry Patmore sort. Given the captive's imprisonment in a life ruled by abject submission, she can only hope for deliverance—a return to personal freedom—beyond life: "A messenger of Hope comes every night to me, / and offers, for short life, eternal liberty" (238).

Her creative energies stifled by the narrow demands inherent in the middle-class concept of feminine passivity and innocence, her physical being locked into self-negation, this woman-prisoner comes to see the vision of death as a promise of new possibility:

> Winds take a pensive tone, and stars a tender fire,
> And visions rise and change that kill me with desire—
>
> Desire for nothing known in my maturer years
> When joy grew mad with awe at counting future tears;
> When, if my spirit's sky was full of flashes warm,
> I knew not whence they came, from sun or thunderstorm (238).

But as yet that hope of release is a mere dream—reality still encircles her. The prison house of life reasserts itself each time her narrow, manmade senses reassert their dominion:

> Oh, dreadful is the check—intense the agony—
> When the ear begins to hear and the eye begins to see;
> When the pulse begins to throb, the brain to think again;
> The soul to feel the flesh and the flesh to feel the chain! (239)

The visitors to the prison house, although unsympathetic to their captive's plight, recognize that in her very longing for death she is trying to defy the world's demand that she live a life of intellectual paralysis merely to maintain the semblance of saintly virtue:

> Her cheek, her gleaming eye, declared that man had given
> A sentence unapproved, and overruled by Heaven (242).

Brontë's poem helps to explain the cult of invalidism and death which began to preoccupy middle-class women during the second half of the nineteenth century. The prevailing masculine psychosadistic insistence that women prove their virtue through extreme feats of saintly self-negation created a multiplicity of responses, but all were related to women's attempts to make the best out of an intolerable situation, to construct alternatives, to gain a sense of control over the psychological obliteration they were being asked to undergo. Thus, immersion in illness and even the escape into death came to be seen as creative options, a way of stirring to life their anesthetized senses: Brontë's masochistic ecstasy in "The Prisoner" effectively expresses the nature of this attempt at turning a process of passive suffering ("woman is born to suffer" said Michelet contentedly) into an outlet for creative energy:

> Yet I would lose no sting, would wish no torture less;
> The more that anguish racks the earlier it will bless;
> And robed in fires of Hell, or bright with heavenly shine,
> If it but herald Death, the vision is divine (239).

However, as women everywhere tried hard to become the household nuns they were supposed to be, the act of suffering as a defiant, if passive, form of self-identification lost the rebellious element which had been its main attraction for Emily Brontë and became instead a sign of passive compliance with the cultural image of extreme virtue. What better guarantee of purity, after all, than a woman's pale, consumptive face, fading, in a paroxism of self-negation, into nothingness? Thus, the image of the angel in the house, on her way to becoming an angel in earnest, stirred the hearts of the late nineteenth-century male with comforting thoughts of seraphic servants being sent ahead to heaven to secure there for their masters a prime seat at the great spiritual stock exchange which they saw as operating in God's garden.

Of course, even among the angels there were many gradations of virtue, and if woman's true nature was to be angelic, there were still many household nuns who were all too easily tempted by the delicious enticements of worldly pleasure. Even the most willingly subservient women were required to engage in a constant, necessary, tortured self-assessment, as they tried to determine whether they were sufficiently self-effacing to merit being companions to the white-winged, white-gowned, white lily–toting angel of exemplary virtue, or whether in the dark recesses of their souls there might not still stir the vestiges of worldly desire.

This fundamental, dualistic conflict and its health-wrecking effects upon woman were a favorite theme of late nineteenth-century painters. Like Arthur Hacker in "The Cloister or the World" [I, 10], these painters preferred to depict their household nuns in the throes of that struggle. The effects of a conflict of the magnitude delineated by Hacker inevitably had extremely deleterious effects upon the physical condition of the household nun. The more intense the struggle, the more impressive the triumph of true virtue. A healthy woman was therefore regarded with suspicion. In consequence, the late nineteenth-century painters generally liked to paint their paragons of virtue in an advanced state of physical debility and illness. Artists ransacked literature and history to come up with affecting instances of women in states of terminal illness. ("At Death's Door" was the title of a popular painting by Gaston La Touche exhibited at the Salon des Beaux-Arts in 1893.) As we shall see, the sort of condition represented by a woman in a state of sickness unto death—a woman of whom, as in La Touche's painting, we see nothing but her hollow-cheeked face sunk into a massive pillow, staring hopelessly into nothingness—came to represent the ultimate icon of virtuous feminity. Images such as these constituted a further step in the marginalization of woman, her removal from all meaningful, creative activity. As George Eliot remarked pointedly, ''Men say of women, let them be idols, useless absorbents of previous things, provided we are not obliged to admit them to be strictly fellow-beings.''

CHAPTER II

The Cult of Invalidism;
Ophelia and Folly;
Dead Ladies and the Fetish of Sleep

Throughout the second half of the nineteenth century, parents, sisters, daughters, and loving friends were kept busy on canvases everywhere, anxiously nursing wan, hollow-eyed beauties who were on the verge of death. For many a Victorian husband his wife's physical weakness came to be evidence to the world and to God of her physical and mental purity—that precious commodity which would ultimately secure for him spiritual succor from the world of sordid business affairs and rescue his soul from perdition. Late nineteenth-century painters were eager to oblige his sense of virtue with affecting images of feminine weakness bearing such titles as "A Shadow," "A Lull," "Anxiety," "The Dying Mother," "In Excelsis," and so on.

The French painter Louis Ridel's contribution to the Salon of 1900 is a typical example of this genre. Titled "Last Flowers" [II, 1], it shows a young woman in the loose-fitting clothing of a terminal consumptive, leaning exhaustedly against her best friend's comforting shoulder. The latter's strong and healthy presence contrasts strikingly with the helpless inanition of her companion. The healthy woman, it is clear, has taken her friend on a last outing to the lake along whose secluded banks the two had drifted so often in the past, picking flowers and exchanging the gentle confidences of close friends. Once again, as if to deny the course of time and the ravages of nature, they have gone back to haunt their favorite spots, but as was fated, destiny has overtaken them: Among the flowers plucked this time, among these *dernières fleurs*, is the life of one of them—a woman who will fall into the waters of time as surely as the flowers in her hand must fall into the silent waters of the pond.

Ridel's painting is a striking, indeed, a genuinely moving image, but like its many counterparts of the fin de siècle, it exploits and romanticizes the notion of woman as a permanent, a necessary, even a "natural" invalid. It was an image which in the second half of the nineteenth century came to control and not infrequently destroy the lives of countless European and American women. More and more the mythology of the day began to associate even normal health—let alone "unusual" physical vigor in women—with dangerous, masculinizing attitudes. A healthy woman, it was often thought, was likely to be an "unnatural" woman. Proper human angels were weak, helpless, ill. Michelet's formula for the treatment of women read like set of instructions for the treatment of the terminally ill: "More fragile than a child, woman absolutely requires that we love her for herself alone, that we guard her carefully, that we be every moment sensible that in urging her too far we are sure of nothing. Our angel, though smiling, and blooming with life, often touches the earth with but the tip of one wing; the other would already waft her elsewhere" (169).

Michelet's writings on women and on love had a huge readership in England and America. His book *Love,* for instance, had a circulation of over two hundred thousand copies in the United States alone during the 1860s. By 1870 Nicholas Francis Cooke, himself certainly not a friend of women's emancipation, could in his *Satan in Society* refer contemptuously to "the dwarfed, miserable, sickly specimens of feminine humanity" which he saw as constituting "the rule rather than the exception" among the middle classes. Indeed, Coventry Patmore's sketch of the nubile young woman's virtuous desperation—

II, 1. Louis Ridel (b. 1866), "Last Flowers" (1900)

"She wearies with an ill unknown," while in her sleep "she sobs and seems to float, / A water-lily, all alone, / Within a lonely castle-moat" (131)—was only too close to reality.

The "fearful collapse of female health" (*Satan in Society,* 35) was not an imaginary state of ideal feminine weakness dreamt up by poets. With more than a touch of irony Abba Goold Woolson, a good friend of John Greenleaf Whittier and a wonderfully level-headed observer of contemporary mores, entitled a chapter in her book *Woman in American Society* (1873) "Invalidism as a Pursuit." "This invalidism," said Woolson, "is apparent on every hand." American women everywhere were "afflicted with weakness and disease." To be ill was actually thought to be a sign of delicacy and breeding: "With us, to be ladylike is to be lifeless, inane and dawdling. Since people who are ill must necessarily possess these qualities of manner, from a lack of vital energy and spirits, it follows that they are the ones studiously copied as models of female attractiveness." As a result, Woolson pointed out, feminine invalidism had become a veritable cult among the women of the leisure class. Instead of "being properly ashamed of physical infirmities, our fine ladies aspire to be called *invalides;* and the long, French accent with which they roll off the last syllable of this word seems to give it a peculiar charm. If you happen upon a group of them conversing together on a bright summer afternoon, you will be sure to find them endeavoring to outshine each other in the recital of past illnesses." It was indeed clear that the cult of invalidism was closely tied to the requirements of Veblen's category of "conspicuous leisure." These demanded that a gentleman of means should put "in evidence his ability to sustain large pecuniary damage without impairing his superior opulence" (*Theory of the Leisure Class,* 58). Woolson identified this connection quite bluntly: "We have even heard a husband wind off a glowing enumeration of his wife's accomplishments with the remark that she was an invalid, as though that revealed at one stroke her fine manners and high social position, as well as the ample means which could allow him to support such a helpless elegance" (192–94).

In addition, Woolson pointed out, this pursuit of invalidism had come to prevail only in relatively recent years: "Girls had formerly some out-door life and jollity; but nearly all the active sports in which they were wont to indulge seem abandoned" (203). Instead middle-class women everywhere now shunned "the outer air and the sunlight as if they would harm us" (208). If they did go out they would "balance along on their high heels, holding parasols carefully in glove-cramped hands lest a drop of sunshine should touch their pallid cheeks" (203). To dare to be active and energetic was considered a social faux pas. When a young woman Woolson observed at a summer hotel showed disconcerting tendencies toward "surmounting stone walls and clambering upon stage-coaches," she caused consternation among the women present. "Everybody apologized for her, and assured her dismayed mother that these dreadful ways she would certainly outgrow, and that a judicious application of corsets and juvenile parties would soon make her presentable" (210–11).

Things were hardly different in Britain. In one of her widely read commentaries on contemporary feminine mores gathered in *Modern Women* (1889), Mrs. E. Lynn Linton described the manner in which early strictures on movement rapidly turned healthy children into sickly young women: "Less and less every year are the nerves and muscles, the restless activities of arms and legs, exercised and made to purvey new vigour to the life. The body is allowed to grow stagnant. The life of the woman, even as mere animal, becomes poor and morbid and artificial" (38–39). Already in 1848, in *Mary Barton,* Elizabeth Gaskell had described the manner in which the cult of invalidism was becoming ingrained among women of the privileged class. Mrs. Carson, the wife of a wealthy mill owner,

> was (as was usual with her, when no particular excitement was going on), very poorly, and sitting up-stairs in her dressing-room, indulging in the luxury of a head-ache. She was not well, certainly. "Wind in the head," the servants called it. But it was but the natural consequence of the state of mental and bodily idleness in which she was placed. Without education enough to value the resources of wealth and leisure, she was so circumstanced as to command both (254).

There was thus for late nineteenth-century audiences a strikingly insistent, but rarely admitted, suggestion of social status and economic privilege embedded in paintings dealing with feminine illness. This element helps explain, at least partially, why the art-buying public of the period should have been so eager to pay large sums of money to acquire images of women in stages of abject physical degeneration, painted by the highest-paid artists of the day.

Nor did the women thus depicted of necessity have to be members of the ruling class. Images of women whose demise was taking place amid the trappings of "decent poverty" provided middle-class women—who perhaps felt that they ought to but did not have the actual will to be decorously ill—with excellent surrogates: To hang such a picture was to acknowledge at least one's consciousness of virtue. The cult of feminine invalidism, then, was both among men and women inextricably associated with suggestions of wealth and success. In addition, such images were attractive because in an environment which valued self-negation as the principal evidence of woman's "moral value," women enveloped by illness were the visual equivalents of spiritual purity. They went Florence Nightingale one better: They did not merely pursue self-sacrifice to demonstrate their virtue; they personified its virtues through their self-obliteration by means of illness.

How directly, and often tragically, the late nineteenth-century cult of invalidism affected the lives of even the apparently "best"-situated women can be seen in the case of Alice James, the sister of Henry and William James, who spent most of her short life fighting and succumbing to the effects of a variety of increasingly oppressive, shadowy ailments which both fascinated and perplexed her. That her condition was directly related to her desperate efforts to repress her own active, inquisitive nature in order to be more perfectly the infinitely tolerant, pliant woman she was expected to be is movingly reflected in her (probably unconscious) choice of words when, in desperation, she confessed in her diary: "How sick one gets of being 'good,' how much I should respect myself if I could burst out and make everyone wretched for 24 hours . . ." (64). She might have gained in health, as she would have in self-respect, had she let herself go, but she could not overcome her environment and slipped deeper and deeper into invalidism, until every time she tried to pull herself out of her malaise her will became weaker. Finally, the realm of what she called "my suffocation" came to seem to her "the natural one" (56), even while "every fibre" of her being protested "against being taken simply as a sick carcass" (183). She died in 1892, a year after she wrote those words.

As the sad case of Alice James all too graphically demonstrates, there was no reason for the cult of invalidism to stop at illness. Such relative reticence was not favored by the extremist temper of time. Poets and artists everywhere drove women to go even further in their acts of sacrifice. As every properly trained, self-denying woman knew, true sacrifice found its logical apotheosis in death. The absence of exercise, the dusty housebound environment, and the deliberate shunning of sunlight and fresh air had created a new breed of women who could virtually not exist unattended by nurses and servants. Inevitably, many of these women contracted actual diseases, such as tuberculosis—consumption—and were doomed, like Alice James, to a life which must have seemed to them to be no more than a long, protracted process of dying.

The many more young women whose resilience was such that they remained relatively healthy even under the conditions to which they were subjected soon came to harbor feelings of guilt and, not infrequently, an undefined sense of not being as virtuous as their sisters, who were languishing so delicately among them. To put a good front on their continued but suspect feelings of physical well-being, they therefore began to affect illness even when they did not really feel ill.

Many, realizing that a consumptive look in women was thought to be evidence of a saintly disposition, began to cultivate that look of tubercular virtue by starving themselves. Women everywhere, said Woolson, had acquired a taste for a variety of forms of "slow suicide" (209). Indeed, it is very likely that the psychological antecedents of our twentieth-century disease of anorexia nervosa, which gives the sufferer a false sense of virtuous self-control, are to be found in the fad of sublime tubercular emaciation which, as we have seen, began to take on epidemic proportions in the 1860s and has continued to serve as a model of what is considered "truly feminine."

Thus, what may be termed the cult of the "consumptive sublime"—which was characterized, as Woolson pointed out, by "the undeveloped forms, the pallid, flabby faces, the languid air, of the prevailing type of young ladyhood" (209)—was a direct result of the mid-century middle-class woman's cultivated inability to fend for herself in the manner which had still been common among the women who had lived in Defoe's time. "Society," according to Woolson, was "ever doing its best to crush out of them every trace of healthy instincts and vigorous life, and to reduce them to the condition of the enfeebled young ladies that meet us on every side, who are all modelled after one wretched pattern, and as much alike as so many peas" (210).

Art, which rarely shapes but nearly always helps to consolidate and entrench prevailing cultural prejudices, inevitably stepped in to celebrate the cult of the consumptive sublime. Woolson, with her fine, levelheaded capacity to state things as they were, once again accurately described the prevailing fashion and its effect on her contemporaries:

> The familiar heroines of our books, particularly if described by masculine pens, are petite and fragile, with lily fingers and taper waists; and they are supposed to subsist on air and moonlight, and never to commit the unpardonable sin of eating in the presence of man. Longfellow, Tennyson, and the whole tuneful throng, immortalize the maidens of their verse as slender and wand-like, with a step so light that the flowers scarcely nod beneath it. Evidently, fine constitutions, strength of muscle, and hearty appetites, are becoming only in washerwomen and amazons. A sweet-tempered dyspeptic, a little too spiritual for this world and a little too material for the next, and who, therefore, seems always hovering between the two, is the accepted type of female loveliness. No wonder, then, that boarding schools hold the tradition that it is interesting to be pale and languishing and consumptive; that the Venus of Milo spoiled her form by not wearing stays in her youth, and that Hebe's complexion would have been greatly improved by a judicious course of slate pencils and pickles (136).

Death became a woman's ultimate sacrifice of her being to the males she had been born to serve. To withhold from them this last gesture of her exalted servility was, in a sense, an act of insubordination, of "self-will." Woolson poignantly described the fate

II, 2. Frank Dicksee (1853–1928), "The Crisis" (ca. 1891)

which awaited the young middle-class woman who had been so "lucky" as to have found a male:

> Nothing is more sad than to note the rapid fading of young women after marriage. The friends we loved at school, who were full of life and jollity, and seemed born for a splendid, triumphant womanhood, are scarcely recognizable when we meet them a few years after their bright wedding. The abundant wishes for health and happiness showered upon them on that festive day appear to have blessed only the husband. He has grown handsomer and heartier with the lapse of time; but the poor little wife, with her sunken cheeks, lost color and wasted smiles, looks like some heart-sick wraith of the girl we remember (190–91).

The cultural apotheosis of the consumptive sublime, as Woolson clearly indicates, represented the socially ritualized acceptance by the middle-class woman of the prevailing concept that she must transfer the essence of her well-being, symbolically her "jewel," the fragile lily of her virtue, to her chosen mate to help revivify his moral energies. This principle of "spiritual" transference came more and more to be "validated" by the wife's physical, and hence visible (again, in Veblen's terms, "conspicuous"), degeneration.

Nor did women have much choice in the matter. When Tennie Claflin, in her book *Constitutional Equality* (1871), remarked that "at present marriage is the all in all for woman," and saw it as the "end of woman's individual existence" (78), she was not voicing the exaggerated opinion of a confirmed radical but the general consensus of men and women alike. And, as Claflin also pointed out, any attempt on the part of a woman to break out of this straightjacket existence was countered with ostracism: "When wives are brought into active contact with the world, it has been, and still is, to a great extent, the rule to consider her as 'abandoned.' In fact, men make it their special duty to attempt to stigmatize all women who move outside of the specific circle of the wife as 'common women' " (79).

If in the 1870s women such as Claflin and Woolson began to find a receptive audience for their objections to the manner in which women's energies were, in a sense, being aggressively "vampirized" by the males around them, the cultural-ideological counterof-fensive of artists and writers gained in momentum as well. Thus, just when, toward the final years of the century, feminists had become quite vociferous and daring, ideologically charged counterimages of women ill, dying, or already safely dead proliferated.

Sir Frank Dicksee, for one, specialized in these affecting themes. In "The Crisis" [II, 2], exhibited at the Royal Academy in 1891 an older gentleman—a kindly physician, no doubt—thoughtfully contemplates a woman propped up lifelessly against heavy pillows in her sickbed, her features feverish, her body clearly in the throes of a final, probably fruitless, struggle with illness. In another painting, exhibited in 1895, Dicksee presents the viewer with a gentleman seated in the comfortable surroundings of his well-appointed home, listening intently while his daughter sings him the song sung "in the years fled, [by] lips that are dead," as the couplet accompanying the painting goes, while the memory of his dear, departed wife looms in ghostly form behind the seated figure of the daughter.

Leopoldo Romanach's "The Convalescent" [II, 3] had the merit of depicting virtuous poverty as well as a young girl who, given the painting's title, offered the promise to live another day, presumably to try again, at some later date, to be a proper martyr. The young girl as convalescent was, as a matter of fact, a popular subtheme of the genre. Carl Larsson's image of a wan young thing, variously titled "The Invalid" or "Convalescence" [II, 4], is a Scandinavian opus which basked in international popularity toward the end of the first decade of the twentieth century. The British painter A. Chevallier Tayler's "In Sickness and In Health," his contribution to the Royal Academy exhibition of 1900, shows us a husband bending anxiously over his wife, who, flat on her back in bed, her head heavy as a stone among soft pillows, struggles with her disease as best she can. One of the sick woman's hands hangs limply over the side of the bed, while her fingers cling desperately to a rose. Will the flower drop?

Alfred-Philippe Roll's "Sick Woman" [II, 5], in her exhausted, spent, near-death condition, would seem to have felt the full effects of Auguste Comte's positivist "wor-ship" of the feminine. Roll has placed his sufferer against a brick wall, in a brick-paved courtyard, as if the material world, in its barren monotony, had nothing more to offer her.

II, 3. Leopoldo Romanach (1862–1951), "The Convalescent" (ca. 1911)

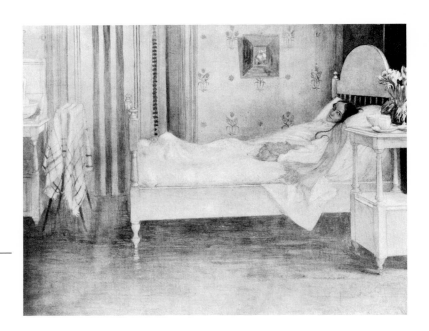

II, 4. Carl Larsson (1853–1919), "The Invalid" (or "Convalescence") (1899)

A fading flower, as surely uprooted from the fertile soil of life as the flowers in and around the decorative marble vase she faces, she can only wait patiently for death to deliver her from the stony prison of reality. In her helpless inanition she is an ideal constituent member of the feminine trinity Comte saw as necessary to attend to the male's soul in health.

For, indeed, in our positivist's "perfect world" the notion of woman as the savior of man's soul had suffered considerable inflation since the early years of Richardson's eighteenth-century "one-on-one" soul exchange, which merely focused on the soul's life beyond life. Comte's system for the recuperation and proper care of the male soul required (because the survival and growth of the human over-soul depended on it) not just one but three feminine victims to work its fructifying magic properly at the husband's "age of full maturity (aet. 42). By that time the mother is generally removed by death; the daughter is alive, her type therefore is objective; the wife may be equally either one or the other." Certainly Roll's "Sick Woman" is not yet quite dead ("subjective" in Comte's terms), but she can't really count as being alive ("objective") either; she is thus a perfect example of Comte's notion of the wife's transitional function between the "objective" and "subjective" types of womanly inspiration. These, if properly cultivated, could effectively "strengthen" the male brain, especially in the particularly pleasing admixture of deadness and aliveness among the trinity of women who were to feed this process, "the subjective element purifying, the objective vivifying it." (*System of Positive Polity*, IV, 98) In any case, Roll's woman gives every indication of having been used up completely by the devastating effects of what one might call the principle of "energy vampirism" underlying Comte's "worship of woman" and her cleansing passivity, which was to take the place of metaphysics.

Woman, Comte had insisted, in yielding up the active impulses within her to her mate, provided him with additional power to triumph in the supreme realm of Force. Thus, as the emissary of "goodness," she contributed to man's power to dominate even as she endeavored to "modify the harshness with which men exercise their authority" (II, 170).

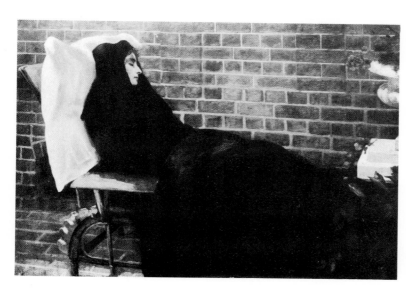

II, 5. Alfred-Philippe
Roll (1846–1919), "The
Sick Woman" (1880s)

Women, being the "originators of spiritual power," should therefore "abstain altogether from the practical pursuits of the stronger sex" (II, 197), especially since "equality in the position of the two sexes is contrary to their nature" (II, 198). Instead, given the "essentially domestic character of female life" (II, 199), women should quietly and unobtrusively "conduct the moral education of Humanity" and do so "free in the sacred retirement of their homes" (II, 204).

Thus, as the "spontaneous priestesses of Humanity" (II, 208) they would assume the role now given to a figure such as Christ in organized religion and save not just man's soul but the soul of humanity from perdition, even while the male blithely continued to wreak his power-seeking havoc in the realm of public life. As a substitute Christ figure, then, sympathy and sacrifice were the realm of "woman's peculiar vocation" (I, 212). In this respect, one could regard Roll's "Sick Woman" as the perfect representative of woman as Christ figure: Her sacrifice was not necessarily to suffer on the cross of metaphysics; instead, it was to "suffer and be still" and to die of self-effacing inanition, so that Comte's males might not feel threatened by any representative of the sex "with whom, from their position in society, he is in no danger of rivalry in the affairs of life" (I, 208).

By 1900, in the real world the cult of feminine invalidism had already begun to give way before the angry denunciations of feminists everywhere. Many women were once again beginning to realize the beneficial effects of exercise and sports. But for painters such as Roll, or Giovanni Segantini, perpetually in search of symbols of the "ideal," the efforts of living women were no more than crude shadows of the fevered light of sacrifice which shone in the eyes of those on the verge of forever. Segantini saw in the image of his sick wife the essence of a rose leaf lying wilting on its pillow [II, 6]. In this painting sentiments of genuine pity and concern merge with evidence of the artist's morbid and erotic fascination with woman as invalid to create what could be considered the ultimate icon of this cult of the suffering, diseased woman as Christ figure. "Her disposition is devout, / Her countenance angelical," Segantini might have explained in Coventry Patmore's words, "the faithless, seeing her, conceive / Not only heaven, but hope of it" (*Poems*, 83).

II, 6. Giovanni Segantini (1858–1899),
"Rose Leaf"

II, 7. Albert von Keller (1844–1920),
"Moonlight" (or "Martyr") (1894)

The images of self-sacrifice provided by painters such as Roll and Segantini were apparently not emphatic enough for others. The Munich-based Albert von Keller, for instance, felt called upon to paint piously prurient images of women who fulfilled the demands of self-sacrifice to an extent beyond even August Comte's most aggressive dreams of feminine submission. The German painter, following in the footsteps of his compatriot Gabriel Max, depicted the Christlike attributes of sainted women in such paintings as "Moonlight" [II, 7], taking an obvious sadistic pleasure in the representation of a vulnerable, naked woman tied to a cross. Religious sentiment is likely to be very far from the mind of anyone seeing this image. Others, such as the Belgian painter Fernand Khnopff, were content to adopt such simple descriptive titles as "A Martyr" for their paeans to female sacrificial submission. Indeed, in a number of works Khnopff combined the fetishized emaciation of the tubercular woman with the theme of crucifixion to obtain double evidence of the condition of true feminine perfection. Already in 1869 Horace Bushnell had asserted that "there is more expectancy of truth and sacrifice in the semi-christly, subject state of women than is likely to be looked for in the forward self-asserting leadership of men" (142).

The esthetic, psychological, and ideological fascination which the theme of the dying or physically spent woman as martyr held for males of the late nineteenth century—and for women who failed to question its validity—is perhaps nowhere expressed more strikingly than in George du Maurier's novel *Trilby*, (1894), in which a purification of male brains through feminine sacrifice analogous to that envisaged by Comte takes place. *Trilby* was without doubt among the most universally admired best-sellers of the 1890s. This novel, which introduced the reading public to the arch-villain Svengali, one of the key creations of anti-Semitism, codified for young women everywhere additional extremes of self-destructive behavior. Its heroine was, as James D. Hart has pointed out,

> much admired by young girls, who made Trilby their model in all but a few of the more questionable aspects of her behavior. Girls by the thousands yearned for feet as graceful as hers, spoke of their own as "Trilbies," wore Trilby slippers, cultivated a so-called "Trilby-type" of beauty, and dressed themselves in Trilby hats and coats (decorated with costume jewelry shaped like Trilby's own foot). They nibbled Trilby chocolates, played Trilby waltzes, and during many a summer evening of 1894 and 1895 lifted their voices to the sad strain of "Ben Bolt," the fifty-year-old lyric revived by its introduction into du Maurier's novel (*The Popular Book*, 194).

The fictional heroine who caused this furor was a young woman of little moral training (she had been brought up among the bohemians of Paris) but with a transcendent instinct for cleanliness and "natural" virtue, or, as du Maurier put it, "a virginal heart," as befitted the daughter of a British gentleman who was himself "the son of a Dublin physician and friend of George the Fourth." However, what made Trilby attractive to the readers of her story, notwithstanding the dubious quality of her virtue, was that she was the epitome of the passive, yielding woman whose only identity came from what the men around her made of her, whether they were imaginative artists, such as the "three nice clean Englishmen" whose adventures dominate the narrative, or the evil Svengali, whose hypnotic powers made her into a conduit for heavenly music. At the end Trilby pays for her sins of the flesh, as inevitably she had to, by dying as the personification of the consumptive sublime.

Trilby is originally gifted with a heavenly voice. At the same time she is—as a woman, as a matter of course, is du Maurier's implication—tone deaf. Eventually she falls into the clutches of Svengali. He hypnotizes her to turn her into a passive nightingale through whom he can express his artistic talents. Trilby soon becomes "weak," "ill," and "languid," to use du Maurier's own terms, as she is physically drained by the effort involved in being the conduit of Svengali's heartless but extraordinary musical insights. When, through a belated and indirect intervention of the novel's trio of well-scrubbed British heroes, Svengali's spell over Trilby is broken, all energy and life have already been drained out of her. Only at this point does she become of more than merely prurient interest to our clean-cut trio. Her deadly illness, in fact, brings virtuous inaccessibility to the woman whom they once feared as a "tall, straight, flat-backed, square-shouldered, deep-chested, full-bosomed young grisette" (32), a woman who, with a frankness never permitted mid-century British women, had bestowed warm favors on the men who loved her "for love's sake." Given this habit, Trilby's heart was inevitably originally wrapped in "a thin slimy layer of sorrow and shame" (34). But under the eyes of our trio of masculine heroes the cleansing power of dying instantly transforms her from a sinner caught in the depths of vice into the very woman of their "clean" British dreams.

As she slowly fades and ages beyond her twenty-three years, they notice with delight how her hands become "almost transparent in their waxen whiteness; delicate little frosty wrinkles had gathered round her eyes; there were grey streaks in her hair; all strength and straightness and elasticity seemed to have gone out of her . . . .'' Physically, du Maurier stresses, Trilby may have become a "wreck," but, he insists, "tuneless and insane, she was more of a siren than ever—a quite unconscious siren—without any guile, who appealed to the heart all the more directly and irresistibly that she could no longer stir the passions" (310).

Thus, now that she has ceased to be a direct sexual threat, Trilby is free to become the true sexless, high-Victorian feminine ideal: the woman who, in her very physical helplessness, makes no further overt erotic demands upon the male, guaranteeing him a restful respite from the energy-draining requirements of sexual involvement and thereby proving to him that even in the 1890s, the heyday of the dreadful "new woman," she could still be the same as she used to be in her mother's day: a comforting emissary from the spiritual realm rather than a dangerous, competing inhabitant of the world of aggression and exchange.

Du Maurier tells us that as she "seemed to be wasting and fading away from sheer atrophy," the insane and helpless Trilby became an object of admiration for the British trio: "Day by day she grew more beautiful in their eyes, in spite of her increasing pallor and emaciation—her skin was so pure and white and delicate, and the bones of her face so admirable!" (317) In *Trilby* we see the apotheosis of an ideal of feminine passivity and helplessness whose tubercular or anorexic presence is still with us virtually unchanged in the fashion pages of magazines and newspapers throughout the "civilized" world. A few years after the publication of *Trilby* Maurice Greiffenhagen offered his admiring public an ecstatic version of the mystical, purified, feverishly sacrificial woman in a late Pre-Raphaelite–inspired "Annunciation" [II, 8]. Carlos Schwabe, in an 1894 concert poster [II, 9] gave an equally ghoulishly accurate "ideal" interpretation of this long-lived creature of the misogynist imagination.

When we are told that Trilby became "tuneless and insane" as she deteriorated into the Victorian ideal of the dying woman, her link with numerous other figures in the pantheon of Victorian virginal heroines on their way to death and transfiguration becomes especially apparent. Indeed, with nineteenth-century opinion loudly clamoring for any "decent" woman to conceive of herself, in or out of marriage, as the vestal maiden lovingly portrayed by Arnold Böcklin in one of his paintings—a young woman with her mouth hermetically sealed in cloth wrappings and her body hidden as in a shroud—it became hard for women not to conceive of themselves as permanently scarred.

In her story "The Yellow Wallpaper" Charlotte Perkins Gilman was to demonstrate the immediate link which existed between the male creation of (and many women's compliance with) the principles of the cult of invalidism, the physicians' encouragement of that cult, and the increasing incidence of madness in women. The male world's resolute refusal to recognize the creative intelligence of women was leading many to desperate attempts to break through the "pointless pattern" of the wallpaper of social constriction, only to find themselves enmeshed ever more tightly in its design:

> Sometimes I think there are a great many women behind, and sometimes only one, and she crawls around fast, and her crawling shakes it all over.
> Then in the very bright spots she keeps still, and in the very shady spots she just takes hold of the bars and shakes them hard.

II, 8. Maurice Greiffenhagen (1862–1931), ''The Annunciation'' (ca. 1897)

II, 9. Carlos Schwabe (1866–1926), concert poster (1894)

And she is all the time trying to climb through. But nobody could climb through that pattern—it strangles so; I think that is why it has so many heads.

They get through, and then the pattern strangles them off and turns them upside down, and makes their eyes white! (30).

Many late nineteenth-century women felt themselves being strangled, felt as if they were losing their minds, caught in the patterns of a society which had come to see even expressions of insanity as representative of feminine devotion to the male. Under the influence of poets such as Tennyson and Patmore, the reading public was being swept up into an ever more attenuated glorification of those women who demonstrated such an intense need to be allowed to sacrifice themselves to chivalrous males that the knowledge of failure, or their male's lack of interest in them, was more than enough to drive them mad. Indeed, such transcendentally insane heroines as Tennyson's Lady of Shalott, Elaine, and Mariana kept the Pre-Raphaelites and other British painters of the later nineteenth century busy finding new ways to depict the madness of a woman in need of a man to whom she might sacrifice herself.

In ''The Lady of Shalott'' Tennyson had, as early as 1832, pointed out that insanity and death were the likely outcome for a woman whose sacrifical urge remained unrequited. In Tennyson there is always an (unexpressed) element of desperate erotic longing in the women who do not get their men. To the poet it was clear that the sacrificial impulse in woman was what he might have called, in the language of the time, the ''sex-impulse,'' turned to civilized use and made subservient to woman's role as housekeeper and resident polisher of the male soul tarnished by the ''struggle for existence.'' Woman's attempts to

II, 10. William Holman Hunt (1827–1910), "The Lady of Shalott," preparatory design (ca. 1857)

fulfill her sacrificial duty became all the more pleasingly pathetic in Tennyson's eyes because, basically, except for her ceremonial duty as the nominal keeper of virtue, she was obviously expendable in the male world of "serious" struggle and achievement. Thus, in "The Lady of Shalott" the nameless lady of the title "weaves by night and day" on her "silent isle," and as she looks "through a mirror blue," symbolic of her static, passive existence, upon the world of Camelot outside, she sees "young lovers lately wed" and realizes that "she hath no loyal knight and true, / The Lady of Shalott." This realization stirs "the curse" of sensual awakening in her, making her aware that she has no male to "melt into," in the fashion of Wagner's Elsa.

As a result, she easily succumbs to the manly charms of Sir Lancelot, whom she sees in her mirror as he rides by on the mainland like "some bearded meteor." The effect of her glimpsing this paragon of masculinity is instantaneous: "The mirror cracked from side to side; / 'The curse is come upon me,' cried / The Lady of Shalott." Unfortunately for our lady, as Elaine was also to discover to her misfortune, she happened to have picked the most inaccessible of Tennyson's Arthurian heroes, a man who already had his hands full with Guinevere. But, undaunted and crazed with longing, our lady hopped into a boat, trying to get to her hero before "her blood was frozen" by the curse of passion which had come upon her. However, without an immediate opportunity to let herself "fuse" with her man and gain vital energy from his male "flame," she was not likely to survive long— something she obviously intuited right away, for "down she lay" even while getting started on her watery voyage to Camelot. And, indeed, "ere she reached upon the tide / The first house by the waterside, / Singing in her song she died," another tender victim to woman's need to be absorbed by a man. Her expendability in the world of the male is thereupon nicely expressed by Lancelot, who, when as "a gleaming shape she floated by / Dead-pale between the houses high,'' gives her a passing glance and appreciatively comments, ''She has a lovely face,'' and goes about his knightly business (*Poems and Plays,* 26-28).

Late nineteenth-century painters loved Tennyson's combination of incipient madness, self-destructive, passive yearning, and a beautiful dead woman floating downstream. William Holman Hunt, for one, did all he could to catch the nuances of the poet's affecting narrative. Working diligently on various versions of this theme between 1850 and 1905, he depicted the Lady of Shalott at the moment her ''mirror crack'd'' and the ''curse'' of passion came upon her, a mad longing to merge with the image of Lancelot whipping her body into a frenzy and causing her hair to stand on end as if charged by the electric shock waves of her need [II, 10]. John William Waterhouse, too, showed the lady in the throes of her curse, although he made her look rather more glowering and threatening, ready to pounce on the viewer—something Tennyson clearly had not intended. A few years earlier, in 1888, Waterhouse had still shown her floating in her boat [II, 11], far more decorously wan and helplessly insane, even though he had painted her sitting up, not lying down as she is described in Tennyson's poem.

Much as the painters loved ''The Lady of Shalott,'' they loved Tennyson's ''Elaine'' more. In ''Elaine'' Tennyson had created a nearly perfect verson of the ''tuneless and insane'' feminine martyr. Like the Lady of Shalott, ''the lily maid'' Elaine, having laid eyes on Lancelot, at once becomes a devoted, yearning follower of this honorable but not overly constant paragon of male muscular virtue. In *The Idylls of the King,* where Tennyson tells her story, Elaine is bent on keeping ''the one-day-seen Sir Lancelot in her heart.'' Wishing, as she says to her father, ''to be sweet and serviceable'' (379) to the knight of her dreams as he lies wounded, she goes to him, ostensibly to bring him, in a neat bit of symbolism right out of *Pamela,* the diamond which he had won at the tournament in which he was hurt (while wearing her ''favour at the tilt''). Lancelot, receiving the diamond from Elaine, ''kiss'd her face, / as we kiss the child / That does the task assign'd.'' Elaine, however, mistaking this fatherly gesture for something more adult, ''slipt like water to the

II, 11. John William Waterhouse (1849–1917), ''The Lady of Shalott'' (1888)

floor,'' no doubt weak-kneed in anticipation of the glorious opportunities of ''service'' to follow. And she soon gets her chance, for while Lancelot, who is in truth badly wounded, waxes ''brain-feverous in his heat and agony'' and seems ''uncourteous'' to her in that state,

> . . . the meek maid
> Sweetly forebore him ever, being to him
> Meeker than any child to a rough nurse,
> Milder than any mother to a sick child,
> And never woman yet, since man's first fall,
> Did kindlier unto man, but her deep love
> Upbore her; till the hermit, skill'd in all
> The simples and the science of that time,
> Told him that her fine care had saved his life (380–81).

Lancelot, however, remains unimpressed by Elaine's to-do over him, and when, bursting with self-sacrificial energy, she confesses to him, ''I have gone mad. I love you: let me die,'' thus succinctly expressing the range of amorous sentiments proper to a nineteenth-century ''lily-maid,'' Lancelot in effect replies that he doesn't much feel like getting married and that if he did, he certainly wouldn't think of her: ''Had I chosen to wed, / I had been wedded earlier, sweet Elaine: / But now there never will be wife of mine'' (382). Elaine, ''deathly pale'' thereupon proceeds to faint impressively, crying, ''I needs must follow death, who calls for me,'' and is carted off to her father's house, where she makes careful arrangements for her self-evidently imminent demise. She writes Lancelot a letter and tells her father to ''lay the letter in my hand / A little ere I die, and close the hand / Upon it; I shall guard it even in death'' (384). Much of the rest of Elaine's story is a

---

II, 12. Toby Rosenthal (1848–1917), ''Elaine'' (1874)

repeat performance of the Lady of Shalott's doings, except that Elaine instructs her father to hoist her, deathbed and all, on a barge with a "dumb old man" as oarsman, and to let her float down to Camelot. Once properly dead, she is indeed sent on her way in this fashion, with "in her right hand the lily, in her left / The letter—all her bright hair streaming down" (385).

Lancelot, a little world-weary from the fuss, sees the dead young woman floating toward him, is handed the letter to him which she clutches in her lifeless hand, and reads her last words: "I loved you, and my love had no return, / And therefore my true love has been my death." Lancelot thereupon acknowledges that she did what she could, remarking, appreciatively for once, "good she was and true, / But loved me with a love beyond all love / In women, whomsoever I have known" (387). Still, he concludes, those things can't be forced, for "to be loved makes not to love again." But this time, it must be said, Lancelot is a bit uncomfortable about his own role in the matter, and his world-weariness deepens—only to have Tennyson rush in to assure the reader that Elaine's self-sacrifice was not in vain, for as our hero laments his own illicit liaison with the queen, "these bonds that so defame me," the poet hastens to tell us: "So groan'd Sir Lancelot in remorseful pain, / Not knowing he should die a holy man" (389). Obviously, Tennyson would have us believe that the passionate love-deaths of a few desperate sacrificial maidens can only enhance a man's standing in heaven.

The part of Elaine's pathetic story which inevitably attracted the painters most was the dead lily-maiden's journey downstream for her last meeting with the lover in whose service she had sought death. Of the innumerable renditions of the scene, that by the American Munich-based painter Toby Rosenthal, painted in 1874 [II, 12] was one of the most popular. It was widely reproduced both in the United States and in Europe, for its deft visualization of a woman, pale and lifeless, floating on water, surrounded by flowers, and guided by an oarsman who looked like death personified, could not fail to thrill the artistic sensibilities of art lovers everywhere. The public rejoiced in the sight of a beautiful woman in love, safely dead, and hence not likely—in her perfectly self-evident state of extreme self-sacrifice—to complicate the emotional life of the viewer any more than Elaine had complicated that of Sir Lancelot. When the painting was brought to San Francisco and exhibited, it created something of an "Elaine" cult, itself becoming the subject of a sensational art theft, only to be speedily recovered from the thieves, who had hoped to get twenty-five thousand dollars for the work—not an insignificant sum even now, and certainly a king's ransom in 1875.

Inevitably a subject which could stir so much public enthusiasm was to remain a favorite with painters. Pale and dead Elaines floated toward oblivion year after year at the Royal Academy—and not infrequently at Continental exhibitions as well. Gustave Courtois omitted the oarsman and made it look as if his Elaine was merely sleeping; others, such as Blair Leighton in his Royal Academy painting of 1899, tried to give the scene a "genuine" medieval look, making the oarsman an old, bearded man of the people; but most, like Ernest Normand in his entry of 1904 to the Royal Academy, preferred to follow Rosenthal's lead and milk the theme of a dead maiden accompanied by an oarsman, who was death personified, for all it was worth. John Atkinson Grimshaw's version of 1877 is one of the few which varied markedly from the iconography established by Rosenthal. In his painting, a peculiar fantasmagorical proliferation on the horizon of needlelike pseudo-medieval towers at nightfall, the equally angular dragon decoration of Elaine's boat, and the shadowy, devilish outline of the oarsman together create the impression that this Elaine

II, 13. Arthur Hughes
(1832–1915), "Ophelia"
(1852)

II, 14. Sir John Everett
Millais (1829–1896),
"Ophelia" (1851)

is not so much the victim of self-immolating fervor as the helplessly sacrificial victim of
some unspeakable, satanic rite. In this painting, in other words, the sadistic-aggressive
impulses underlying Tennyson's narrative, and his evident delight in Elaine's passionate
self-sacrifice, were brought to a level of visual equivalence unmatched by any other major
late nineteenth-century rendition of the theme.

However, even Tennyson's heroines had to yield in popularity to Shakespeare's Ophelia,
the later nineteenth-century's all-time favorite example of the love-crazed self-sacrificial
woman who most perfectly demonstrated her devotion to her man by descending into
madness, who surrounded herself with flowers to show her equivalence to them, and who
in the end committed herself to a watery grave, thereby fulfilling the nineteenth-century
male's fondest fantasies of feminine dependency. Though a significant secondary character
in Shakespeare, Ophelia was far more important than Hamlet to the nineteenth-century
ideologues of female sacrifice. She was first brought center stage by the Pre-Raphaelites,
who saw in her the same feminine qualities they so admired in Tennyson's heroines.
Spurred by their renditions of Ophelia's flowery madness and watery death, her story soon
became the one theme no self-respecting turn-of-the-century painter could avoid depicting
at least once. Arthur Hughes and Sir John Everett Millais created the principal iconographic details for her portrayal at mid-century.

We find Arthur Hughes' Ophelia [II, 13], painted in 1852, at the edge of the brook where Shakespeare placed her. In a state of madness and anguish, she has crowned herself with reeds as she watches the flowers she drops into the water float away in anticipation of her own imminent fate. She is emaciated and tubercular and therefore has all the requisite attributes of the icons of illness. Consumptive fever has heightened the contrast between the pallor of her skin and her red lips and the deathlike shadows around her eyes. In the issue of *The Art Journal* in which the engraving here reproduced was first published, an enthusiastic commentator remarked on Hughes' singular success in bringing a "look of vacancy" into Ophelia's "sweet, child-like face" (332).

Millais' even more famous "Ophelia" of 1851 follows her journey into death: woman and water united forever in a passive voyage to eternity among the reeds [II, 14]. Millais' dead Ophelia, floating prettily but uselessly in the water, and Hughes' mad Ophelia, showing her weakness and lack of control over her own fate, her touching expendability and her inherent debility, gradually became the models for a host of other analogous treatments of the theme by British painters such as Richard Redgrave, Henrietta Rae, and Louise Jopling, the latter basing herself, in her 1892 version, on Mrs. Tree's crazed-eyed, emaciated-looking stage Ophelia. John William Waterhouse was especially drawn to the Ophelia theme. He returned to it time and again, depicting her variously as rolling madly in a field, a flower toppled off her stem and seeking to regain the balance of nature, as in his 1889 version; strolling weirdly near the water's edge, having made her skirt into a basket piled high with picked flowers (1894); or, finally, sitting among the reeds, staring vacantly into herself just before the stream will claim her and carry her away (1910). George Frederick Watts, too, placed his Ophelia, a pale and strangely lurking hollow-eyed creature, among the rushes, staring vacuously into the water; and in 1909 Alice Pike Barney put her, Millais-like, among water lilies, only her head still above the water level, eyes glaring, and a strange, crazed grin playing about her mouth.

On the Continent, Tony Robert-Fleury sent his profile of a much better fed, almost Germanic Ophelia to the Paris Salon of 1887, while Elie Delaunay had our wide-eyed heroine wear an intricate crown of flowers, nestled like a still life in her long, flowing hair. Ernest Hébert's portrayal of Ophelia was closer to the tradition established by the British Pre-Raphaelites [II, 15]. To accompany a reproduction of this work, the editors of the popular French magazine *Je Sais Tout* wrote an especially telling caption: "This is truly that helplessly abandoned ideal creature, whose hallucinating eyes see nothing more than what is within, and who, hair loosened and streaming down, will in a few moments enter gently into the stream which will carry her—a cut flower among other cut flowers—away to that world beyond whereof her madness is already an expression" (vol. 3 [1907], 2nd sem., 673).

In 1910 Adolphe Dagnan-Bouveret, anticipating Hollywood's sense of melodrama (for which this painter's imagery was a source of inspiration), exhibited at the Salon des Beaux-Arts an Ophelia who seemed to have wandered out of a provincial dramatization of a medieval costume-drama into an unfamiliar part of the forest, and who now found herself, flowers in hand, terrified at her dilemma and unable to figure out how to get back to the village by curtain time [II, 16].

Madeleine Lemaire, considered by many of her contemporaries to be one of the most remarkable women painters of all time (obviously not least because she had quite effectively and uncritically adopted the thematic preoccupations of her male colleagues), managed to combine in her "Ophelia" [II, 17] a number of popular elements of the theme.

II, 15. Ernest Hébert (1817–1908), "Ophelia" (1890s)

II, 16. Adolphe Dagnan-Bouveret (1852–1919), "Ophelia" (ca. 1910)

She depicted Shakespeare's heroine in the precarious, tottering stance introduced a century earlier by Sir Joshua Reynolds in certain of his full-length portraits of society ladies. In addition, she placed her, as was generally customary, among the reeds and flowers at the water's edge. But what was far from customary was that she made Ophelia leer with the glowering light of a vampire in her eyes, thus emphasizing the sexual origin of her madness—an aspect further accentuated by the very undecorous fashion in which her dress has slipped off her shoulders to reveal her breasts. Male painters, in contrast, preferred to show Ophelia fully clothed to emphasize the heroic nature of her choice of madness and death over a state of dangerous arousal.

Germanic artists were also fond of the mad Ophelia. In 1871 Hans Makart gave his rendition a stagey quality analogous to Henrietta Rae's medievalized "There's Rue for You." The mysterious druid priestess of Fritz Erler (1900) and Georg Richard Falkenberg's depiction of a modern woman wracked with nervous self-doubt and obsessive introspection [II, 18] were both promptly reproduced in leading magazines. Back in France, Georges Clairin, Sarah Bernhardt's favorite painter and her constant mythologizer, painted an "Ophelia Among the Nettles" in which hair, nettles, and grayish leaves surround a dead-eyed head which seems to be growing out of dusty soil. The inspiration for this work was, as always, the "Divine Sarah," whose sharp, pinched features were exactly those of Clairin's Ophelia. Among her many talents, Bernhardt could also pride herself on being a fine sculptor: Witness her beautifully executed bas-relief, exhibited in the Women's Building at the Chicago World's Fair of 1893, which presents an overhead view of the ubiquitous Ophelia as she floats downstream, flowers still tangled in her hair and trailing along her breast, which like Lemaire, she had chosen to reveal. Bernhardt made Ophelia's hair weave itself into a pattern that echoes the ripples of water which surround her dying flesh [II, 19].

Sarah Bernhardt, of course, as this relief clearly indicates, was all too well aware of the fascination which the theme of the weak-witted, expiring woman exerted over the males of her time, and she used this knowledge wherever she could to cultivate her own eccentric image, not least by carting about with her on her *tournées* what she described as her own coffin, and by letting it be known that she was not averse to sleeping in this rather narrow cot to emphasize the (to her admirers evidently delicious) fact of her own mortality. Photographs of her lying thus in near-excelsis did the rounds of afficionados and appropriately thrilled men everywhere. [II, 20] In his autobiography *Time Was,* Walford Graham Robertson describes a scene which took place about 1887 at Count Robert de Montesquiou's Paris rooms and which gave him the thrill of a lifetime:

> One day, while exploring amongst the curious objects strewn about the rooms, I came upon a tiny photograph in a black frame, the photograph of a dead girl in her coffin. There was nothing of horror in the picture, only pathetic loveliness, and I felt that I had surprised some tragic secret as I hastily replaced the frame and Comte Robert looked up quickly.

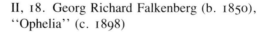

II, 17. Madeleine Lemaire (1845–1928), "Ophelia" (1880s)

II, 18. Georg Richard Falkenberg (b. 1850), "Ophelia" (c. 1898)

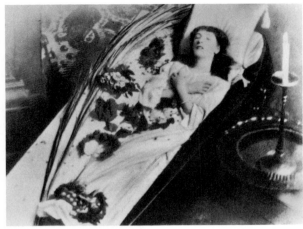

II, 19. Sarah Bernhardt (1844–1923), "Ophelia," bronze bas-relief (c. 1890)

II, 20. Sarah Bernhardt, the coffin portrait, photograph (1870s)

> "You know her, don't you?" he asked.
> "No," I whispered. Did I ever see her? Could I have known her when she was alive?
> "She isn't dead," he said smiling. "That is Sarah Bernhardt."(107).

Bernhardt herself had, in her youth, displayed all the symptoms of emaciation and terminal illness characteristic of the cult of invalidism and tubercular weakness, even to the point of coughing up blood, and at fifteen she had been given only a few years to live. Bernhardt's case shows how much the very nature of the cult of invalidism was psychologically linked to women's desire to be given some individual attention, some personal recognition as individual entities, rather than be seen as generalized "flowers of virtue." Her sense of drama clearly found an early outlet in her youthful illness (which, for all that, was no doubt quite as real as the illness from which Alice James ultimately did die). It was most likely Bernhardt's ability to deflect her desire to be recognized as a person into her theatrical aptitude which saved her from Alice James' fate. Her dramatic doings with that infamous coffin were, given the late nineteenth-century's fascination with the notion of feminine insanity and the cult of the woman as corpse, truly flawless responses to the cultural taste of the times. What was a woman who slept in her own coffin if not insane? The emaciated features she carried with her, like her coffin, virtually throughout her life could have been modeled on those of Arthur Hughes' Ophelia, and what was Ophelia if not the late nineteenth-century's favorite madwoman? Nothing is more indicative of the singular popularity of Shakespeare's self-effacing heroine than the fact that around 1890 the Parisian cosmetics firm of Houbigant sought to create massive interest in its latest facial powder by calling it "Poudre Ophélia." The new product was widely advertised as a true "talisman of beauty." Presumably a less than self-effacing woman might, by making use of the powder, create at least the outward appearance of being as decorously pale and fragile as any true Ophelia.

It was, of course, not merely Ophelia's mental and physical debility that kept the painters busy. Other women's weaknesses and insanity could serve quite as well. "Isabella and the Pot of Basil" was another favorite theme, pursued diligently by Pre-Raphaelites such as William Holman Hunt, who were fascinated with the (to them apparently delight-

II, 21. John White Alexander (1856–1915), "Isabella and the Pot of Basil" (1897)

II, 22. Noé Bordignon (1843–1920), "Matelda" (ca. 1890)

fully morbid) notion—taken from Boccaccio via Keats—of a love-crazed woman who, in the words of a late nineteenth-century critic, "nourished her basil-plant from the brain of the lover assassinated by her merciless brothers," and had planted this very fertile source of the male imagination for that very purpose in the pot she is always seen hugging in late nineteenth century paintings. There were, as we shall subsequently see, some very unvirtuous reasons for a woman to covet a man's severed head, but in Isabella's case, since it was widely understood that she would have preferred the whole man to the fetishized part-object, there was good reason to sympathize with her plight. Still, toward the end of the century such a painter as John White Alexander, in his 1897 version of the theme [II, 21], could show that he was very well informed about suggestions fashionable at the time concerning the sort of fertilizer that was represented by a man's brain. In depicting Isabella as a hollow-eyed, emaciated snakelike vampire-creature, Alexander seems to have equated her quite directly with such turn-of-the-century headhunters as Salome and Judith.

In the wake of Ophelia's success as a vehicle for the representation of female madness, artists and writes diligently combed world literature and classical mythology for other

II, 23. Frederick Sandys (1829–1904), ''Cassandra'' (ca. 1895)

satisfying examples of the decorously self-sacrificial insane woman. Noé Bordignon found her in Matelda, from Dante's *Inferno,* although from the looks of it she would merely seem to be Ophelia in disguise [II, 22]. Cassandra was another favorite. She was usually shown, as she is in Frederick Sandys' rendition, stalking about wide-eyed and desperate, presumably because no one would listen to her—itself a by no means unusual experience for a woman in the nineteenth century [II, 23].

A literary pretext, however, was, with the advent of the twentieth century, not considered necessary any longer. Pierre-Georges Jeanniot's ''The Madwoman'' [II, 24] is quite straightforwardly the portrayal of a deranged woman, although the ''literary'' connection is maintained by the artist by having this woman, Ophelia-like, hold a rose. The painting is one of those extraordinary ''documentary'' images characteristic of turn-of-the-

II, 24. Pierre-Georges Jeanniot (1848–1934), ''The Madwoman'' (1899)

century official art whose narrative power is undeniable but whose emotional impact is made ambiguous by the morbidity and exploitative prurience of its focus.

The image of the dead, floating Ophelia, the self-sacrificial female corpse, was often adapted to other appropriate narratives of woman's taste for martyrdom. In Paul Delaroche's "The Christian Martyr," for instance, she became a sainted woman, floating elegantly into sacrificial oblivion, her hands tied very prominently to emphasize her absolute helplessness.

Others adapted the theme of the floating woman to the fashionable mode of fake historical orientalism, as Federico Faruffini did in his "The Virgin of the Nile." Ingeniously combining the Ophelia theme with that of the female crucifixion, he showed his tantalizingly bare-breasted virgin floating downstream, draped supinely on a cross and surrounded by flowers. All this was presented against a soon-to-be-seen-in-Hollywood architectural backdrop of monumental angularity, consisting of vaguely antique riverside ramparts populated with obviously no longer virginal odalisques and bevies of industriously harp-playing houris.

A major artistic principle at the turn of the century was that if something seemed to catch the viewers' attention, more of that something in a single image was unquestionably better. Pursuing this aesthetic theory with admirable consistency, Henry A. Payne, in "The Enchanted Sea," offered his viewers an obscure combination of the themes of the Lady of Shalott, Elaine, and Ophelia in a single image by painting a sea which was literally clogged

II, 25. Henry A. Payne (1868–1940), "The Enchanted Sea" (ca. 1898)

with floating women. Navigating this sea, the artist clearly demonstrated, was an extremely hazardous, and certainly excruciatingly morbid, endeavor [II, 25].

Always, however, whatever the narrative excuse, representations of beautiful women safely dead remained the late nineteenth-century painter's favorite way of depicting the transcendent spiritual value of passive feminine sacrifice. Once a woman was dead she became a figure of heroic proportions, and for such heroines the nameless bourgeois suffering of the consumptive housewife so characteristically depicted by Roll and Dicksee was no longer appropriate. Thus, paintings of dead women continued to be most usually associated with the portrayal of famous sacrificial heroines from literature and classical mythology, even when it was clear that the principal object of the painters' fascination was the generalized subject of "the death of a beautiful woman."

Aside from Ophelia, other Shakespearean heroines (especially Juliet) and operatic heart throbs such as the terminally consumptive Mimi (from Murger via Puccini's *La Bohème*) and Manon Lescaut (according to Massenet) were favorite subjects, making their appearance year after year on the walls of the great exhibition halls. Americans such as William Deleftwith Dodge and Eanger Irving Couse gave the theme a sense of the picturesque and of local color by treating Indian subjects such as "The Death of Minnehaha." In Germany these themes were matched by painters such as Hans Thoma in patriotic renditions of the thoroughly Aryan death of Brünnhilde. In England Lord Leighton had Hercules, in an imposing display of rippling muscles, struggle with Death for the body of Alcestis, who was all the while lying demurely helpless and oblivious in the center of the painter's oversized conception of the scene. Chateaubriand's Atala was buried by a wide range of painters, from the Brazilian Rodriguez Duarte to patriotic Frenchmen such as Gustave Courtois.

James Bertrand was among the many who showed Bernardin de Saint-Pierre's excruciatingly modest heroine Virginie washed ashore [II, 26], and he did such an affecting job of it that when the original was snatched up by the French state for display at the Luxembourg, the artist was importuned to paint several replicas for admiring American art lovers. The Austrian painter Maximilian Lenz, editor of *Ver Sacrum*, sent a whole series

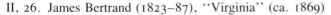

II, 26. James Bertrand (1823–87), "Virginia" (ca. 1869)

of emaciated feminine corpses in search of eternity, and Jean-Jacques Henner had "The Levite of Ephraim" study the prominently displayed prostrate nude body of his dead wife on a wall of the Paris Salon of 1898.

As the fin-de-siècle painters pursued the theme of the dead woman with ever greater determination, commentators on their work also became more deeply involved in descriptions of a dubious nature. For instance, Cecil van Haanen's portrayal of the dead Juliet, exhibited at the Grosvenor Gallery in 1885, inspired the following comment by a reviewer for *The Magazine of Art:* "It is a sombre and powerful conception, firmly handled, solidly painted, and impressive in effect. The dusky flesh of the swooning face is broadly and powerfully modelled, and the tone-values of the varying textures of the flesh are admirably rendered. Amid the white swathed draperies and the great coils of lustrous black hair, the massive, powerful face is wrought in telling relief, and full of solemn significance" (466).

Owen Meredith's poem "The Earl's Return" tells of a sad mismatch between a grim aristocrat and another of the author's famed paragons of feminine virtue. This young lady is so virtuous that the very return of her earl after a long absence and the potential carnal implications of his imminent presence scare the poor girl quite literally to death: "The staggering light did wax and wane, / Till there came a snap of the heavy brain; / And a slow-subsiding pulse of pain; / And the whole world darkened into rest" (350). The scene of a fragile young woman "darkly filled with sleepy death," as Meredith put it, was simply irresistible to painters. Sir Joseph Noël Paton composed "The Dead Lady" [II, 27] even before Millais sent Ophelia floating downstream. Paton's work, combining moon, stars, and a dead woman in an elegantly elongated pose of terminal rest, brought together a set of iconographic details which was to be exploited endlessly throughout the rest of the century. It is, in this respect, interesting to juxtapose Paton's composition, and its very traditional narrative methodology, with such a painting as Romaine Brooks' "Le Trajet" of 1911. [II, 28] The stylistic differences between the two paintings are certainly remarkable. Brooks' brilliantly modernist gestures of planar simplification allowed her to reduce the visual constituents of her composition to their most dramatic essentials. The bed on which her subject lies becomes like a raft, its sheets like the fins of a strange, airborne, sharklike fish. In terms of style Paton's and Brooks' works are worlds apart. Yet in terms of basic compositional structure they share far more similarities than differences. Brooks' work is, in a sense, a version of Paton's painting stripped of inessential detail. Less cluttered and, to the modern viewer, much more efficient in conveying its emotional impact, it is nonetheless morbidly similar in its frame of reference: the dead woman as object of desire.

The viewer of Paton's work, it is true, has two narrative options, while the viewer of Brooks' work has only a single choice. For, in identifying his or her narrative place in Paton's painting, the viewer can take either extreme of the mid-century dualistic concept of male-female relationships: One can take the male's position and experience the aggressive-sadistic sentimentality of the dominator for whose sake the woman has died, or one can opt for the passive-masochistic pleasure of self-sacrifice centered in the figure of the dead lady. In Brooks' work, however, the viewer is invited to become quite single-mindedly the prostrate subject: Its impact is virtually exclusively centered in the masochistic response of the viewer. Indeed, the very absence of any possibility to move beyond the visual enclosure of Brooks' figure in darkness, as well as the loss of any suggestion of narrative alternatives to the viewer's identification with the subject in the painting, forces that subject into the center of the viewer's masochistic self-perception as a passive entity

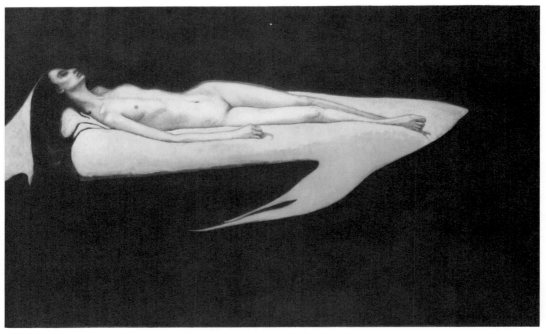

II, 27. Joseph Noël Paton (1821–1901), "The Dead Lady" (c. 1850)

II, 28. Romaine Brooks (1874–1970), "Le Trajet" (or "Dead Woman") (ca. 1911)

waiting to be acted upon, as a willing victim in a world in which aggression is the privilege of the other.

Thus, Brooks' painting becomes not a critique of the turn-of-the-century's victimization of woman as passive object but an emotionally charged expression of the manner in which women attempted to transform their passive position in this society as manipulated objects into the illusion of an active participation in their domination through a supposedly self-elected ideal of physical invalidism and consumptive fragility. To Brooks and to the viewer, her subject, Ida Rubinstein, is an object of desire because she represents an inverted ideal of personal control. The hollow eyes and the anorexic emaciation were well-known and well-publicized features of Rubinstein's dramatic personality. Through her ability to personify the cultural ideal of woman's subjection, Rubinstein was seen by her contemporaries, especially by her women friends, to have gained a position of extraordinary power as, one could say, an ambulant fetish expressive of the ideologically manipulated desires of that society. By depicting Rubenstein prostrate, floating whitely in a field of self-referential darkness, Brooks appropriated Rubinstein's identity as cultural fetish to herself and to the viewers of her painting.

But by doing so Brooks, in effect, also helped feed rather than combat a male fantasy whose precise implications were neatly expressed by the critic Ezra Tharp as he waxed lyrical in the March 1914 issue of *Art and Progress* about the whispy, ghostlike women of Thomas Wilmer Dewing [see V, 1]. "The thinness of Dewing's women," declared Tharp, "is part of their modernness—thinness being a modern and an American ideal. There's no other animal one wants to see thin, except a dog." After having thus indicated succinctly where he placed woman on the evolutionary scale, the critic continued with sensorious arrogance toward the general public: "One is impatient when people,—even when people who ought not to know any better, lament the thinness of his women, for after all there is nothing so handsome as a skeleton, as the drop and set and hang of the bones" (156). Clearly, the fetishized emaciation of iconic figures such as Rubinstein made it possible for males to respond to them in either a sadistic or masochistic fashion, depending on whether they were seen as subjects in control of their own destiny (and hence a threat to the aggressive self-identity of the men observing them) or as ultrapassive objects of aggressive desire. Tharp, for his part, left no doubt concerning the message he read in Dewing's paintings: "Among women and flowers, he cares for one type—puts one long-stemmed graceful flower in a beautiful vase" (159).

However, the principal progression of suggestive effect from Paton's mid-nineteenth-century narrative painting of a "Dead Lady" to Brooks' early twentieth-century version of the same subject resides in the manner in which the range of subjective options had narrowed for the latter painting's audience. As we shall see, this is a characteristic pattern in the development of late nineteenth-century art, a development which, to a large degree, was a reflection of the middle-class male's changing perception of the nature of his position in the arena of economic domination.

The aesthetic pleasures attendant upon the contemplation of feminine mortality continued even so to fascinate painters throughout the historical period under scrutiny. The German painter Herman Moest's "The Fate of Beauty" [II, 29] occupies, in both date (1898) and narrative focus, something of a middle ground between the works of Paton and Brooks. The straightforwardly male-sadistic narrative point of view, a virtually polar opposite to Brooks' image, was still operating largely unchanged in the work of most turn-of-the-century painters, although it may be true that such a work as Paul-Albert Besnard's

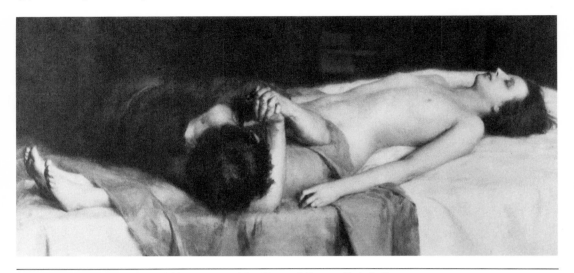

II, 29. Hermann Moest (1868–1945), ''The Fate of Beauty'' (1898)

''The Dead Woman'' [II, 30] represents a stage of development in which the aesthetic pleasure in female suffering of earlier generations of artists had begun to mingle with a morbid appeal to the viewer's participatory self-projection into that suffering. But if in Besnard's image the agonized, contorted face of the dead woman speaks of extreme pain and not of the fetishized feminine ''beauty in extremity'' so dear to the nineteenth-century ideologues of male domination, the image nonetheless represents the aggressive appropriation of a woman's suffering. Under the guise of documentary concern, Besnard has robbed this woman of even the dignity of privacy on her death bed. In consequence, she must forever remain ''Besnard's 'Dead Woman,' '' not an individuated person. The painter, in coolly documenting her at this moment, has effectively obliterated her personal identity, turning her into another crushed Ophelia, helplessly and anonymously fading into her pillows, as if these were the waters of oblivion.

In a number of other works, moreover, Besnard made it all too clear that his fascination with dead and dying women was mingled with a morbid eroticism. For instance, an etching of his called ''The Dying Woman'' [II, 31] combines the expressionist/documentary approach with a realistic study of the woman's still voluptuous torso to create an image full of sadistic ambiguity. In this respect Besnard's works are Gallic counterparts to those of Albert von Keller, who, with characteristic Germanic insistence on accuracy of detail, visited the morgue to paint the bodies and facial expressions of dead women [II, 32] so that he might more successfully depict the extreme religious joy of the nuns he liked to portray at the very moment of their ecstatic self-sacrifical demise. Hans Rosenhagen, in a monograph on Keller published in 1912, reported that the artist had determined, as a result of his field research, that at least ''in the case of the girls and women who had died a natural death,'' if one ''studied their faces intensively one could see them take on an expression of pain made so noble and almost so sympathetic by their suffering that it allowed an otherworldly happiness to shine through which could often only be compared to the miraculous expression of a woman who is in love to the point of ecstasy'' (*A. von Keller*, 55). Keller's repulsive observation shows how literal an equation late nineteenth-century males made between virtuous passivity, sacrificial ecstasy, and erotic

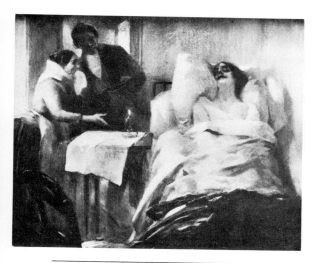

II, 30. Paul-Albert Besnard (1849–1934), "The Dead Woman" (1880s)

II, 31. Paul-Albert Besnard (1849–1934), "The Dying Woman," etching (1885)

II, 32. Albert von Keller (1844–1920), "Study of a Dead Woman" (1885)

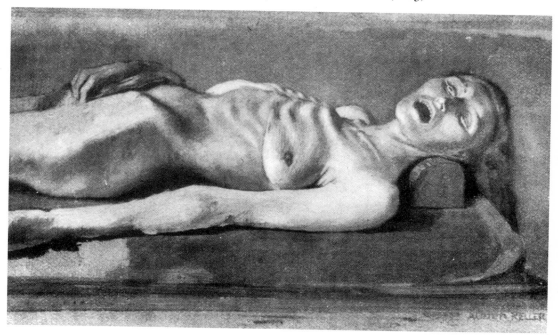

55

death as indicative of "feminine fulfillment." Even those women whose "animal energies" made them threatening, active forces while alive could be brought back into the realm of passive erotic appeal by painters who chose to depict them safely dead.

Most painters of the late nineteenth century, however, did not bother much with the pseudo-expressionist ambiguities of Besnard and Keller. If they were not illustrating how death rewarded the saintly passivity of literary heroines, painters everywhere liked to portray the deadly virtue of sainted women. "St. Eulalia" [II, 33] John William Waterhouse's contribution to the Royal Academy exhibition of 1885, filled *The Magazine of Art*'s reviewer with lugubrious admiration: "There are many ways of presenting martyrdom," he stressed, "the act and its significance, the literal scene with its brutality of fact, and the final consummation in all its religious and poetic glory." After a short history of martyrdom in art, he waxed lyrical over Waterhouse's achievement, lingering lovingly on the painter's treatment of the dead body of the female saint:

> The artist's conception is full of power and originality. Its whole force is centered in the pathetic dignity of the outstretched figure, so beautiful in its helplessness and pure serenity, so affecting in its forlorn and wintry shroud, so noble in the grace and strength of its presentment. The tone of the dark, almost livid flesh is finely realized, and the drawing of the foreshortened figure displays masterly skill; the disposition of the body and the curves of the lower limbs are circumstances of real subtlety of design in this beautiful composition (388–90).

St. Cecilia, St. Elizabeth, St. Catherine of Siena, and various other female saints provided further opportunities to depict woman's virtuous "helplessness and pure serenity." But literary heroines remained the artist's sacrificial victims of choice. John Collier's version of "The Death of Albine" [II, 34], a huge success when first exhibited at the Royal Academy in 1895 and equally popular at the Venice Biennale two years later—is a case in point. Emile Zola's Albine, the Eve of his version of paradise and the cause of *The Sin of Father Mouret* in his novel by that title (1874), is "like a huge, strong-smelling bouquet" (39), "a great rose" (123). In the novel she turns out to be a hothouse version of the flower-gathering Ophelia of the British. Albine has all the virginal, desperate yearning for the man who must "complete" her which characterizes Ophelia, but unlike Ophelia she does not let virginity rule her behavior. Like Ophelia she has an intimate relationship with flowers. When she moves, we are told, one can smell "the sharp odor of the vegetation she carried on her body" (39). And she emits a "clear laughter that died gradually, as if it came from the motion of an insane animal released in the grass" (40).

Awakened to passion by a hopeless—though not fruitless, since she is expecting a child toward the end of the novel—love for Serge Mouret, a young priest whom she nursed back to health (and into "manhood") after he had fallen into a fever, she is Zola's version of the Madonna. Given Zola's skepticism about any woman's willingness to remain a genuine virgin when offered a chance to be otherwise, she is as close to the late nineteenth-century ideal woman one is likely to get in this writer's voluminous oeuvre. Zola even makes the priest's sister, Désirée, recognize in Albine a resemblance to the statue of the Virgin Mary her brother has in his room: "She was all white, like you; she had big curls on her neck, and she showed her red heart here, where I feel yours beating" (231). But Father Mouret, having reverted to his priestly ways after a short Edenic convalescence among the flowers while nestled in Albine's arms, is now beyond the latter's passionate reach. At the same time "the flame of fever was in her eyes" (230).

Inevitably Albine falls victim to Zola's version of what one might designate as the

II, 33. John William Waterhouse (1849–1917), "St. Eulalia" (ca. 1885)

II, 34. John Collier (1850–1935), "The Death of Albine" (ca. 1895)

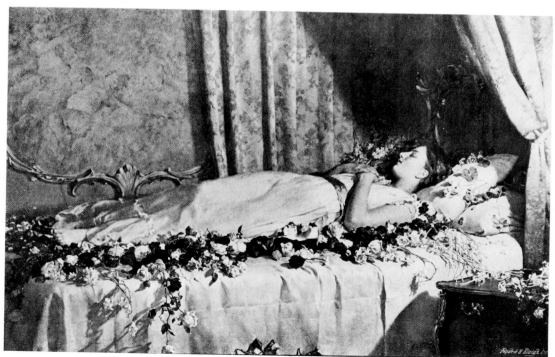

nineteenth-century's Florence Nightingale transfer principle, which held that a good woman who nurses a man back to health after a serious illness must complete her self-effacing duties by falling ill in a similar manner—assuming the man's illness, as it were, to bear it aloft with her into the heaven of sacrificial women. Faithful to this principle, Albine falls ill as Serge Mouret returns to health and to his priestly duties. Thus, feverish and lacking the male for whom her nature yearns, Albine, like Ophelia, a flower herself, turns to the flowers. Hearing "the voice of the plants saying goodbye, wishing one another a happy death" (284) at the end of the season, she realizes that she, too, is destined to die: "The garden was undoubtedly preparing her death as her supreme delight. It was toward death that it had guided her so tenderly. After love there was only death" (285). Albine consequently decides that, being nature's flower, she will die like a flower among the flowers. Filling a room with all the flowers she can find in her primitive paradise, she commits what was for Zola clearly the ultimate erotic suicide: She lets the exhalations of the flowers suffocate her: "The marriage was prepared; the roses' fanfare announced the awesome moment. Swooning, dying, her hands pressed tighter and tighter to her heart, she was gasping, opening her mouth, searching for the kiss which was to smother her" (290).

John Collier deliberately painted Albine as if she were still merely sleeping rather than dead. Henry Blackburn, in his *Academy Notes* for 1895, wrote enthusiastically that "a warm light suffuses the picture and falls upon the red flowers and satin draperies of the bed." The peculiar, somewhat boudoirlike atmosphere created by these colors and the rich draperies, joined with the subtle modulations of the woman's body, her right knee raised ever so slightly to hint at the soft curvature of her belly and loins, are ample indications of the dubious sincerity of the painter's interest in the spiritual and inspirational aspects of his subject. Instead he made the painting a striking expression of the erotic ambiguity of the Victorian ideal of passive womanhood—the dead woman—indicating how easily a painterly homage to feminine self-sacrifice could shift toward a necrophiliac preoccupation with the erotic potential of woman when in a state of virtually guaranteed passivity. Late nineteenth-century men were fascinated by woman's ability to be, in du Maurier's words, an "unconscious siren" even when on the verge of death.

It is necessary to point out that just as many women willingly—or resignedly, because they had no other choice—became camp followers of the Trilby fad, so many of the fine women painters of the period were as adept at depicting the "sensuously dead heroine" as their male colleagues. An excellent example of this is a pastel of the dead Albine by Lucy Hartmann [II, 35], which was exhibited at the salon of the Société des Beaux-Arts in Paris in 1899. A strong and technically accomplished composition, it focuses on the theme of the erotic dead woman by showing Albine's nude body in a pose resembling sensuous inanition more than death. Hartmann, in fact, emphasizes the deadly nature of this inanition by nearly blotting out the woman's head with a strangely localized shadow, thereby calling our attention to both the macabre and erotic implications of the scene.

One of the most explicit celebrations of the dead woman as an object of erotic desire had occurred relatively early in the century. Théophile Gautier in his story "La Morte Amoureuse" ("A Dead Woman in Love," or "Clarimonde," 1836), describes the forbidden passion of his priestly narrator Romuald for the dead Clarimonde—in life a courtesan with expensive tastes. Romuald studies the body of his sensuous temptress, posed, much like Collier's Albine, "at full length, with hands joined upon her bosom," her "graceful corpse" covered with a linen cloth of dazzling whiteness which "concealed nothing of her

II, 35. Lucy Hartmann
(active 1890s), "Albine,"
pastel (ca. 1899)

body's charming form, and allowed the eye to follow those beautiful outlines—undulating like the neck of a swan—which even death had not robbed of their supple grace. She seemed an alabaster statue executed by some skillful sculptor to place upon the tomb of a queen, or rather, perhaps, like a slumbering maiden over whom the silent snow had woven a spotless veil.''

It soon becomes clear to the reader that for a man of religious bent such as Romuald, death had served to cleanse even the brazen courtesan Clarimonde, returning her once again to the purity of adolescence and virginity. For, being a proper nineteenth-century male, nothing stirs Romuald's fervor so much as the intimation of virginal innocence in a woman: ''That exquisite perfection of bodily form, although purified and made sacred by the shadow of death, affected me more voluptuously than it should have done,'' Romuald admits, adding, in self-justification, that Clarimonde's repose in death ''so closely resembled slumber, that one might well have mistaken it for such.'' The recently deceased woman's ideally passive material presence finally makes it possible for Romuald to mold her attitude into that of the perfect woman, to ''sculpt'' her into what she should have been all along: For now the dead woman's arms could be made to cross ''on her bosom in an attitude of pious rest and silent prayer.'' Such a pose clearly gave her, once and for all, that combination of qualities the nineteenth-century male craved most in woman: ''an unspeakably seductive aspect of melancholy chastity and mental suffering'' (114–17).

In the progression of Gautier's story Romuald was still to experience, to his costly enlightenment, that to seek the embrace of a virgin in the dead body of a woman of pleasure can only yield a passion for the works of the devil. Soon, however, the appetite of the mid-century bourgeois male for the chaste woman of death was to lead him to seek the devil incarnate in the bodies of those women who had the gall to remain in the world of the living. Zola's tale of Albine, for instance, was a feverishly erotic fantasy concerning Eve's primordial ability to lead sainted manhood astray. In its presentation of an unabashedly allegorical dream of Eden, it might seem unlike Zola's other ''naturalistic'' works. But the high flush of desire with which Zola focuses on Albine as an innocent child of nature, willing and even eager to die an Ophelian death all for love, presents us with an excellent explanation for the extreme ferocity and anything but "realistic" extravagance of his negative characterizations of most of the other women in his work. Zola, like most of his contemporaries, was in love with the idea of woman as the personification of compliance—as the bed of flowers into which man could sink to be caressed by the petals of submissive admiration. The women who died—of yearning, preferably—were the icons of this religious-erotic fantasy. The women—most women, fortunately—who refused to comply with this ideal usually got their comeuppance by being typecast, like Nana, as evil flies rising from the dunghill of degeneration.

Zola's tale of Albine is a version of the familiar European legend of "The Revenge of the Flowers," itself the subject of numerous paintings. In this legend, to quote a commentator for *Famous Paintings of the World* (1897), a young girl, "lying down to sleep surrounded by the many-hued plunder of the garden, is overcome by the influence of the air surcharged with their baneful odor, and passes from the slumber of perfect health into the endless sleep of death" (202). As this description and numerous others in the vein of Owen Meredith's phrase "sleepy death" indicate, death as the ideal state of submissive womanhood had become such a staple of the later nineteenth-century imagination that many males could barely look upon a sleeping women without imagining her to be virtuously dead.

Dante Gabriel Rossetti's poem "My Sister's Sleep" (1850) is a characteristic early expression of this sleep-death equation. Rossetti, to boot, throws in a welter of suggestions which create equivalences with Mary, Mother of God, the household nun, and woman as Christlike martyr. The poem is a fantasy concerning an event Rossetti was quite obviously looking forward to with a great deal of eagerness as an occasion for virtuous rejoicing: his sister's death. Since she had as yet been irritatingly uncooperative in providing her brother with the wealth of artistic inspiration her actual death could have provided, Rossetti, as it were, took matters into his own hands. He consequently invented a narrator who imagines that he is contemplating his sister while she is sleeping, when, in fact, she has died. "She fell asleep," Rossetti tells his readers, "on Christmas Eve"—a perfect time certainly, for religious analogies. "Without, there was a cold moon up, / Of winter radiance sheer and thin; / The hollow halo it was in / Was like an icy crystal cup." Unaware, as yet, that her daughter has died, the mother, who has been caring for her in her illness, hails the birth of Christ: "Glory unto the Newly Born!" In fact, as she will soon realize, she is hailing the Christlike sacrifice of the daughter's passing from life into death, her "birth into eternity." Once the mother discovers the truth concerning the sister's ultimate self-effacement in a death so meek it seems like sleep to everyone else, the brother remarks, in a tone of considerable satisfaction, "God knows I knew that she was dead. / And there, all the while, my sister slept" (96–98).

With this sort of moral suasion, it is a wonder that women did not up and die quite as willingly as Rossetti might have wished. Still, Christina Rossetti could take a hint, and while wisely not quite practicing what he preached (she outlived her brother by twelve years), she did live an exemplary life of self-effacement while writing poems in which she portrayed herself affectingly in the extreme throes of the sleep-death equation, as in "After Death" (1862), in which she fantasizes virtuously about the meager rewards self-effacement brought to an exemplary woman: "He leaned above me, thinking that I slept / And could not hear him . . ." Discovering that she is dead, the unidentified "he" of this poem, for whom the female speaker has sacrificed herself, mutters, "Poor child, poor child"—a response which is not exactly overwhelming in its agonized intensity. The dead woman explains the man's coldness as follows: "He did not love me living; but once dead / He pitied me; and very sweet it is / To know he still is warm though I am cold" (106). In this fashion Christina Rossetti tried to convince herself that a good woman should be eager to sacrifice herself for love; the condescending sentiment of pity should be reward enough for any self-effacing female, since her death, in its cold reality, represented an important aspect of her function if it served as a validation of the warm life of masculine achievement.

The self-sacrificial sleep-death of a woman thus came to symbolize the extreme form of woman's compliance with the dualistic notion that made male-female relationships a simple matter of dominance and submission in an arena in which the nineteenth-century male could live out and realize the dreams of power which might have escaped his grasp in the actual realm of worldly affairs. The sleep-death equation had clearly become charged with morbid erotic implications, presenting the male with at least the fantasy of conquest without battle, of a life of power without constraints. The less demanding his mate, then, the greater his conquest.

Tennyson, champion of the woman who gave and gave and asked nothing in return, played countless variations on the sleep-death equation throughout his career as a poet. When, for instance, the dead Elaine is set afloat in her barge on her voyage to Camelot and Lancelot, Tennyson tells us that "she did not seem as dead, / But fast asleep, and lay as tho' she smiled" (385). The very fact that she asks to be sent downstream, bed and all, is also part of the sleep-death equation. A maiden's bed, after all, was supposed to be her "bridal-bed"—and a virgin was therefore thought to be immersed in a perpetual sleep of innocence until her chosen knight came along and awakened her with a kiss into the useful state of matrimonial service. Thus, a virgin who died before her virginity had been put to its appointed use—that of rescuing a male soul from worldly damnation—merely slipped from the sleep of innocence into the sleep of death.

The fairy tale of the sleeping beauty, too, inevitably came to be seen as symbolic of woman in her virginal state of sleep—her state of suspended animation and, as it were, death-in-life. In late nineteenth-century art representations of the sleeping beauty prolifer-

II, 36. Frances MacDonald (1874–1921), "The Sleeping Princess" (ca. 1897)

II, 37. George W. Joy (1844–1925), "Joan of Arc" (ca. 1895)

II, 38. Madeleine Lemaire (1845–1928), "Sleep" (c. 1890)

ated. Often the artists would make it a point to show the "deathlike" quality of the virgin's sleep. Frances MacDonald's "The Sleeping Princess" [II, 36] could just as easily have been labeled "Ophelia," since this princess seems to be floating in a watery grave rather than to be merely sleeping. An "Ophelia" painted by Odilon Redon in 1905, on the other hand, could just as well have been given the title of MacDonald's work, since the French painter's Ophelia seems to be sleeping comfortably rather than to be dead.

Indeed, portrayals of women whose obvious inanition seemed to prove that sleep was death and death was sleep became a source of endless delight among late nineteenth-century painters. Images of women who were so fast asleep that they looked as if they were dead became legion. The trick was that they could be portrayed in this ultimate stage of passive sensuality with all the more impunity since, after all, they were only sleeping and not actually dead; hence no one could accuse the artist of morbidity. At the same time, however, nothing could prevent the male viewer from indulging in the sleep-death equation and immerse himself, to virtually any degree of pleasurable morbidity, in thoughts of sensual arousal by a woman who appeared to be safely dead, and therefore also safely beyond actual temptation, even while that viewer could continue to tell himself that he was merely looking at a harmless image of a beautiful woman sleeping. George W. Joy, for instance, in his "Joan of Arc" [II, 37], exhibited at the Royal Academy in 1895, depicted this heroine—who, of all women, is certainly most immediately associated with the active life, with battle and with sundry martial doings—as being asleep, so much so that the viewer could not be blamed for thinking her dead. In fact, we see her experiencing the ideal sleep of feminine innocence: her virginity clothing her, as it were, with armor; her sword laid across her loins, a weapon against all men who might dare to approach her with impure thoughts, while the angel of saintly purity and militant maidenhood crouches protectively at her feet.

Everything that the late nineteenth-century art lover had learned to associate with the iconographic representation of a beautiful woman safely dead also came to be used to depict women sleeping. For instance, Madeleine Lemaire's ''Sleep'' [II, 38], which was exhibited at the Paris salon of the Société des Beaux-Arts in 1890, could just as well have been titled ''The Death of Albine,'' while the German sculptor Hans Dammann created what is perhaps the ultimate visual representation of the sleep-death equation in a tomb-stone sculpture exhibited in Berlin in 1899 and simply titled ''Sleep.'' His stately figure of a woman standing, her eyes closed, her hands full of flowers, and her thoughts on death, came as close to a representation of the turn-of-the-century's concept of the ideal woman as one is likely to encounter anywhere.

Unfortunately for the devotees of this ideal of woman as sexless sacrificial virgin, the likelihood that the object of their affection was able or, for that matter, willing to further the cause of man by applying herself consistently to such magnificent feats of self-negation had diminished drastically by this time. Instead of having learned to shrink and die in response to a century of indoctrination, women seemed more and more determined to stand up and fight, to make themselves heard. Sexuality, moreover, once more insisted on rearing its bestial head.

CHAPTER III

The Collapsing Woman: Solitary Vice and Restful Detumescence

No matter how eagerly the late nineteenth-century middle-class male wanted to believe in the sexless, virginal purity of the holy trinity of womanhood—mother, wife, and daughter—the facts of life kept getting in the way. For one thing, many of the women in his life seemed perversely unwilling to take the strains of the wedding march from *Lohengrin* as their cue to efface themselves forever in the bosom of matrimonial subservience. The ideological campaign of the Michelet contingent between 1840 and 1860, which had been so intense that it might have turned Venus herself into a paragon of virtue, had, as early as 1848, spawned such "unnatural" aberrations as the Women's Rights Convention of Seneca Falls and a rising tide of belligerence among women. Instead of peacefully scurrying into the walled garden of domesticity, women—who, as everyone knew, should, like well-behaved children, be seen and adored but not heard—were beginning to challenge their rightful rulers. They seemed to misunderstand not only their need to submit to the greater worldly powers of the male but were even challenging the notion that a good woman had no sexual impulses of her own but was content to mold herself into the vessel of male desire.

Moreover, the scientific investigation of feminine behavior patterns, which had gotten under way in the 1860s, was beginning to take a rather ominous turn, and by 1870 authorities such as Nicholas Francis Cooke, author of a cautionary handbook on human sexuality called, significantly, *Satan in Society,* were making the most shocking revelations about the private habits of young girls. What they had to say was enough to shake the faith of the most determined worshipers of woman as the sexless household nun. Cooke, who, certainly not without justification, called Michelet's writings on women "false, sentimental, and pernicious" (41), was only one among a host of experts who were beginning to push the dualistic pendulum of male notions concerning the essence of femininity back on its return swing from the extremes of the mid-century ideal of woman as household nun.

These writers almost invariably saw the women's rights movement as indicative of a perverse strain of delusion in the female mind. For them feminine self-assertion represented a reversion to earlier conditions of human life, a return to a more primitive stage of human civilization. Woman needed to be guided by the male, toward the proper apprehension of humane motivation since the male had a more "intellectual" comprehension of right and wrong than woman, given her greater intuitive being and her natural need to be "absorbed" by the male, which in her case made intellectual acuity basically an unnecessary attribute. Intensely concerned about the terrible effects of notions related to women's rights on the fragile feminine mind, Cooke warned that "the bare toleration of the idea, is sufficient in itself to injure the mind and operate powerfully upon the imagination of these impressionable creatures—to excite in them feelings of indignation and dissatisfaction with their present condition." Such terrible sentiments of incipient independence could only lead to "a sort of sentimental rebellion dangerous to tranquil repose and to feminine modesty." The mind of woman, once excited, would, as Cooke knew all too well, have only one outlet—and that outlet was not one of constructive participation in the governance of the world.

What was likely to happen instead was that man would be deprived of the gentle oasis of household peace presided over by his modest mate. She would "cease to be the gentle mother, and become the Amazonian brawler" (86). Man's fear was that woman, who, in the course of time, had learned to lift herself out of the bog of physical excitations—somewhat, one might suppose, like Holman Hunt's young woman, whose "awakening conscience" had caused her to lift herself from her lover's lap—would instead sit down again and, coasting on the comfort of man's loins, slide back into that realm of sensuality she had only just escaped. This would be catastrophic, for Cooke was certain that the absence of desire for sexual excitation in woman was the cornerstone of a healthy society. He concluded that "what remains of the family is only held together by the graces and virtue of woman" (87).

As early as 1858 the British Pre-Raphaelite painter William Lindsay Windus, in his painting "Too Late" [III, 1], had presented the public with a striking cautionary tale of what would happen to a woman—indeed, would happen to the family—if her sexual impulses were not nipped in the bud of her adolescent innocence: Solicitously embraced and supported by a friend, a young woman wracked by tuberculosis feebly holds herself upright by leaning on a cane, her boney fingers gripping its handle while she searches in desperation for something solid that may keep her anchored in the material world a few moments longer. Her eyes burning with disbelief, she stares at the man she once hoped would be her husband. Her face reflects the fevered, agonizing realization that even his belated and remorseful return to her side cannot stay the fatal course of her illness. Her lover, seeing her again after his years of disappearance, covers his eyes with his arm, shocked and pained at the ravages incurred by his irresponsibility and the illness it provoked, to her who once was his bride-to-be. So much for the family-rated part of the narrative, accurately read and related by Victorian as well as more recent critics.

What seems to have escaped twentieth-century commentators, however, is that the element in the painting which caused Ruskin to criticize it as unwholesome was not at all the fact that it dealt graphically with a woman's illness. What upset Ruskin and his contemporaries, but was left unspoken, was that sex had reared its ugly head in this affecting tale, its presence announced by the figure of the innocent child standing between the woman and her onetime lover. The child's face—with its searching eyes, sharply etched nose and

III, 1. William Lindsay Windus (1822–1907), "Too Late" (1858)

mouth—echoes the features of the failing woman, who, we soon come to realize, must be her mother. At the same time, the manner in which the child regards the man leaves no doubt that she has instinctively recognized him as her father. What disturbed Windus' contemporaries was that the painter had dared to soil the melancholy final hours of a beautiful sufferer by impugning her maidenly virtue and by implying that a decent, modern young woman could actually be driven by sexual passion.

A decade later, doubts about feminine continence such as those subtly suggested by Windus' painting and fostered by the less subtle statements and actions of advocates of women's sexual rights, such as Victoria Woodhull and her sister Tennie Claflin, were forcing the nineteenth-century male to recognize that even "civilized" women still showed vestiges of the sexual impulse. In response to this realization, numerous writers appointed themselves society's guardians of feminine purity, determined to warn their contemporaries about the ravages caused by the festering vestiges of sexual feeling among otherwise wholesome young women. What most attracted the attention of Cooke and his companions was the heinous solitary "crime" of masturbation.

Apologizing profusely to his sensitive readers for daring to broach such a ghastly subject, and leaning heavily on the very recent researches of a group of mostly French scientists, Cooke, for instance, sounded the tocsin of preparedness, urging his readers to awaken to the scientific fact that the ranks of female masturbators were "enormous, and the dangers are all the greater, that their very existence is so generally ignored" (106). He was specifically worried about "the female boarding-school," which he saw as "the arena wherein it is most widely acquired and practiced" (137). Cooke's concern itself represented a significant shift in the assumptions which had characterized earlier generations of nineteenth-century educators, who had posited the existence of close friendships among women as one of the principal sources of the development of humane sentiment in society. William Rounseville Alger, writing in 1867, had set out to present an encyclopedic record of the salutary evidence to be found in literature regarding *The Friendships of Women*, as he had titled his book. Alger was still able to come to the conclusion that if

> one-tenth of the efforts which women now make to fill their time with amusements, or to gratify outward ambition, were devoted to personal improvement, and to the cultivation of high-toned friendships with each other, it would do more than anything to enrich and embellish their lives and to crown them with contentment. Their characters would thus be elevated, their hearts warmed, their minds stored, their manners refined, and kindness and courtesy infused into their intercourse (364).

However, only a few years after the publication of these high-toned words about the beauties of feminine friendship, Cooke was already finding snakes in the garden: "Under the guise of friendship," Cooke warned his readers, girls often sink to the very depths of degradation: "The most intimate *liaisons* are formed under this specious pretext; the same bed often receives the two friends" (107). The strikingly abrupt change in focus represented by Cooke's book can be explained, at least partially, by the rapid rise to public prominence of the women's rights movement. In 1869 Horace Bushnell had warned direly that "women, having once gotten the polls will have them to the end, and if we precipitate our American society down this abyss, and make a final wreck of our public virtue in it, that is the end of our new-born, more beneficent civilization" (31). And Tennie Claflin, in her book of 1871, *Constitutional Equality,* pointed to a fact of very recent history: "Certain it is, that, where five years ago *one paper in a hundred* only, contained something about the progress of the Woman Question, now only *one in a hundred* can be found that has not a very considerable space devoted to it; and this has only become true to this extent within the present year." Claflin saw this rapid change as an outgrowth of the conclusion of the Civil War and a consequent "bold advance" of at least some women "into the heat and strife of active business life" (66). Claflin also emphasized each woman's right to practical equality—indeed, even to equal sexual privileges. She argued that if women

> are forever to be under the ban of society for one false step, they are determined their partners who accompany them shall be held equally culpable. Nor can man evade the point at issue. He must be willing to conform to the same rules he compels women to do, or admit her to those he practices. The extent this condition has actually reached, without his consent, is little dreamed of by the unlearned in the ways of the times. Public prostitution is but nothing to that practiced under the cloak of marriage (16).

Claflin's position, certainly, was held only by the extremist faction of the women's movement, but, then as now, it was the extremist position which received all the press. The average male consequently saw Woodhull and Claflin, with their appetite for business success and their ambitions for sexual liberation, as perfectly representative of the "new" woman. For their perceived economic and sexual voracity there had to be an explanation. Cooke and his cohorts thought they had found that explanation in the dangerous—one might even say incestuous—energies released by women who consorted with each other: The duplication of feminine characteristics which could take place among women friends was based on their lack of individual differences. Woman's high position in society had, in the past, been a factor contributing to her isolation: "The chivalric veneration with which man now regards woman, arises from the distance, as well as the difference, between them.''

On the other hand, in public life woman would team up with other women: She would lose her dependence on the male and hence the positive direction she gained from his guidance. Before long society's dissolution, the plunge into the abyss, would have been set in motion—as Bushnell had already pointed out. In other words, the nineteenth-century middle-class male's rediscovery of feminine sexuality, as well as his discovery of the apparently fearful fact that women could actually ''awaken'' sexual feelings in each other, was, to a large extent, a metaphoric expression of the late nineteenth-century male's unstated awareness that only by dividing women, by keeping them from working together, they could be kept in a state of economic and social submission.

Thus, the very sight of two women together, which only a few years earlier had conjured up cozy images of virtue, now came to stand for a double dose of vice, for, warned Cooke, ''extreme in good, [woman was] also extreme in evil'' (280). No wonder, then, that this writer had come to see girls' boarding schools as hotbeds of vice, where girls, as it were, took lessons from each other in masturbation. Quoting a French researcher whose recent investigations had uncovered a similar proclivity among French girls, Cooke emphasized that ''there is no young girl who should not be considered as already addicted to or liable to become addicted to this habit'' (108–9).

Cooke's cautions are characteristic of the change in the male's perception—the public rediscovery—of female sexuality which was taking place throughout Western culture. This recognition that women did, after all, have sexual drives caused a great deal of consternation among the men who had thought that in developing the myth of the self-effacing sacrificial woman, the undemanding household nun, they had been able to escape from the all too insistent demands related to sexual performance which had figured so largely in their grandfathers' and even in their fathers' smoking-room and tavern conversations. Driven by the ever-increasing pressures of economic survival, they were relieved to be able to remove from their shoulders the heavy burden of male potency, of this form of domestic competition which, they thought, had obstructed the ability of earlier generations of males to concentrate on the much more continuously rigid demands of their ''pecuniary reputability'' as ''an honorific evidence of the owner's prepotence'' (34), to use one of Veblen's very carefully chosen phrases about the pressures placed daily upon the successful businessman.

The weak and dying women of nineteenth-century art, and especially the women who, instead of insisting on their sexual fulfillment, had meekly died of unrequited passion— the Elaines, those tragic ladies of Shalott, even Zola's Albine—were so attractive to painters and writers because their actions indicated their absolute need for a man, even while their actual demands upon man remained virtually nil. Women who died while clutching the lily of virginity in their hands but who were nonetheless in the throes of passion, had allowed the Victorian male to indulge in the pleasurable sense of being desired by woman while yet not having to engage in the cumbersome fulfillment of any of the ''virile'' responsibilities such a desire might otherwise have entailed.

Many middle-class males of the second half of the nineteenth century had come to regard their marital duty as a burden and their own occasional spasms of erotic need as an indication of personal weakness. That this was so can be documented virtually ad infinitum, from the not infrequent incidence of actual celibacy in or out of wedlock (often accompanied by a rampant sexual imagination) to the learned disquisitions of scientific researchers and the newly formed tribe of sexologists which was rapidly coming into prominence. The latter talked with concern about the awkward phenomenon of tumescence and, more specifically, about the male's need to find appropriate relief from that desperate condition. Bernard S. Talmey, a gynecologist whose book *Woman* (1904) though supposedly written for the enlightenment of physicians only, went through numerous editions in a few years—described the problem of male excitability with scientific precision as a condition of tumescence, invariably accompanied by "the impulse of detumescence" or the desire to "cause a relaxation and discharge of the nervous and material genital congestion." Although, Talmey pointed out, this impulse was generally "most imperious," he insisted somewhat cryptically that "the real purpose of the instinct has been hidden to men" (68).

Anthony Ludovici—a man who was the son and grandson of well-known British nineteenth-century painters and who lived and wrote until the middle of the twentieth century yet had his head firmly planted in Queen Victoria's pot of basil—elaborated upon the problem in *Enemies of Women,* one of his numerous volumes of advice to men and women concerning the "woman question." Pointing to the physical frustrations attendant upon middle-class males who had come to take their duty to be virtuous seriously even in marriage, he identified "the state in which men are most prone to sentimentalize over women, especially young and attractive women, and to exalt them unduly, [as] one of sex-starvation or tumescence, acute or subacute." This peculiar state of uncomfortable congestion represented "the subjective momentum, driving a tumescent and chaste male to transfigure a sexual object."

Ludovici found this "tumescent tenderness" of men in thinking about women a disaster which affected their masculinity. "A tumescent man," he intoned direly, "no matter what his age, must ultimately fall under the empire of women. If after twenty-five years of marriage, when his wife has long ceased to stimulate him adequately, he is chronically tumescent, this merely places him chronically under the empire of women" (50). Ludovici, who was careful to point out that he was twenty-two in 1904, professed to have had frequent eyewitness experiences of this dire phenomenon, which he identified as a specific characteristic of late nineteenth-century males, and which he insisted had been "almost endemic in England" during this time.

It is clear that the conflict which pitted the call of desire against the ideal of continence, which was seen as the mark of a man of superior breeding and culture, had created a crucial dilemma for many late nineteenth-century males. Demands for sexual continence easily turned into doubts about sexual competence. Therefore the Tennysonian heroines and Ophelias of poetry who chastely panted after males became all the more reassuring when they died, for as long as they were still alive they might afflict the fantasizing male with the embarrassing state of "chronic tumescence," to use Ludovici's picturesque term. However, safely dead and ready for teary contemplation, they must have produced in their male admirers a restful sense of detumescence.

With the widespread rediscovery of feminine sexuality in the 1870s, and especially with the identification of female masturbation as an urgent problem, a new, strangely voyeuristic form of masculine self-reassurance of noninvolvement in woman's sexuality found expression in the visual arts—and, appropriately, given Ludovici's assertions, in British art in particular. This new depiction of woman generally showed her to be at once an object of erotic desire and a creature of peculiar self-containment, not really interested in, and hence not making any demands upon, the viewer's participation in her personal

III, 2. James Abbott McNeill Whistler (1843–1903), "Symphony in White, No. 3" (1867)

erotic gratification. Such images permitted the late nineteenth-century male to have his voyeuristic peek into the world of women and yet leave him with what one might term a soothing sense of "restful detumescence." These images formed the constituent elements of the late nineteenth-century artistic phenomenon of the "collapsing woman."

It is not difficult to trace the fad of the collapsing woman to its origins in the Victorian male's demand that his wife be both a magnificent ornament to his worldly success and the safekeeper of his household's collective spiritual virtues. We have seen that in order to maintain those dual functions, she must not—at least not publicly—engage in productive labor, for that would imply her husband's inability to maintain her. She must also under no condition let the ruddy texture and sunburnt glow of the peasant girl's skin impugn her husband's ability to keep her. The high Victorian bourgeois wife, myth would have it, was the true personification of the *dolce far niente,* the sweet indolence appropriate to the existence of a creature who, if it were not for her decorative and reproductive tasks, would have no function at all in this world.

As featured in the art of the turn of the century, the collapsing woman was primarily characterized—quite appropriately, given her spineless nature—by an overwhelming aura of lassitude. She was often portrayed as being on the very brink of sleep or otherwise already asleep. She originally clearly shared features with the goddesses of the consumptive sublime; as a matter of fact, she most likely originated from these withered flowers of lassitude, as can be seen in James Abbott McNeill Whistler's "Symphony in White, No. 3" of 1867 [III, 2], in which elements of the cult of invalidism, stylish boredom, and a new, slightly troubling undercurrent of erotic languor mingle to form an early example of the genre. In fact, while sometimes depicted alone, the collapsing woman was most frequently, as in Whistler's painting, shown in the company of a woman friend or friends. Often they held hands or leaned against each other languorously, sleepily, expressions of a rather contented exhaustion lingering on their faces. Unlike the self-destructive heroines of the consumptive sublime, they were generally portrayed as being in excellent physical health. Nonetheless, as if weighted down by the sheer rotund volume of her material existence, the collapsing woman simply could not seem to manage to keep herself upright. The very effort of living seemed to exhaust her. It is instructive to follow the trajectory of her collapse.

As the women in Edward Coley Burne-Jones' familiar, and in its time very popular painting entitled "The Golden Stairs" [III, 3] descend along a spiral staircase, they already display a peculiarly disequilibrious air. The painter has deliberately given them an only vaguely angelic character—they seem like the inhabitants of a somewhat underpatrolled classical boarding school for girls. In his depiction of the movement of their descent, he has suggested an extraordinary sense of downward pressure, as if the women, propelled by their own weight and lack of strength, were being gathered in clusters, curving weakly or leaning into one another rather helplessly and clumsily, their volume inexorably dominated by the earth's gravitational pull.

Clearly, the burden of their materiality was pulling these women toward the earth upon which, in a minimally defensive measure, they would soon be forced to sit down to accommodate their own weight. Lord Leighton, for instance, in his painting "Solitude," placed a hefty, fleshy woman squarely on a pedestal of rock, but imposed upon her so much body weight and so little apparent muscular control that the image gives the viewer the impression that the woman portrayed is in imminent danger of collapsing into herself. In this painting elements of existential lassitude and a faint touch of the consumptive sublime in the woman's face are given a common anchor in the enormous material presence provided by the grotesquely exaggerated, cocoonlike bulk of her cloth-wrapped hips and thighs. In studying paintings such as these, one cannot but applaud the accurate perception of an anonymous contributor to *The Magazine of Art,* who in 1881 characterized

III, 3. Sir Edward Coley Burne-Jones (1833–98), "The Golden Stairs," (1880)

III, 4. Albert Moore (1841–1893), "The Fan" (ca. 1874)

III, 5. Sir Edward Coley Burne-Jones (1833–1898), Study for the Garden Court in "The Legend of the Briar Rose" (ca. 1887)

the "modern taste" as given to a "tendency to long lines and limpness and a severity of shape" in its depiction of feminine beauty.

When the American critic Earl Shinn, writing his monumental multivolume *The Art Treasures of America* (1879) under the name of Edward Strahan, came to discuss a display of languorous female flesh by the French painter Alexandre Cabanel (whose work had a considerable stylistic influence on his British contemporaries), he pointed to the "nerveless weakness" characteristic of Cabanel's women, and then added that this artist seemed to have a predilection for the depiction of women who represented "a melting of human flesh into glutinous deliquescence, unbraced by bone" (I, 120). Shinn's astute comment is certainly equally appropriate for the women of such painters as Lord Leighton and Albert Moore. It was, in fact, inevitable that women of such fleshy and exhausted materiality as these artists insisted on painting would end up collapsed into themselves by the sheer pressure of their own weight, cradled, as best they might be, by sofas and pillows. Indeed, both Moore and Leighton made the depiction of such women a specialty eagerly sought after by connoisseurs. A painting by Moore of the mid-1870s, of which he produced several replicas for the delectation of wealthy contemporaries [III, 4], shows us two comely, diaphanously clad maidens so overcome with exhaustion, and so little in control of their own muscular capacities, that their arms and legs appear to have been stilled by sleep even while they were in the process of trying to find a comfortable position, causing them to be trapped in peculiar twists which emphasize the extent of their languorous passivity.

Burne-Jones, in a series of paintings illustrative of "The Legend of the Briar Rose," itself a version of the sleeping princess theme, combined elements of the dead woman with the collapsing woman theme in a sequence of "piled up" female bodies all shown in various stages of collapse, as if they were the girls from that imaginary boarding school of "The Golden Stairs" in an advanced state of dubiously acquired physical exhaustion [III, 5].

Lord Leighton's "Flaming June," currently unquestionably his most often reproduced work, is also a characteristic example of the genre of the collapsing woman. Near the waters of a golden, moonlit sea a young woman reposes in a flame-red, diaphanous gown. She is clearly utterly exhausted. Curled up and, as it were, sunk in upon itself, her body has become a carefully composed, solid mass of limbs and flesh. The parts of her body appear to have only a tenuous relation to one another, but in their substantial materiality

III, 6. Frederic, Lord Leighton (1830–1896), "Summer Moon" (ca. 1872)

they suggest an entity of enormous weight slowly continuing to compact under downward pressure. Describing the painting as representing "a sleeping maiden sunk on a marble couch in utter abandon of repose," Henry Blackburn, in his *Academy Notes* for 1895, accurately indicated the work's underlying theme of passive feminine eroticism and its function as a true icon of restful detumescence.

In an 1897 issue of *The Magazine of Art,* whose editors were indefatigable champions of Leighton's art, an answer to a reader's query provides a telling insight into Leighton's "innocent" intentions in painting this work and others like it. The composition of "Flaming June," we are told,

> was the adaptation of a chance pose assumed by the tired model during the period of a "rest." So the present writer was informed by the artist himself. Lord Leighton was charmed by the unusual attitude, expressing as it did the utter lassitude of an exceptionally supple figure. He at once made a sketch of it and used it as decoration in the small bas-relief painted in the lower right-hand corner of the bath in "Summer Slumber." He stated at the time that he proposed enlarging the scheme into an important picture for the following year. He kept to his intention, and "Flaming June" was the result (336).

The same artist's "Summer Moon" [III, 6] compounds the sense of enormous weight so notable in "Flaming June," as Leighton conjoins the bodies of two women in an image which serves truly as the apotheosis of feminine lassitude, the definitive statement of the collapsing woman. The two women, as always with Leighton, have hefty, substantial bodies, but they seem to be "unbraced by bone," to have no spinal columns and no musculature with which to support the weight of their flesh. Leighton has stacked them rather disdainfully on a ledge, almost as if they were no more than sacks of potatoes to be disposed of. Weak, exhausted, and dangerously heavy, these sleeping women reflect, in the apparent sterility of their passive existences, an erotic conception virtually excluding any invitation to the viewer to become actively involved in the experience of the figures portrayed. The women's lassitude, in its overbearing emphasis on their physical being, unmistakably posits a sexual cause for their exhaustion—but one far removed from masculine participation. Self-contained and imploded, as it were, these women are part of an endlessly repetitive series of turn-of-the-century depictions of women whose energy has apparently been sapped by excessive indulgence in solitary pleasures.

In *Satan In Society* Cooke had given a stirring summary of the latest "scientific data" gathered by French and British researchers which would cause an alert observer to detect the chronic female masturbator. His comments can therefore be regarded as representative of the late nineteenth-century's wisdom concerning the practice. As we have seen, the offender was most likely a boarding-school girl who had picked up the practice from friends, since it was "a contagious vice." It should be realized, Cooke declared ominously, that "the example of a single masturbator never fails to bear its fruit." This, he was convinced, furnished "the explanation for its frequency in establishments where a great number of young subjects are gathered together—schools, boarding houses, colleges; in short all places where education is in common" (113). Then he got down to brass tacks:

> The symptoms which enable us to recognize or suspect this crime are the following: A general condition of languor, weakness, and loss of flesh; the absence of freshness and beauty, of color from the complexion, of the vermillion from the lips, and whiteness from the teeth, which are replaced by a pale, lean, puffy, flabby, livid physiognomy; a bluish circle around the eyes, which are sunken, dull, and spiritless; a sad expression, dry cough, oppression and panting on the least exertion, the appearance of incipient consumption. (111).

In addition there were such moral symptoms as "sadness or melancholy, solitude or indifference, an aversion to legitimate pleasures" (112). Cooke was convinced that "the broken health, the prostration, the great debility, the remarkable derangements of the gastric and uterine functions, for which the physician is consulted, too often have this origin" (109).

Clearly, the males of the late nineteenth century—faced with a generation of women who, having grown up under the strictures and (absence of) expectations fostered by the purveyors of the cult of the household nun in the 1850s and 1860s, had become virtual invalids—were trying to assign blame for these women's apparent debility to their constitutionally inverted identity. As such, even the sad psychophysiological condition of many of the women of this time—which, as we have seen, was a fact universally attested to by feminists and antifeminists alike—was made into a subject of voyeuristic, clinical fascination. Males, afraid of actual tactile involvement with these by now almost mythical females, and fearing the loss of self-control implied by such tumescent involvement, could now recognize the women's invalidism as a factor of an inherent feminine tendency to "criminal self-abuse" and hence in no way to be regarded as a condition for which the male could be held responsible.

In consequence, as has been pointed out by numerous recent researchers, the turn-of-the-century male establishment became obsessed by the degenerative effect sexual stimulation of any kind was likely to have on women. Only complete absorption in the practice of motherhood was considered a fit activity for women. A virtuous woman might, following an appropriate display of distaste, submit reluctantly to the sexual advances of her husband, but only in the joyful anticipation of immediate impregnation. None but fallen women were supposed to find actual pleasure in sexual intercourse. Husbands who continued to detect vestiges of apparent enjoyment in their wives' response to physical stimulation suspected their spouses of indulging surreptitiously in "the crime." Many men had come to fear that by engaging in the act which was necessary for the survival of the nuclear family, their wives might actually come to enjoy sexual activity and begin to seek it out.

There was considerable suspicion that even decent women—given woman's presumably much feebler capacity for self-control, which was seen as primarily a manly virtue—might ultimately succumb to the call of the flesh and secretly become addicted to "solitary vice." Thus, the Cooke formula for the detection of female masturbators came to be seen as expressive of a condition charged with unspoken implications of autoerotic transgression. Solitary sin had left these women physically and mentally exhausted. In effect, the Cooke formula was little more than a description of the characteristic condition of the late nineteenth-century woman, suffering from the psychophysiologically based invalidism induced by her compliance with the conditions of the cult of the household nun. Still, the formula conveniently served to identify a newly discovered form of undignified and unwise feminine behavior. Physicians explained to horrified husbands that as blood drained from their wives' brains to rush to their excited reproductive organs, their minds as well as their bodies weakened, and soul and body alike would trail off into a sleep induced by erotic self-stimulation.

Such a singular condition of passivity and crime, such fascinating new evidence for woman's preoccupation with graceful self-abuse, could not but hypnotize the painters of the time. They rushed in to document the stages of woman's autoerotic physical collapse. The master painter of the exhausted, masturbatory woman, was Albert Moore, who invited his viewers to become voyeuristic observers of orgies of solitary feminine sensuality. He exploited the theme of passive feminine eroticism in a seemingly unending series of paint-

III, 7. Albert Moore (1841–1893), "Yellow Marguerites" (ca. 1880)

ings of groups of women (in an apparent response to the theory that the "vice" was contagious) as well as of individual figures lost in a fevered world of their own, far removed from all men, or even, it would seem, the thought of men.

In works with deceptively harmless titles such as "A Workbasket," "Yellow Marguerites" [III, 7], or "Rose Leaves," as well as in a work combining elements from all these images, a truly exhaustive display of exhausted femininity entitled "The Dreamers" [III, 8], he pursued the theme in a manner which allowed his contemporaries to enjoy their voyeurism with composure, since, in the words of Cosmo Monkhouse, his art was "unsuggestive, undomestic, and unemotional for the most part" (*The Magazine of Art*, 1895, 191). It is, of course, precisely this flaccid, singleminded focus on an absolutely passive and hence unthreateningly "feminine," sensuality that made Moore's paintings true blueprints of the Victorian upper-middle-class male's fantasies about "woman's nature." At the same time, his paintings helped to rationalize and "estheticize" the invalidism of the Victorian wife.

No well-informed late nineteenth-century man of the world could have mistaken the "scientific" implications of the peculiarly languid sensuality of the women depicted by Leighton and Moore. These painters and their host of followers, which included William Reynolds Stephens, John William Waterhouse, William Godward, Harold Speed, and Evelyn Pickering de Morgan, found it easy to justify their representation of such women by putting them in classical garb, thereby making their works "historical documents" which depicted the uncontrolled, "animalistic" permissivenenss of ancient times, "when the world was young," to use the title of a painting by Edward Poynter exhibited in 1892 at the Royal Academy, representing a trio of this painter's favorite collapsing women. The decorative and mildly—but certainly unthreateningly—erotic implications purveyed by such

III, 8. Albert Moore (1841–1893), "The Dreamers" (1882)

III, 9. William Reynolds Stephens (1862–1943), "Summer" (1891)

bevies of pleasantly exhausted painted ladies made these works ideal for clubs and public buildings. Stephens' monumental "Summer" [III, 9], painted in 1891, was designed as a mural decoration to be hung in the "Refreshment Room" (!) of the exhibition galleries of the Royal Academy.

Although British painters unquestionably pursued the theme of the collapsing woman with the most singleminded intensity, the subject was a favorite with painters everywhere. In France, for example, Pierre Puvis de Chavannes specialized in the depiction of women's stately listlessness. His work, in turn, helped give direction to the wilting females of the symbolists. The more orthodox academic painter Charles Chaplin produced extraordinary examples of sensuous feminine bone- and spinelessness. His magnum opus, "Sleep" [III, 10], painted in 1886, is probably the most exaggerated image of eroticized feminine passivity ever produced. In a bed of drapery as fluid in texture as the waves of the sea nearby, a woman sleeps in self-satisfied exhaustion. Her body, as soft and unstable as an undercooked pudding, would appear to be in the process of dissolving into the textures of cloth and wave with which she is surrounded. That this woman could ever awaken, let alone lift herself out of the throes of her self-dissolving languor, would appear to be a physical impossibility.

The Americans tended to emphasize the mysterious self-involvement of the collapsing woman. John White Alexander, in his haunting work "Memories" of 1903 [III, 11], brilliantly managed to combine suggestions of virginal purity and mystical transport with hints of a hothouse attachment between the two young ladies he had painted: "Manifestations of spurious spirituality are often induced by some perversion," Bernard Talmey warned his considerable audience at about the same time Alexander painted his "Memories." Were the cryptic "memories" of this pair perhaps those of unmentionable doings during their boarding-school days together—or was the book held by one of them an example of their "clandestine reading of certain books, in which abject authors have traced, in the liveliest colors, the deplorable deviations of the senses?" According to the anonymous

III, 10. Charles Chaplin (1825–1891), ''Sleep'' (1886)

specialist of the French *Dictionary of Medical Sciences* faithfully quoted by Cooke, that was, among such young persons as those depicted by Alexander, ''one of the most active causes of their depravation'' (107). Perhaps what we are witnessing in Alexander's painting are the catlike incursions of an already depraved young woman upon the still lingering innocence of her younger friend. The strength of the painting is that its meaning must remain ambiguous, but it is certainly not an expression of that ''naive, unconscious sexuality'' the twentieth century has so fondly—and wrongly—attributed to the work of Victorian painters.

Woman's indulgence in solitary vices inevitably led to exhaustion and hence to sleep. Turn-of-the-century painters had a veritable field day exploring the autoerotic ramifications of the representation of the sleeping woman. This theme was obviously closely allied to the more decorous representations of the sleep-death equation favored by the Pre-Raphaelites. However, rather than appealing to the viewer's appreciation of the woman as self-sacrificial martyr, this variation based itself on a generalized assumption of the ability of woman to satisfy her own physical needs, thus clearly removing the male from sexual responsibility and allowing him once more to enter into a voyeuristic, passive erotic titilation within a soothing, undemanding context conducive to a state of restful detumescence.

Even if there was no immediate intimation of solitary activity on the woman's part, and even if she had clearly been content to lend herself to the submissive pleasures of a ritual *a deux*, the contemplation of her erotic exhaustion seemed to some nineteenth-century males to promise greater ecstasies than those which could be expected as part of the process which was to bring her to that condition. As early as 1831, in his novel *The Wild Ass's Skin*, Balzac had made his doomed intellectual hero Raphael look upon the pleasantly somnolent body of his love, Pauline, and construct an enthusiastic version of the sleep-death equation: ''To watch your mistress as she sleeps, smiling in a peaceful dream as you mount guard over her, loving you even as she dreams, at a time when she seems not to exist as a sensate being, offering you silently her lips which in sleep speak to you of the last kiss; . . . such joys have surely no name?'' (242)

The painters who explored the erotic potential of the image of the sleeping woman usually gave their paintings neutral descriptive titles such as ''Sleep,'' ''Dreamland,'' ''The Sleeping Beauty,'' ''The Tired Dancer,'' ''A Summer Afternoon,'' ''The Enchanted Hour,'' ''Repose,'' and so on. Even a painter such as John Singer Sargent, whose principal occupation as a portraitist required from him a more ''upright'' standard of representation in his women, occasionally indulged, as in his 1894 painting ''Rest,'' in the depiction of women in a state of utter collapse—no doubt the result of similar opportunities for the observation of tired models as the one encountered by Leighton in conceiving his ''Flaming June.''

Obviously, it is foolish to seek behind every depiction of the collapsed or sleeping woman of this period a deliberate depiction by the artist of a woman in the throes of overindulgence in autoerotic pleasures, but it is certain that the average, well-educated turn-of-the-century viewer of these paintings was, by and large, by no means unfamiliar with the current medical wisdom. The gist of that wisdom was conveniently stated by Talmey, who, many years after Cooke, still continued to claim that the outward signs of the cult of invalidism were, in fact, proof of sexual excess. Overindulgence, insisted Talmey, led ''directly to anemia, malnutrition, asthenia of the muscles and nerves, and mental exhaustion. Immoderate persons are pale and have long, flabby, sometimes tense features. They are melancholic and not fit for any difficult and continued corporal or mental work'' (170). Cooke had already warned that even the most innocent-seeming young women

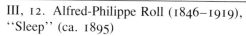

III, 11. John White Alexander (1856–1915), ''Memories'' (1903)

III, 12. Alfred-Philippe Roll (1846–1919), ''Sleep'' (ca. 1895)

might become very adept at concealing their vice under the guise of sleep. Hence, those who wanted to catch the young female masturbator in the act should sneak up on her in her retirement: "Her hands are never outside the bed, and generally she prefers to hide her head under the coverlet. She has scarcely gone to bed ere she appears plunged in a profound sleep." To Cooke a young woman's sleep was, as a matter of fact, a "circumstance, which to a practiced observer is always suspicious." Therefore "the affectation that the young person carries into her pretended sleep, the marked exaggeration with which she pretends to sleep, may often serve to betray her" (115).

Thus, even the sight of a woman sleeping came to suggest very specific voyeuristic erotic fantasies to the turn-of-the-century art lover, ones which the artists of the time did their best to foster—whenever, that is, they did not get involved in the even more morbid pursuit of the sleep-death equation. The sleeper in Alfred-Philippe Roll's "Sleep" [III, 12] for instance, presents a very close approximation of the state of the erotic sleeper of Cooke's description, whose breathing is "more precipitate, the pulse more developed, harder and quicker, the blood-vessels fuller, and the heat greater than in the natural condition" (115–16). The Russian artist Konstantin Somoff's "Sleeping Woman" [III, 13], given the extremely pleased expression on her face and her strategically and suggestively placed hand (deliberately snakelike in its modeling), would appear to be a similarly direct expression of the artist's suspicions concerning the lady's activities before she lapsed into sleep. Somoff's painting could have served as an illustration to Swinburne's well-informed lines from "Laus Veneris": "Her eyelids are so peaceable, no doubt / Deep sleep has warmed her blood through all its ways" (*Collected Poetical Works*, I, 11). Arthur B. Davies' "Sleep," which won him honorable mention at Pittsburgh's Carnegie Institute in 1913, brings together in nature a bevy of very exhausted young ladies who seem to be overcome by a contented languor not fully attributable to their excursion into the fresh outdoors [III, 14].

Like Davies, who made this sort of image a speciality, many of the turn-of-the-century painters delighted in combining the restful and pleasantly uninvolving depiction of

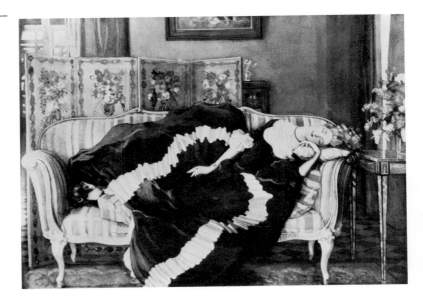

III, 13. Konstantin Somoff (1869–1939), "Sleeping Woman" (ca. 1911)

III, 14. Arthur B. Davies (1862–1928), "Sleep" (ca. 1905)

sleeping women with a natural setting. Joseph Engelhardt, a member of the Viennese *Ver Sacrum* contingent, placed a trio of tired young women under the dappled shade of a tolerant tree. Out of the upper-right corner of this painting, as usual modestly entitled "Sleep" [III, 15], crawls a peculiar little monster, half-businessman and half-slug, with the sunken eyes of the sort of male Cooke had also found prone to self-pollution, the type who "acquires a dull, silly, listless, embarrassed, sad, effeminate exterior." In fact, the peculiar creature which Engelhardt has portrayed here only seems capable of contemplating the beautiful, self-sufficient women under the tree from a safe distance. The lower part of his body would seem to have melted away, as if he were a symbolic representation of the essence of voyeurism. Clearly, this timid little half-male shares numerous characteristics

III, 15. Josef Engelhardt (1864–1941), "Sleep" (ca. 1898)

III, 16. Robert van Vorst Sewell (1860–1924), "The Garden of Persephone" (ca. 1896)

with Cooke's degenerate masturbatory males, who "fall into complete apathy, and, sunken, below those brutes which have the least instinct . . . retain only the figure of their race" (101).

It would indeed seem that turn-of-the-century painters, in their depiction of the sleeping woman, were trying to come to terms with the fact of feminine sexuality without having to confront the concomitant realization that if she ceased to be a household nun woman might also cease to be a passive object for contemplation. By placing woman in nature in a state of utter exhaustion, the painters were trying to indicate that although she was no longer an ideal creature, and instead very much a part of nature, she was still not an active threat to man, who was, after all, in many ways far superior to brute nature. In a sense, she was still part of the flowers, a self-contained "plant-like entity" with all the static indolence of nature's vegetable matter. The women in Robert van Vorst Sewell's "The Garden of Persephone" [III, 16], are intended to be seen as exemplary followers of this goddess of the hollow earth, of fertility, and of death, who was the personification of vegetable life and the never-changing cycle of the seasons. It is obvious that Sewell's women are strong, healthy creatures of nature, but they have been felled, like sheafs of wheat, as if following an almost bestial instinct of lassitude at midday. These women are clearly primal beings; it is as if they are wilting even while still growing out of the soil, like the tree roots which surround them. Passive but fertile, they personify what had come to be a standard conception of woman as the infinitely receptive, seed-sheltering womb of a sweltering earth.

CHAPTER IV

The Weightless Woman; the Nymph with the Broken Back; and the Mythology of Therapeutic Rape

At mid-century Coventry Patmore had warbled excitedly: "Boon Nature to the woman bows; / She walks in earth's whole glory clad" ("The Angel in the House," 85). For the votaries of the household nun it had been important to play down any direct comparison between woman and nature. The earth was her servant, nature an abject slave to her capacity to transcend the call of the flesh. However, with the scientific rediscovery of a sexual drive in woman, such encomiums to woman's sexless virtue became less convincing. Within a relatively short period of time—a decade or two at most—she now fell, in the eyes of many men, quite down to earth from her former perch in the heavens. Soon she was once again to become the very personification of mother earth. Even in this capacity, however, she often continued to be seen as a viable symbol of passivity and sacrifice. Did not the earth, nature herself, meekly permit her body to be plowed, seeded, stripped, and abused by man?

Louis Vauxcelles' description, in *The Salons of 1907*, of a symbolic mural design by Paul-Albert Besnard destined to serve as a decoration for the Petit Palais in Paris, provides a useful delineation of the concept of woman as "matter":

> Under the open sky, amid a whirl of clouds, a faun with a goat's head, foreshortened as Michaelangelo might have drawn him, carries in the close embrace of his strong arms a nude being, unconsciously voluptuous; from her fly the germs of life in the form of little children who soar up, whirl around in the air, and fall back to the earth. Following the curve of their fall, we return to earth with these infants conceived in joy, and we find a woman's corpse lying on the ground, where the soil, fertilized and enriched by its decay, is more abundantly green (11).

Woman, in other words, was fertile—and fertilizer as well. She was the symbol of nature, she was the earth, eager to give—but also voraciously hungry. Vauxcelles called Besnard's symbolism pantheistic, but perhaps it would be more accurate to see the painter's concep-

IV, 1. William Adolphe Bouguereau (1825–1905), "Alma Parens" (1883)

IV, 2. Léon Frédéric (1856–1940), "Nature" (central panel) (1894)

tion as the apotheosis of materialism. Once the sexual nature of woman had been redis-
covered by the scientists of the 1860s, and she therefore could no longer remain the virginal keeper of the soul, her ancient mythological associations with fertility were resurrected
by artists and writers everywhere. With the start of the age of rampant consumerism in the
1880s and the consequent cultural fascination with quantity, woman as nature became
nature as abundance.

The pendulum of dualism was swinging again: The concept of woman as the Virgin
enthroned was easily altered to make her nature enthroned. Her habit of near-immaculate
conception became instead one of indiscriminate fertility, her charity of heart became an
offering of milk-filled breasts. William Adolphe Bouguereau's "Alma Parens" of 1883
[IV, 1], is characteristic of this new image of indiscriminate abundance enthroned. Bouguereau's woman is the personification of intuitive altruism: A symbol of domesticity, of
motherhood, she is at the same time too fertile to be holy. Virtuous in her voluptuous
maternity, she makes men long for her, wishing to nestle, like the babies which surround
her, at her abundant bosom and forget about their duties in the grim world of affairs. In
the same vein, Friedrich von Kaulbach's "Mother Love," a huge favorite with the American public, was, as Earl Shinn pointed out, "a sort of altar-piece of the *cultus* of family

piety'' (*Art Treasures*, III, 75). These images were expressive of men's dreams of generous, unrestrained inclusion; of nature as simultaneously receptacle, fertile soil, and comforting breast. They are odes to mother as the earth's warm womb of passivity, gentleness,
and the promise of repose—three characteristics which, according to Henry Drummond in
The Ascent of Man, typified the role of woman in an evolving world (327). It is thus that
Léon Frédéric portrayed her in ''Nature'' [IV, 2], a painting which unquestionably represents the ultimate apotheosis of the consumerist dream of abundance, fertility, and satiety.

Frédéric's painting represents as grossly incontinent an indulgence in the fantasy of
woman as the all-giving, all-receiving womb as one is likely to encounter anywhere—
except, perhaps, in other works by the same artist, which are characteristic of the tendency
of turn-of-the-century painters to try and top what had already come before by resorting to
a quantitative approach in the means of representation. Not even Otto Greiner's 1911
etching of an ''Earth Goddess'' [IV, 3] can match the dizzying proliferation of symbols of
woman's all-giving fertility in Frédéric's painting. In its turn, the mindless primal mother
in Greiner's etching is an illustration of the pervasive fin-de-siècle notion, succinctly expressed by August Strindberg in his novel *By the Open Sea,* that ''the woman is the man's
root in the earth'' (314).

Woman as earth mother, the fertile enveloper, was sexual, certainly, but she was also
protective. Her role as a giver of succor to the tired businessman coming home from the
economic wars to the protective embrace of his ever-yielding domestic slave was still
intact, but now her flower identity—as, indeed, everything else in late nineteenth-century
bourgeois society—became quantified. She was no longer a pale, wilting flower but a

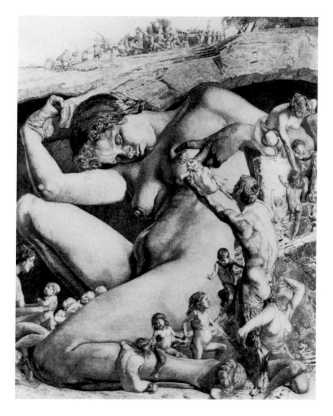

IV, 3. Otto Greiner (1869–1916), ''The
Earth Goddess (Gaea),'' etching (1911)

IV, 4. Charles Courtney Curran (1861–1942), "The Spirit of Roses" (ca. 1899)

IV, 5. Paul-Albert Besnard (1849–1934), "Vision of a Woman" (1890)

IV, 6. Childe Hassam (1859–1935), "The Butterfly" (1905)

sweltering bower of roses, the earth's warm flower-womb, always there for the tottering tot of the masculine ego to wander into whenever he wanted to escape from the realities of the world.

Thus, the eroticized body of woman became the late nineteenth-century male's universal symbol of nature and of all natural phenomena. She sat, a flower among flowers, a warm, receiving womb and body, waiting patiently for man, the very incarnation of the spirit of the rose. This is how Charles Courtney Curran saw her in his contribution to the American section of the Universal Exposition of 1900 in Paris [IV, 4]. Paul-Albert Besnard took his eyes off his designs of woman as matter, or woman in the throes of death, just long enough to see her, for once, before she wilted again, growing like a swollen, tuberous bud out of the oceanic flower beds of French suburbia [IV, 5]. She also watched—and was—Childe Hassam's careless, fragile "Butterfly" [IV, 6], among the leaves and flowers of all-encompassing nature. In numerous paintings bearing such titles as "Summer," "Wintry Spring," "Morning," "The Dew," and "Falling Leaves," she personified every month, every season, every time of day, and a gamut of meteorological conditions as well. She tripped naked through the woods, roamed carelessly among the bushes, and grew like a white-breasted human waterlily among the flora of the sea, as in Ary Renan's rootlike "Sappho".

But most often, persistently and relentlessly, she floated through the air. She floated, because to walk is to act, and to beckon a form of invitation, a way of taking charge. And to the late nineteenth-century male nothing was as unwelcome as the thought of woman—even woman as the embodiment of nature—taking charge. *He* wanted to be in charge, it was his *right* to be in charge. To him, Henry Drummond assured him, had been "mainly assigned the fulfilment of the first great function—the Struggle for Life" (329). It was woman's appointed role, even as the personification of nature, to float weightlessly in the breeze.

To float in the air was the eroticized alternative to Ophelia's watery voyage. Woman's weightlessness was still a sign of her willing—or helpless—submission, still allowed the male to remain uninvolved, still permitted him to maintain his voyeur's distance from this

IV, 7. Henry Mosler (1841–1920), "The Spirit of the Rainbow" (ca. 1912)

IV, 8. William Adolphe Bouguereau (1825–1905), "The Oreads" (1902)

IV, 9. Martin Brandenburg (1870–1919), "Whirling Sands" (ca. 1905)

creature of nature, this creature that *was* nature, who both fascinated and frightened him. So he made her tumble like a brown leaf through the air, using her identity as nature personified as his excuse for making her do so. She floated, preferably in the nude, the enticing "Spirit of the Rainbow," as in Henry Mosler's painting of that title [IV, 7]. She might personify the rushing hours or, as in the eyes of the aging Bouguereau, a whirling group of very fleshy "Oreads" [IV, 8], those otherwise flighty nymphs of the mountains who, in Bouguereau's fantasy, were clearly capable of daunting even the satyrs. Sometimes woman was blown and tossed about like the "Whirling Sands," as in Martin Brandenburg's popular painting [IV, 9]. The American Louis Eilshemius' continued fame as an inspired modernist-primitive rests, to a large extent, upon his well-known image of a cluster of women afloat in "The Afternoon Wind" (1899). The painting is thought to represent his eccentric and precociously surrealist imagination. In fact, it is no more than a perfectly orthodox and ordinary version of the turn-of-the-century academic artists' floating women, who could be seen everywhere drifting along like clusters of dandelion seed. With many companions she swirled, undaunted by the cold, through mountainous regions, as in Paul Legrand's "The Snow" of 1902 [IV, 10]. As a gentle drift of that same cold, moist matter woman took to the air in Carlo Fornara's "Alpine Legend" [IV, 11] to form, hand in hand with other paragons of feminine insubstantiality, a chain of slightly eerie, mildly underclad, and definitely flaky nymphs.

It is clear that the weightless woman has a great deal in common with the collapsing

IV, 10. Paul Legrand (b. 1860), "The Snow" (ca. 1902)

IV, 11. Carlo Fornara
(1871–1968), "Alpine
Legend" (ca. 1905)

woman, for she was both passive and sexual, but once again she was sexual in a manner
that was likely to discourage any sort of participation on the part of the observer of her
airy trajectory. It is decidedly difficult to imagine engaging in any sort of sustained bodily
contact with a woman who is lighter than air. At the same time, even as a weightless
floater she was most often "a nymph in the full flower of womanhood, with low brow,
languid eyes, and glorious locks" (154), as a commentator for *The Art of the World*
described this ideal of fertility in an appreciative note concerning a "Pomona" by George
Maynard.

Moreover, she often tended to lift off into her heavenly trajectory from amid signs of
a rather decadent, even oriental, hypnotic somnolence, as in the case of Jean-Louis Hamon's "Twilight" of 1867 [IV, 12], whose symbolic significance was effectively explicated by Earl Shinn:

> She represents the "Opium Dream," seemingly. Beneath stretches a field of poppies, lifting
> up their stems and their shapely seed-pods, chiseled like Indian capitals; from among them,
> her feet disentangling themselves from their cold stems, floats up the Vision, a dim figure in
> human shape, her filmed eyes lifted, her arms crossed in Oriental adoration, and all her faint,
> smoky figure ready to blend with the clouds and fumes that overweave the unsubstantial heaven
> (*Art Treasures*, II, 24).

Alone or in clusters, the weightless, floating woman was thus to be suspected of indulging
in a similar sort of secretive, self-involved, luxuriant yielding to the call of nature characteristic of the collapsing woman. This was made abundantly clear by Giovanni Segantini
in his well-known painting of 1891 entitled "The Punishment of Luxury." Segantini's
women float weightlessly through a northern Italian landscape which has abruptly turned
into an undiscovered circle of Dante's *Inferno*. They seem to be curiously unconcerned
about their wintry environment, and appear instead, to be warmed by a languorous, self-
satisfied inner glow. And, no wonder, for these are women whose fate as perpetual floaters
had, according to Segantini's own indications, been sealed by their luxurious self-indulgence in the pleasures of the flesh. Their egotism was symbolized by their refusal to
become doting mothers and, in so doing, renounce the sensuous side of their nature. Any

woman, after all, must know instinctively what the indefatigable Cesare Lombroso had determined scientifically, in association with his compatriot Gugliemo Ferrero, that "in the ordinary run of mothers the sexual instinct is in abeyance" (*The Female Offender*, 153).

Given the antimaternal, luxuriantly autoerotic implications of the theme of the weightless woman, it was not surprising that she most often took to the air during the twilight times of day, the times most frequently associated with indulgence in the pleasures of the flesh. Bouguereau's "Twilight" was among his entries to the Paris Salon of 1882. The painter showed a pale-skinned woman rising, like the moon, out of the waters of the sea, surrounded by a rather undecorous but certainly very decorative flutter of drapery, her naked body taut with anticipation of the pleasures of a weightless evening.

Bouguereau's ascending and cautiously suggestive personification of twilight found its early-morning counterpart in the American painter Walter Shirlaw's "Dawn," executed a few years later in 1886 [IV, 13]. In brisk, bold strokes of paint he has conjured up a somnolent beauty, enveloped by little more than her radiant and opulent auburn hair and a hint of veils. Shirlaw has suspended her in air forever as she floats down in blissful languor toward the surface of a pond partly overgrown with leaves and lilies. His figure has the half-parted, softly rosy lips, the velvet-lidded eyes, and the butter-cream complexion of an extremely well-satisfied young lady. The slightest hint of a smile plays around the edges of her mouth, and the soft pinks and beige edging into rose of her luxuriantly curving body play against the rust-shaded proliferation of her long, free-floating hair, whipped into flamelike swirls by the fluttering wings of the amorous doves who accompany her on her downward trajectory toward the mirroring universal waters of undifferentiated femininity.

IV, 12. Jean-Louis Hamon (1821–1874), "Twilight" (1867)

IV, 13. Walter Shirlaw (1838–1909), ''Dawn'' (1886)

Shirlaw's "Dawn" has a bravura, improvisational, very modern, broadly painted quality which, although virtually unmatched by other American painters at this time, nonetheless establishes the sophisticated international frame of reference available to American painters by the mid-1880s. The painting represents a fascinating stylistic and thematic synthesis of British Pre-Raphaelite and French academic elements which, in their juxtaposition, led to a remarkable prefiguration of the stylized organic linearity that was to characterize European Art Nouveau productions of the 1890s. At the same time, Shirlaw combined knowledgeable hints of Rossetti and Leighton-like elements of modulation in this "stout and boneless nude," as Oliver Larkin, in his *Art and Life in America,* described her. The painter, furthermore, added lessons learned from Bouguereau to make this American "Dawn" into a striking document of the New World's vanguard participation in the increasingly heterogeneous stylistic explorations of the supposedly so homogeneous ranks of the academic painters of the late nineteenth century. The welter of influences, synthesized into a very personal style, which characterizes Shirlaw's painting also demonstrates dramatically how thoroughly internationalized the culture of Europe and the United States was becoming by the mid-1880s.

Around 1890 another American painter, Sarah Paxton Ball Dodson, tried to bring the theme of the weightless, airborne woman back to a more spiritually focused, pantheistic concept—stressing the link between woman and nature as the material representation of a fundamental celestial harmony, the chain of universal feminine beauty connecting humanity with the Ideal—in her processional work entitled "The Morning Stars" [IV, 14]. "Each of these floating nymphs," commented an anonymous critic in *The Art of the World,* when the painting was exhibited at the World's Columbian Exposition in 1893, "bears aloft her glittering star, while the leader waves a torch and beckons onward her endless throng of followers" (196). And, indeed, Dodson's floating women seem determined to regain the heights of transcendence from which they had been forced to descend in Burne-Jones' "The Golden Stairs."

Not surprisingly, there were, among the many weightless women who populated the

IV, 14. Sarah Paxton Ball Dodson (1847–1906), "The Morning Stars" (ca. 1890)

turn-of-the-century art world like clouds of drifting seeds, quite a few who ended up caught in the branches of trees. These young ladies who found themselves so very literally and uncomfortably out on a limb were, inevitably, part of the woman-as-nature theme. Anna Klumpke, an American painter who was closely associated with Rosa Bonheur but who, quite unlike Bonheur, allowed herself to indulge in what turn-of-the-century critics were wont to call "ideal" imagery, made the connection between the weightless woman and the woman-up-a-tree theme quite clear when she titled a 1911 painting, showing a young girl gracefully pendent from the branches of a tree loaded with blossoms, "The Breeze Flutters Among the Trees of the Forest." And Giovanni Segantini followed his depiction of women's self-indulgent luxuriance with his 1894 painting "The Evil Mothers" [IV, 15], in which, like withered, dry leaves, these nasty women, with nonexistent children gnawing like an evil conscience at their uselessly voluptuous breasts, are seen twisting and turning in the wind, caught in the sterile branches of barren trees, mere empty shells of what they might have been had they not forsaken the sacred duties of motherhood to pursue their lascivious private pleasures. However, as if to make certain that his audience would understand that the dualistic conception of reality demands consistency in its portrayal of extremes, and that it was hence every woman's business to be in the trees, he also painted his ideal representation of woman as the maternal household nun, "The Angel of Life," ensconced in a tree. This time, however, he gave her a comfortable perch, so that she might hold and hug her healthy baby safely and securely.

For Dante Gabriel Rossetti it was woman's failure to keep her mind on the serious matters of the intellect which made her end up in the trees, as in his painting "The Day-Dream" of 1880 [IV, 16], in which one of the painter-poet's characteristic large-lipped, droopy-eyed ladies of languorous pallor finds herself weighted down among the branches of "reverie," where, as he explained in his poem of the same title, unheedful of "her forgotten book," she fantasizes her "woman's budding day-dream spirit-fann'd" (260).

Zola's story of Albine, of woman as Eve, contains a wealth of intimations concerning the reason for, and the focus of, woman's affinity with the trees. The tree, of course, is a principal symbol of nature—it had for woman, or so the nineteenth century male mind conceived it, an irresistible attraction since it grew out of the earth in a state symbolizing that chronic tumescence which, if it were to afflict the male, would make him inevitably fall, in Ludovici's words, "under the empire of women" (50). Thus the tree symbolized fertility and primal sexual excitement in its most archetypal form. In *The Sin of Father Mouret,* and at the center of Albine's paradise, stands the tempter—a colossal tree, the tree of sensual knowledge: "From its green vault fell all the joy of creation: flower smells, bird songs, drops of light, cool, awakening dawns, warm, sleepy dusks. Its sap was so powerful that it flowed through its bark, bathing the tree in a vapor of fertility, making it the very manhood of the earth" (182). No wonder, then, that Albine is irresistibly drawn to this tree, indeed, to all trees, for even when we first meet her—this daughter of middle-class parents who had once been a model young lady, "educated, reading, knitting, chatting, banging on pianos" (40)—she has a very determined love for trees. When the Abbé, after encountering Albine, takes his departure, he hears "the noise of branches being shaken" and the voice of our young lady telling him, "I'm kissing the tree, the tree's sending you my kisses" (41).

But even if, in its perpetual turgidity, the tree symbolized "the very manhood of the earth" it represented that manhood only in its primeval state. Since a tree is static, immobile, and cannot "evolve" but only grow in size, it could not represent true spiritual

IV, 15. Giovanni Segantini (1858–1899), ''The Evil Mothers'' (1894)

IV, 16. Dante Gabriel Rossetti (1828–1882), ''The Day-Dream'' (1880)

masculinity. Rooted in the earth and fundamentally passive, it represented only static energy; that is, it stood for the feminine mentality in nineteenth-century thinking. Albine not only loves the tree, and the condition of being a tree, she *is* the tree. Her guardian, when questioned concerning the manner in which he has permitted her to grow up, remarks that it is ''bad to keep trees from growing the way they want'' (41). Women's relationship to nature is as the tree's relationship to the soil in which it grows. Désirée, Serge Mouret's sister, who truly is nature personified, seems, Zola tells his readers, ''to draw strength from the barnyard, to suck life up through her strong legs, as white and solid as saplings'' (48).

Woman's desire to hop into the trees, then, was not merely symbolic of her desire to be ''fertilized'' and of her urge for physical pleasure but also of her ''static'' qualities, her physical, erotic self-sufficiency. As always, classical mythology was a fine source for artists as they developed their iconography for the representation of woman's connection with the trees. The painters simply posed their models as tree ''nymphs,'' as supposedly historical representations of such young, unmarried women as were designated by the original Greek meaning of that word—even though the women they depicted in this manner, aside from having been draped with some usually rather skimpy ''classical'' garb, tended, given their usually very stylish modern hairdos and facial features, to look just like contemporary women of marriageable age. Classical mythology, then, became once again a useful source for appropriate narrative pegs upon which to hang suggestive contemporary images.

The stories of Daphne and of Phyllis were especially useful to this end. Not surprisingly, given his passion for the Ophelia theme, John William Waterhouse portrayed the story of Phyllis and Demophoön in terms of the nymph's affecting passive desire to sacrifice all for love—for, after all, this poor girl had hanged herself when her Trojan warrior failed to return to her side. In Waterhouse's painting [IV, 17], we see her to our full voyeuristic delectation as the almond tree into which the gods had turned her to reward her for her not quite virginal virtue as a masochistic martyr. Daphne, in her unwillingness to get involved with men at all, was, of course, an even more soothing tale for devotees of images conducive to restful detumescence. Unwilling to requite Apollo's love for her, she fled from him, praying to her father for help, and for her effort she was turned into a laurel tree—an incident that gave turn-of-the-century painters a chance to combine a pleasantly titillating show of shapely feminine limbs in elegant conjunction with a wealth of verdant branches.

In general, however, the artists needed little mythological excuse to send their women into the trees. It usually sufficed to call their arboreal explorers dryads or, better yet, hamadryads to designate them as those lowly, simple, spontaneous creatures of nature, the nymphs of the forests and the trees, over whom, judging from their prolific presence in the works of these painters, one was bound to stumble every time one ventured into the nineteenth-century woods.

For at least some of the many women artists of the turn of the century who showed themselves as intrigued as their male counterparts with the perched position of the tree nymph, the latter's presence among the branches clearly had a different meaning. Anne Brigman's remarkable photograph ''The Dryad'' [IV, 18], for instance, is an expression of this superb artist's pervasive pantheistic fascination with woman's body as a source of creative energy. Brigman's dryad is a young woman whose body, rather than being mired in lassitude, is supple and strong; she does not, as in the images produced by men, seem

IV, 17. John William Waterhouse (1849–1917), "Phyllis and Demophoön" (1907)

IV, 18. Anne Brigman (1869–1949), "The Dryad," photograph (ca. 1906)

to teeter precariously on a limb. She is not limp and passive but forceful, independent, and free. Brigman's image represents, in this sense, a woman's wish for liberation from the maudlin manipulative visual mythology constructed by the men around her. The photograph represents a subtle transformation of the means of visual expression from a form of domination into a statement of liberation. Few of her fellow women artists were able to do what she did, and it would seem reasonable to attribute her ability to transcend tradition to her choice of the—in her time still relatively novel—medium of photography to express herself. Since photography was as yet hardly recognized as a medium for significant artistic expression, she was less constrained than the women painters of her time in the development of her individual vision. The women painters, to prove themselves in their predominantly male enclave of expression, were forced to become more conservative than the men, were forced, in order to be taken seriously, to take fewer stylistic and compositional risks. Hence, the sort of independent vision we see in Brigman's work was much less prominent in the work of women painters.

While the women painters tended to sidestep the implications of the ever-louder rumblings of women's dissatisfaction with the stifling effects of the roles into which they had been cast, the predominant male response to these rumblings was to wallow in ever-cruder representations of their favorite antifeminine themes rather than to scotch the mythologized representation of women as passively sexual, self-involved creatures.

IV, 19. Franz von Stuck
(1863–1928), "The See-
saw" (1898)

IV, 20. Paul Klee
(1879–1940), "Woman
in a Tree," etching
(1903)

With the arrival of the 1890s, in consequence, the painters' representation of the self-
directed sexual energy of the dryads became even more explicit. Such artists as the Belgian
Félicien Rops—a favorite among the writers of the French symbolist movement, who never
made a secret of his suspicions concerning the erotic inclinations of women—showed very
contemporary-looking nude women riding ecstatically, but utterly self-involved, on the
branches of trees as if they were horses. He gave these works such innocuous titles as "A
Hama-Dryad," thereby escaping, for the sake of decorum, the otherwise inevitable accusation of pornography. A.-J. Chantron, a perpetual purveyor of naked nymphs, showed
one at the Paris Salon of 1905 positioned slightly more modestly than those of Rops, but

hugging the main trunk of her tree very closely, all the while looking knowingly and with amusement at the viewer from the corner of her eye. This opus he coyly entitled "Ivy." Franz von Stuck, like Rops never one to dawdle excessively in metaphoric subtleties, showed his dryads having a great deal of fun on a seesaw of remarkably organic construction in a work dating from 1898 [IV, 19]. In 1903, Stuck's youthful pupil Paul Klee produced an etching of a "Young Woman in a Tree" in which he demonstrated both his acute fear of women and the characteristic turn-of-the-century sources of his symbolism [IV, 20].

Inevitably, the solitary or collective but always self-directed arboreal ecstasies of the dryads would lead them to lose their grip—women, after all, were prone to such incautious behavior—and to fall out of the trees in which they had pursued their pleasure. If we are to believe the turn-of-the-century painters, the trees of every forest were not only full of dryads but at the foot of every tree one could find some of the former inhabitants of those trees, who, having lost their balance, were now sprawled among the leaves and flowers, in the throes of terminal satiety. Gabriel Guay's tactfully titled "Poem of the Woods" of 1889 [IV, 21] is one of these artistic documentations of feminine recklessness, and Arthur Hacker's "Leaf-drift" [IV, 22] is another. We have stumbled, almost accidentally, it would seem, upon woman in her natural habitat. The season is autumn, the leaves are falling, and among the leaves we encounter a pack of women, blending—for they are themselves nothing more than leaves—with ease into their world of roots and branches. Indeed, the very roots of the trees are echoed in the women's flanks and legs, which seem to dig themselves into the earth.

What is especially striking is that, prostrate and naked, these women of nature remain intensely vulnerable. At the same time, they seem in desperate need of sexual fulfillment. We can almost hear them call to us like animals waiting to be fed. They are passive, but

IV, 21. Gabriel Guay (b. 1848), "Poem of the Woods" (1889)

IV, 22. Arthur Hacker (1858–1919), "Leaf Drift" (1902)

in the intensity of their primal needs their passivity is the source of aggressive suggestions. The woman in the foreground of Hacker's painting, for example, would seem to be suffering from a paralyzing backward curve of the spine. Indeed, it would seem that by falling out of the trees she had injured, perhaps even broken, her back. Yet in the very vulnerability of her position she seems, in her continued ecstatic passivity, to go beyond an appeal merely to the voyeuristic tendencies of the turn-of-the-century male audience. She, her companions, and all the other endlessly repeated images of prostrate women who were seemingly unable to stand up straight catered to a latent fantasy of aggression. Because woman was incapable of independent action—a creature of the earth who lolled about, doomed to wait helplessly yet ever more eagerly for man as her body tensed further with every minute of unfulfillment—it was only natural to take her by force, since by her very behavior she seemed forever to be pleading to be taken by force.

Thus, the turn-of-the-century painters, in creating these images of helplessly ecstatic women, were playing directly on their audience's fantasies of aggression and "invited" rape by depicting women who were extremely vulnerable and naked, usually sprawled flat on their backs in primarily sylvan surroundings, yet who appeared to be in the last throes of an uncontrollable ecstasy. Inevitably these paintings must suggest a perverse combination of intense sexual need and abject helplessness on the part of the women depicted. There is only one way in which these creatures can be joined: We must descend to their level, become part of their world of earth and trees, part of the preconscious universe, responsive to nature's barbaric invitation to engage in the purely sexual, purely materialistic ritual of reproduction. Both Guay's and Hacker's wood nymphs are part of the enormous output of paintings by turn-of-the-century artists catering to an audience of men who, fed in their youth with fantasies of woman as the all-suffering household nun and constrained in their own sexual development by images created by their fathers, were now seeking relief in daydreams of "invited violence," of an abandonment to aggression for which they could not be held personally responsible.

While the peculiar twisted poses these painters tended to give their prostrate female nudes were no doubt at least partially inspired by a rather prosaic search for novelty in an overcrowded field of representation, these poses soon became standard fare and can therefore not be explained away as attempts at "compositional innovation." Instead, the recurrence of salon nudes lying supine in sharp backward angles must be seen as a deliberate transgression of the voyeuristic focus representative of the collapsing woman and even the woman in the trees. The almost spastic, uncontrollable helplessness of these women—as if their backs had been broken in some violent movement, so that they are now sprawled out as if paralyzed, having been made doubly vulnerable by their nakedness, ready to be taken with impunity by any man who happens to be passing by—is a characteristic visualization of the male preoccupation which developed during the late nineteenth century with the notion that women were born masochists and loved nothing better than to be raped and beaten. The sexologists, with their case studies of women eagerly in search of pain, did everything they could to encourage the spread of this notion, never asking themselves what the social origins of the phenomena they recorded might be.

Bernard S. Talmey, in describing the "uncontrollable sexual passion" he observed in one of his patients, could, in fact, have been describing any one of a thousand salon nudes he might have seen not only in Paris but in any expensive "society" saloon in cities across the United States. When reaching ecstasy, Talmey reported, this woman—a French woman, he was careful to point out—"has spasms of the muscles of the neck and back, by which the entire body is stretched backward, the spinal column forming a convex arc at the anterior aspect, the veritable opisthotonus of hysteria" (*Woman*, 135). The word *opisthotonus*, used in the nineteenth century by pathologists to designate, in the words of the Oxford English Dictionary, a "spasm of muscles of the neck, back, and legs, in which the body is bent backward," might indeed serve as a generic designation for the many paintings of women shown in terminal backward spasms of uncontrollable sexual desire which began to litter the walls of the yearly exhibitions in the 1870s and continued to do so well into the twentieth century.

The term *masochism* was invented by Richard von Krafft-Ebing, who, in his *Psychopathia Sexualis* (1886), linked the phenomenon he defined as "the wish to suffer pain and be subjected to force" (86) to the name of Leopold von Sacher-Masoch, a popular author of the period whose heroes usually spent their time in enthusiastic pursuit of maltreatment. But Krafft-Ebing saw the phenomenon of masochism as being a true "perversion" only in men. "In woman," he contended, "voluntary subjection to the opposite sex is a physiological phenomenon. Owing to her passive *role* in procreation and long-existent social conditions, ideas of subjection are, in woman, normally connected with the idea of sexual relations. They form, so to speak, the harmonics which determine the tone-quality of feminine feeling." Nature itself, Krafft-Ebing insisted, has given woman "an instinctive inclination to voluntary subordination to man; [who] will notice that exaggeration of customary gallantry is very distasteful to women, and that a deviation from it in the direction of masterful behavior, though loudly reprehended, is often accepted with secret satisfaction. Under the veneer of polite society the instinct of feminine servitude is everywhere discernible" (130).

The late nineteenth-century male thus had it from the very highest, most advanced "scientific" authority that women, even if they might seem to indicate otherwise, *wanted* to be beaten and subjected to violence. In addition to being instructed by what Krafft-Ebing was saying, men were by 1893 being reassured by such other eminent—and widely read—authorities as Lombroso and Ferrero, that the "normal woman is naturally less sensitive to pain than a man" (*The Female Offender*, 150), so that there was clearly absolutely no reason to be squeamish about pushing women around a bit. On the basis of the "find-ings" of these and other "scientific" observers, the proponents of dualistic thought thus installed another durable antifeminine myth whose ramifications still echo daily through the popular arts of our own time. In the literature of the late nineteenth and early twentieth centuries an author's adherence to the theory that women just naturally liked to be beaten was a sign of extreme intellectual sophistication. It was an indication that one was truly well informed about matters of scientific interest. Hence, instances of this feminine desire grew to be legion in fiction, especially in the works of devotees of Zola and the symbolist poets.

Zola himself set an example for his followers in his portrayal of the supreme goddess of his celestial harem of perverse women in the novel *Nana* (1880). As a matter of fact, if we are to believe Zola it was for Nana one of the more enjoyable features of life to be beaten by the one she loved. In her affair with the actor Fontan, Nana rapidly grows accustomed to the beatings she is habitually given by the man. Indeed, Zola tells us, Nana comes to feel that "it was even nice getting a slap, provided it came from him" (251). She and her friend Satin begin to trade stories about particularly interesting beatings they have received: "Both of them revelled in these antecdotes about slaps and punches, and delighted in recounting the same stupid incidents a hundred times or more, abandoning themselves to the sort of languorous, pleasurable weariness which followed the thrashings they talked of." They begin to be such connoisseurs of pain that even while talking about past incidents "they secretly preferred the days when thrashings were in the air, for the prospect of a beating was more exciting" (255). In fact, Zola tells us, Nana thrives as never before on the systematic violence administered to her by Fontan: "By dint of being beaten, Nana became as supple as fine linen; her skin grew delicate, all pink and white, so soft to the touch and pleasing to the eye that she looked more beautiful than ever" (265).

Zola's American follower Frank Norris, in *McTeague* (1899) characterizes Trina, the woman whose very presence in the world becomes the cause of McTeague's troubles, as someone who instinctively feels "the desire, the necessity of being conquered by a supe-rior strength." Just the thought alone that McTeague might consent "to crush down her struggle with his enormous strength, to subdue her, conquer her by sheer brute force" gets her all hot and bothered: "Why did it please her?" Norris makes her ask rhetorically, "why had it suddenly thrilled her from head to foot with a quick, terrifying gust of pas-sion, the like of which she had never known?" (65) The cognoscenti among Norris' read-ers needed no footnotes to understand their author's call upon their familiarity with the latest scientific disquisitions on woman's innate desires. And, undoubtedly, these readers also nodded their approval of the young author's understanding of the feminine psyche, when later on in the book they were told that in response to the increasing violence of McTeague's behavior toward her, "this brutality made Trina all the more affectionate; aroused in her a morbid, unwholesome love of submission, a strange, unnatural pleasure in yielding, in surrendering herself to the will of an irresistible, virile power" (227).

For Norris this process was part of a "natural" sequence in the eternal encounter between man and woman. As soon as McTeague and Trina kiss, that chain of events commences: "The instant she allowed him to kiss her, he thought less of her." This, Norris insists, "belonged to the changeless order of things—the man desiring the woman only for what she withholds; the woman worshipping the man for that which she yields up to him. With each concession gained the man's desire cools; with every surrender made the woman's adoration increases. But why should it be so?'' (62–63) Well, frankly, the young American novelist knew perfectly well why. It was so, of course, because Schopenhauer had told him it was so. Norris' remarks are no more than a simple echo of a passage in the German philosopher's essay on "The Metaphysics of the Love of the Sexes," in which he remarked that "the love of the man sinks perceptibly from the moment it has obtained satisfaction; almost every other woman charms him more than the one he already possesses: he longs for variety. The love of the woman, on the other hand, increases just from that moment'' (81).

It was thus perfectly all right to force a woman. She could not help her initial resistance to a man's advances, after all, given her basic lack of knowledge about sexual matters. But as soon as she had been initiated, even if the means had been force, her very nature, her very longing for completion, would make her cleave to her master with ever greater intensity. Mark Twain understood the ways of woman quite as well as his younger contemporary and the old German philosopher. In *Eve's Diary* (1906) he has Eve ask herself why it is that she loves Adam. And the answer is clear enough: *"Merely because he is masculine,* I think. At bottom he is good, and I love him for that, but I could love him without it. If he should beat me and abuse me, I should go on loving him. I know it. It is a matter of sex, I think" (101).

Back in France, Pierre Louÿs, in *Woman and Puppet* (1898), had the perverse heroine of that novel, Concha, respond in spasms of yelping ardor to the narrator's violent attack upon her, during which "for perhaps a quarter of an hour" he "struck her with the regularity of a peasant pounding a flail . . . and always on the same spots, the top of the head and the left shoulder'' (218). In a paroxysm of masochistic ecstasy she cries, "How well you have beaten me, my heart! How sweet it was! How good it felt—" Later Concha confesses to her attacker that if she told him lies, it was specifically "to have you beat me, Mateo. When I feel your strength, I love you, I love you so; you cannot imagine how happy it makes me to weep because of you." And, beguilingly, she asks, "Mateo, will you beat me again? Promise me that you will beat me hard! You will kill me! Tell me that you will kill me!'' (220)

Like Louÿs' heroine, Frank Wedekind's Lulu, the archetypal woman at the center of his plays *Earth Spirit* and *Pandora's Box* (as well as Alban Berg's opera based on Wedekind's plays) does not really become interested in a man until he becomes violent toward her. To one of her early lovers she exults, "How proud I am that you will do anything to humiliate me! You degrade me as deep as a man can degrade a woman . . ." (77). For Lulu, as for Concha, the male's violence toward her is supposed to be proof of her power over her man, and this knowledge presumably makes that violence an erotic stimulus for her. The dictum pronounced by another of the men in Lulu's life, that "beating or lovemaking, it's all one to a woman" (122), had become one of the most common clichés among intellectuals at the turn of the century.

Another of these clichés was pronounced by Sue Bridehead in Thomas Hardy's *Jude the Obscure* (1895) when she declared, "I have not felt about [men] as most women are taught to feel—to be on their guard against attacks on their virtue; for no average man—no man short of a sensual savage—will molest a woman by day or night, at home or abroad, unless she invites him. Until she says by a look 'come on' he is always afraid to, and if you never say it, or look it, he never comes'' (154). In other words, a woman who was raped was raped only because she wanted to be raped. Hardy knew this was true because science had proved it to be true.

IV, 23. Pierre Dupuis (b. 1833), "An Ondine Playing in the Waves" (ca. 1896)

It is clear that few of the anti-feminine clichés which had become institutionalized by the 1890s have had a more immediately destructive influence on the daily lives of women throughout the twentieth century than this particular pair of male wishfulfilling items of late nineteenth century "scientific" knowledge. This is the period in which recourse to scientific truth rather than "faith" became the principal justification for the brutal and widespread oppression of human beings on the basis of race and sex, and for the institutionalization of concepts which ultimately led to the blanket justification of violence done to others because one group had decided that another had "asked for it." The women-want-to-be-raped theory is an integral part of the overall self-serving pattern of the rationalization of aggression which still dominates the world today, and which was crucial to the development of the imperialist mentality at the turn of the century.

It may seem a rather bathetic mismatch of causes to point to the supinely sprawling feminine nudes favored by the painters of the Paris salons as a contributing factor to the spread of the aggressive mentality in late nineteenth-century life. But inevitably the mentality of rape, whether it be personal and physical or cultural and intellectual, requires that

IV, 24. Adolf Hirémy-Hirschl (1860–1933), "Aphrodite" (c. 1898)

IV, 25. Alexandre Cabanel (1823–1889), "The Birth of Venus" (c. 1863)

guilt and temptation, and hence the justification for punishment, are to be seen in the other, in this case the woman. All too often the gestures and expressions of ecstatic transport accompanying the supine posture of these nudes suggest a perverse excess of erotic abandonment as the origin of the women's forced posture, as if somehow, in the midst of an intense spasm of uncontrollable desire, they had succeeded in breaking their own backs, thereby dooming themselves to stay forever paralyzed and helpless in the distorted position in which the artist chose to paint them. The sprawling nymphs' helpless postures, joined with their obvious ecstasy, thus suggested quite deliberately to the viewer that these women were, so to speak, "asking to be raped."

Naiads and woodland nymphs with apparently self-inflicted broken backs became a staple of the Paris salon exhibitions, especially during the period between 1880 and 1914. It did not matter whether such women were portrayed as being carried "on the wings of a dream" or, like Pierre Dupuis' playful *ondine,* on the crest of a wave [IV, 23], or whether they had already been rudely washed ashore. Even if the artists' excuse for painting their nude and sprawled bodies was to show them as the personification of "the wave," or "the breeze," or as "Aphrodite" [IV, 24], they always seemed to suffer from the same harsh spinal distortion. Most of these paintings are stylistic echoes of Alexandre Cabanel's *succès de scandale* of 1863, "The Birth of Venus" [IV, 25], which in its own way revolutionized the representation of this theme by having Venus rise out of the waves not standing upright and in control of things, as in Botticelli's famous version, but by having her seem to have floated to the surface from the ocean floor in a conveniently horizontal position. Earl Shinn, commenting on the painting in 1879, remarked laconically that "the form of this personage suffers from bonelessness." Noting this Venus' "delicate and seductive" as well as "rather uncelestial" beauty, Shinn pointed out that "the painter has created her with that absence of soul which evades responsibility" [I, 56]. In addition to exhibiting the lassitude characteristic of the collapsing woman, Cabanel's Venus has landed more than flat on her back. A convenient wave has come along to push up her midriff until there clearly is a considerable strain upon her spinal column. In Shinn's words, "She lies arched on the wave from which she is just born, her whole boneless body twisting on it." In presenting a supposedly classical theme—no less than the birth of a powerful goddess— in a decidedly contemporary fashion, Cabanel had broken new stylistic ground—as well as his model's back, to be sure. But it was certainly not concern for Venus' spinal condition which caught the attention of Cabanel's contemporaries. The well-known critic Philip Gilbert Hamerton remembered the goddess' first appearance at the Paris salon. His remarkably overheated prose still conveys a good sense of the erotic impact of the work among the art lovers of its day:

> . . . they hung in the Salon of 1863 Cabanel's dazzling Venus lying naked on the sea. She lay in full light on a soft couch of clear sea-water that heaved under her with gleams of tender azure and pale emerald, wherein her long hair half mingled, as if it were a little rippling stream of golden water losing itself in the azure deep. The form was wildly voluptuous, the utmost extremities participating in a kind of rhythmical, musical motion. The soft, sleepy eyes just opened to the light were beaming with latent passion, and there was a half-childish, half-womanly waywardness in the playful tossing of the white arms. The whole figure was colored with a dazzling delicacy (*Painting in France,* 110).

IV, 26. Daniel Tixier (active 1890s), "Spring" (ca. 1895)

IV, 27. Albert-Joseph Penot (active 1896–1909), "Autumn" (1903)

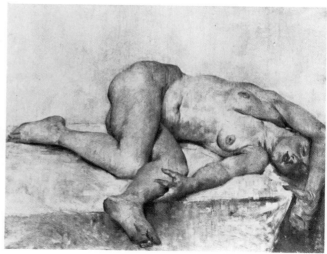

IV, 28. Auguste Rodin (1840–1917), "Springtime," sculpture (1884)

IV, 29. Lovis Corinth (1858–1925), "Reclining Nude" (1907)

Hamerton's description represents a compendium of the elements in the representation of woman which most immediately aroused the aesthetic sensibilities of the late nineteenth-century male. It is therefore not surprising that many acolytes hurried after the master into this tortuous realm of supine feminine beauty. The age of the nymph with the broken back had arrived.

Cabanel's curve was copied by his admirers in what was often a rather slavish fashion, as Daniel Tixier's personification of springtime [IV, 26] and Albert-Joseph Penot's "Autumn" [IV, 27] can attest. Both artists were content to paint a virtual reprise of Cabanel's Venus; only the setting had been changed from sea to land. In addition, Penot had updated his painting by placing in his young lady's listless hand a mandolin, the iconographic sign of *la cigale,* the grasshopper, the fabled unwise woman whose self-destructive exploits were dear to the purveyors of turn-of-the-century culture. In their version of La Fontaine's fable the grasshopper, a beautiful young temptress, played as long as the sun shone, while the domestic ant, the self-effacing housewife, busily cleaned and ordered her middle-class nest. Now autumn had come and the frivolous young lady was getting cold (and old). The domestic ant would not let her into the house which self-effacement had built, and the beautiful, selfish woman's shell was doomed to wither in the ravishing onslaught of nature: "For her, too, as for the gardens, the fields, and the forests, spring has passed," commented Victor Nadal in his *Le Nu au Salon* of 1903, adding, with the grating enjoyment of a man who sees his enemy down, "and though during the next month of May the wild rosebush will regain the adornments of its betrothal, the woman, less favored by nature than the meanest shrub, will have one more disillusionment and one more wrinkle in her brow."

Whatever his undoubted merits as an innovator in the realm of sculptural techniques, in his choice of themes and narrative predilections Auguste Rodin was very much a man of his time, as the truly acrobatic backward curve of the woman in his own version of "Springtime" [IV, 28] clearly demonstrates. In fact, the more often the theme of the nymph with the broken back was repeated, the more extreme became the distortions into

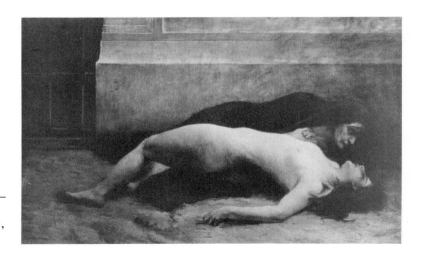

IV, 30. Susanne Bethe-
mont (active 1895–1905),
"The Martyr" (c. 1905)

which the artists tried to force their long-suffering models in order to come up with something new. Lovis Corinth's "Reclining Nude" [IV, 29]—a preposterously understated title—is an example of the "opisthotonic" nude in its most extreme form. Thrown down
forcefully her arms and legs sprawling every which way, as if they were the loose, unattended limbs of an abandoned marionette, this woman is deliberately portrayed as no more
than an insignificant sexual plaything. Corinth's painting represents what is perhaps the
nadir of meretricious representations of sexual abandon at the turn of the century. The
artist is generally regarded as a forerunner of German expressionism, and this fact casts a
sobering light on the extent of the innovations of that movement, whose subject matter all
too often represents no more than a simple continuation of the antifeminine themes characteristic of the work of their stylistically less adventurous contemporaries. Just as Corinth's image does not signal a new beginning in art, but is, on the contrary, merely a
characteristic example of a mode of visual representation driven to an extreme—a factor
of the painter's attempt to "do something new" with an already overworked theme—so
even the stylistic innovations of early twentieth-century expressionism may not unreason-
ably be seen as the logical extension of the practice of the representation of physical
contortions into the realm of visual distortion. The expressionists were, in their perception
of relationships within the world, extremely conventional but, driven by the emphatic
demand for something new, they succeeded in constructing a mode of visual notation
commensurate with the aggression and violence characteristic of the subject matter pre-
dominating in the conventional art of their time. In other words, if the expressionists
succeeded in liberating painting from certain stylistic conventions, their explorations of a
new visual language may actually have been not much more than a logical extension of
the violent content of conventional turn-of-the-century art. The expressionists certainly did
nothing to liberate their work from the vicious antifeminine imagery of the academic painters.

Various forms of what might be termed "academic expressionism"—the pursuit of
melodramatic exaggeration in the production of versions of overused themes—pervaded
the work of many painters active around 1900. For instance, it is tempting, but no doubt
quite inaccurate, to regard Susanne Bethemont's 1905 painting entitled "The Martyr" [IV,
30] as a stark commentary by a woman painter on the plight of the models who had to

accommodate the artists in their dogged pursuit of the theme of the nymph with the broken back. Unfortunately, it is more likely that Bethemont simply accepted the iconographic stereotypes of her time unquestioningly and was here exploiting the then popular themes of nunlike virtue and the "beauty" of feminine paralysis and death in order to make a dramatic—and commercially viable—statement.

Bethemont's obviously exaggerated use of the pose of the nymph with the broken back is, in its extremist appropriation of this theme, analogous to that perpetrated by an unidentified artist in his portrayal of one of art's favorite women in bondage, Andromeda [IV, 31]. This grotesque work was chosen to illustrate the Reverend Ebenezer Cobham Brewer's description of the fate of this heroine in his voluminous *Character Sketches of Romance, Fiction and the Drama,* a work which went through numerous editions in the United States during the 1890s. This Andromeda may well be a work by Paul Merwart, an elusive but often-reproduced Russian artist of this period who specialized in depicting hefty, unclad Teutonic beauties in tortuous poses.

This particular painting's already tenuous links with classical mythology are made even less credible by the absence of either Perseus or the monster he was to slay in the course of the captive's rescue. Clearly, the only monster which threatens this nymph is the distorted libido of the male viewer, who is quite blatantly invited to take violent advantage of the apparent opisthotonic ecstasy exhibited by Andromeda. Paintings such as these help to clarify the direct link between the aggressive "invitation to rape" fantasy represented by the works of painters of the Cabanel curve and a similar "take them by force" notion reflected in the late nineteenth-century vogue for paintings depicting scenes from life in barbarian times or among "barbarian" peoples.

With the arrival of the 1880s and the ever-increasing chest-pounding, "manly" nationalism of the advocates of the "New Struggle for Life Among Nations" (the title of an 1899 article by Brooks Adams for *McClure's Magazine*), there also arose a special fascination with earlier times, when men were men and women were mere property. Many

IV, 31. Unidentified artist (possibly Paul Merwart, 1855–1902) "Andromeda" (ca. 1885)

IV, 32. Fritz Hofmann-Juan (Friedrich Theodor Max Hofmann, b. 1873) "A Raid on the Women" ("Frauenraub") (ca. 1913)

IV, 33. Paul-Joseph Jamin (1853–1903), "Brenn and His Share of the Plunder" (1893)

middle-class men dreamed of those simple times when the sight of a male was enough to make a woman cringe, and when, if you wanted a woman, you simply reached out and took her. By the time the German painter Fritz Hofmann-Juan painted his 1913 version of this fantasy [IV, 32] it had clearly become unnecessary to give its representation a historical flavor, for the protagonists in his drama are thoroughly modern men and women acting out the battle of the sexes. But a decade or two earlier it had still been customary to dream of a return to barbaric times and fantasize about being a ruthless, all-powerful warlord with an ample supply of helpless women always at your beck and call, so that if you felt the urge to have a woman you simply chose one from among your loot, as the lordly barbarian depicted in Paul-Joseph Jamin's 1893 painting "Brenn and His Share of the Plunder" [IV, 33] is about to do. Standing in the doorway of his storage room, our hero, formidably masculine, surveys his booty, his sandals still soaked in the blood of his enemies. Among other trophies and valuables—jars full of gold coins, decorative objects, and decapitated heads—we find a handful of appropriately vulnerable naked women, some with their hands bound behind their backs to make them even more helpless, others in various stages of panic.

The intent of this great-grandfather of Conan the Barbarian is perfectly transparent, and the painting's invitation to the viewer to become an accomplice of the dominant male in his violent attack upon the helpless maidens is equally direct. Jamin, who habitually painted such works, titling them "Rape in the Stone Age" (his entry for the Salon of 1888), and so on, was only one of a legion of painters who took it upon themselves to show their audiences how wonderfully simple matters of sex were in the world they had lost to civilization. Such paintings also helped to remind them, in the words of Arnold Bennett (who insisted that he was "a feminist to the point of passionateness") that "women in the main love to be dominated. They are not entirely happy until they are dominated," that is, "at least in appearance" (105) he added as an afterthought. Even the title of Bennett's book, *Our Women* (1920), in the emphatically possessive designation of its subject, shows that Bennett's soul was still among the barbarians—or at least with the writers of the Napoleonic Code, who had declared (Article 1124) that "woman was given to man so that she could give him children. She is therefore his property, just as a fruit tree is the property of the gardener."

The notion of woman as man's personal property and the sweet dreams of servitude that notion implied were very pleasing to the middle-class male, with his well-developed acquisitive urge and its concomitant aggressive energies. He therefore loved to dwell on scenes which made woman's abject subservience thoroughly explicit. In the midst of a book filled with wisdom concerning woman's proper behavior toward man, Charles Butler, as early as 1836, had described a scene worthy of any late nineteenth-century history painter. Woman, he said, had first shown her true mettle in the age of "gross barbarism," as the just paragon of servitude. It is "she who stands dripping and famished before her husband, while he devours, stretched at ease, the produce of her exertions; waits his tardy permission without a word or a look of impatience, and feeds, with the humblest gratitude, and the shortest intermission of labour, on the scraps and offals which he disdains; she, in a word, who is most tolerant of hardship and of unkindness" (*The American Lady*, 18).

Where the "woman loves to be beaten" theory teamed up with the notion of "woman as personal property," sadistic violence became the nexus of masculine "creativity" in the world of nineteenth-century culture. Not surprisingly, it was to that most staunchly insistent advocate of the "angel in the house" conception of feminine virtue, Coventry Patmore, that thoughts of love among the barbarians came with a special intensity: "Lo," he cried—obviously to make his women readers pleased with the advances of civilization—"how the woman once was woo'd; / Forth leapt the savage from his lair, / And fell'd her, and to nuptials rude / He dragged her, bleeding, by the hair" (186). With scenes such as this one, Patmore and his ilk not only provided late nineteenth-century painters with useful illustrations of barbarian aggression against women but also became, on an even more bathetic level, directly responsible for the endlessly recurring sexist "club-wielding caveman" jokes of twentieth-century cartoonists. In the late nineteenth century, barbarians carrying off naked women in batches could be found roaming through early history at the yearly exhibitions of the academies in numerous paintings with titles such as "The Booty," "In the Grip of the Sea Wolf," and "The Spoils of War."

Specific "historical" events of a catastrophic nature permitting the quantitative display of abjectly posed nudes were considered especially heroic, given the august example of Eugène Delacroix's "The Death of Sardanapalus," that celebrated showpiece of eroticized feminine carnage. Delacroix had provided the following instructive description, which was included in the catalogue for the Salon of 1827–28: "Sardanapalus gives his eunuchs

IV, 34. Georges Roche-
grosse (1859–1938),
"The Death of Babylon"
(1891)

and the officers of his palace orders to cut the throats of his wives, his pages, and even his favorite horses; none of the objects which had served to give him pleasure were to survive him."

Offered this inspiring example by Delacroix, habitual late nineteenth-century purveyors of barbaric carnage such as Georges Rochegrosse clearly felt called upon to increase the body count. In "The Death of Babylon" of 1891 [IV, 34], for instance, he brought the aesthetic of violence to its apotheosis. While the epic of war rages within the portals of iniquity and sin, the painter has taken time off to dwell upon the exhaustion of the flesh which—and he was clearly pleased to be able to emphasize this fact—was the true cause of the fall of Babylon. But in this painting morality hatched a voyeuristic egg. Virtually the entire foreground is taken up by lovingly depicted collapsed and exhausted naked female slaves, most suffering from severe cases of the Cabanel curve.

The sadistic invitation to the viewer presented by this symphony of supine women was strikingly expressed by Armand Silvestre, who described with a connoisseur's enthusiasm what Rochegrosse's painting meant to the painter's contemporaries in his *Le Nu au Salon des Champs-Elysées* of 1891. Rochegrosse's women, said Silvestre,

> are rolled up like serpents in an environment of unhealthy humidity, some of them stretching their limbs in relaxation, while others, in contrast, curl up into themselves as if to savor the touch of their own skin, gathered together in voluptuous folds. Vainly a great cry of terror wells up around them. They do not hear! What is more, why should they care? The instinctive wisdom which rests at the bottom of our unconscious tells them that they will never find a sweeter time to die without even waking from their dreams. The victor's sword can come and plunge into these naked bosoms without finding even a single heart there not long since fallen into torpor. The sacrilegious fury of the soldiers shall bring to these superb cadavers the ultimate profanation—but, once again, why should they care?

In this panoramic opus of rape—and, if one accepts Silvestre's final remark, even necrophiliac violation—as well as in numerous works of a similar nature produced by Rochegrosse and scores of painters like him, the late nineteenth-century mind shaped the imagery which was to give form to the cinematic imagination of early Hollywood film directors such as Cecil B. DeMille and hence became responsible for the visual structure of numerous motion picture spectaculars of the first half of the twentieth century.

But even more popular than these scenes of ancient carnage were scenes of contemporary "oriental" slavery. Painters such as Benjamin Constant, LeComte du Nouy, and Jean-Léon Gérôme in France and stalwart British contributors to the Academy such as Arthur Hacker and Ernest Normand allowed their public to feast their eyes on instances of "barbaric oriental slavery." In scene after scene of slave market doings, as in Normand's huge opus "Bondage" of 1895 [IV, 35], the "timeless cruelty of the East" was brought vividly to life for the many eagerly inquisitive closet anthropologists to be found among the businessmen and intellectuals of Europe and America. In "Vae Victis! The Sack of Morocco by the Almohades" [IV, 36] Arthur Hacker unflinchingly presented the gathering of a considerable cache of helpless captives before the appreciative eyes of their Arab victors—and, in the process, gave his British audience a lesson in oriental ways of dealing with women. A critic for *The Magazine of Art* covering the Royal Academy Exhibition of

IV, 35. Ernest Normand (1859–1923), "Bondage" (ca. 1895)

IV, 36. Arthur Hacker (1858–1919), "Vae Victis! The Sack of Morocco by the Almohades" (ca. 1890)

IV, 37. Stephan Abel Sinding
(1846–1922), ''The Captive
Mother,'' sculpture (1889)

1890 discussed the painting in a manner which showed him to be a true forerunner of the
''let content be damned'' formalist critical mentality which was to rule the twentieth century. Hacker, said this appreciative observer, uses ''the subject merely as a peg on which
to hang the real problem the painter aims at solving, to reproduce the various whites in an
Oriental city—the courtyard, with all its dazzling reflections, and the white skins of the
captive women and children, the whole bound together with the strong colours of the rich
plunder'' (255).

Jean-Léon Gérôme, dean of the Orientalist painters, thrilled his throngs of followers
with scene upon scene of harem doings and the merchandising of women. In his painting
''The Slave Market,'' for instance, he put woman in her place by using a surface narrative
of anthropological documentation to allow his audience to fantasize about the pleasures of
buying or selling a willing woman, a transaction inevitably accompanied by such appropriate horse fair rituals as checking the merchandise's teeth to make sure the beast was
healthy. Artists also liked to dwell on the ability of captive mothers to remain meek,
tolerant, and saintly under all circumstances. The Norwegian sculptor Stephan Abel Sinding impressed the art world in 1889 with his sculpture ''The Captive Mother'' [IV,
37], who is shown crouching and bending forward to suckle a baby she could not touch because
her arms had been bound behind her. Works such as Sinding's were meant to be heartfelt
representations of the glorious virtue of altruism, an estimable quality, but one which the
nineteenth-century philosophers had most magnanimously designated as woman's exclusive possession and principal responsibility. In Auguste Comte's words, woman was by
nature ''more keenly alive to the charm of self-sacrifice'' (*System of Positive Polity*, IV,
100).

Still, in light of the late nineteenth-century male's keen sensitivity to the charm of
domination, it is clear that the elegantly curved, sacrificial posture of Sinding's captive
mother, the ripe fullness of the breast she offers her child, and the casual ''child-Bacchus''
pose of ease of the baby were elements deliberately calculated to play upon his yearning
for the basic dualistic simplicity of male mastery and feminine submission. It is interesting
to note that in Sinding's sculpture the child is the master: He not only is completely at
ease but has the imperially casual air of taking what is his birthright, as if he were a

Roman emperor eating his fill from a bunch of grapes held over his head by a voluptuous and very servile slave. Thus the viewer is invited to take the position of the imperious child and enjoy all the pleasures of dominance while at the same time avoiding any unpleasant sense of adult responsibility.

Undoubtedly it was precisely this mixture of a vicarious feeling of mastery, the pleasing sense of predestined suffering on the part of a beautiful victim, and the comforting knowledge of the absence of any personal responsibility for the misery portrayed which made Orientalist scenes of female slavery so delectable to the art lovers of the fin de siècle. This same conjunction of stimulants also accounts for the popularity of scenes of religious persecution. Under the cloak of a progressive criticism of the intolerance of religious fanatics, painters throughout Europe and America concerned themselves with documenting the historical suffering of religious martyrs. The Briton Herbert Schmalz (no painter had a name more exquisitely descriptive of the general tenor of his work) painted a bevy of very naked, praying, slumping, fainting yet extremely well-fed Christian beauties tied to stakes in a Roman arena, on the verge of being devoured by lions. ''Faithful unto Death'' this tumescent masterpiece was called when the proud creator exhibited it at the Royal Academy in 1888. The French painter Moreau de Tours specialized in martyrs, stigmatics, and ecstatics culled from the annals of church history, whom he depicted variously as nailed to a cross, bare-breasted and supine, or otherwise voluptuously exhibiting their wounds to scientifically interested church officials. In 1891 Philip Hermogenes Calderon's depiction of ''St. Elizabeth of Hungary's Great Act of Renunciation'' showed this sainted woman bending in submission to the church's altar, supremely naked and excruciatingly vulnerable, thus making hearts pound with more than religious fervor at the Royal Academy.

Perhaps most graphically characteristic of this genre was the Portuguese painter José (de) Brito's very exposed, formidably proportioned, and wild-eyed ''Martyr of Fanaticism'' [IV, 38], one of the successes of the Paris World's Fair in 1900. An appeal to the viewer's sadistic enjoyment of the violent torture of a helpless young woman is the real subject of this painting, not the virtuously progressive concern for religious freedom im-

IV, 38. José (de) Brito (1855–1946), ''A Martyr of Fanaticism'' (1895)

plied by the title. Paintings such as this demonstrate the very similar functions of the resignedly submissive Orientalist slaves, the piled-up bodies of the historical massacres, the religious ecstatics, and the supreme sacrifice of Fernand Khnopff's and Albert von Keller's sainted martyrs.

The period's continuing fascination with the self-sacrificial, consumptive, or dying woman and the self-effacement of the household nun found symbolic expression in these historical documents. They were all, in one way or another, a reinforcement of the notion fostered by nineteenth-century culture, both in the social realm and in the realm of sexual relationships, that the only pleasure which was truly meaningful was the pleasure of having power over others, of controlling the lives of others. Life, these paintings proclaimed, echoing the predatory social philosophy of Herbert Spencer and the Social Darwinists, has meaning only in terms of the subjection of the other: The self "lives" only when perched like a vulture upon the supine, lacerated, absolutely submissive body of the other. Few men had an actual opportunity to experience that kind of selfhood in the realm of commerce or politics; therefore many sought it at home, creating the mythology of the naturally submissive woman to justify their demand that their wives be models of abject self-effacement. Men everywhere came to demand that marriage be the continuous enactment of a master-slave ritual that would fulfill in fantasy their search for a power which escaped them in real life.

Nicholas Francis Cooke, the man who was so appalled by the evidence of women's autoerotic practices, became himself erotically aroused when thinking of religious martyrs of the sort depicted by Schmalz and Brito:

> Nothing could stay the transports of these daughters of the Savior—victims to their gratitude and love for Him who had delivered them from bondage. They eagerly poured out their blood upon the scaffold, and gayly reposed their bodies upon the flames. Their delicate limbs were crushed and mangled in the most horrid tortures, but their tranquil and calm souls breathed only of peace and happiness. Kneeling upon the scaffold, mangled by wild beasts, bathed in blood, they seemed, in their sublime courage, in their ineffable sweetness, like veritable angels from heaven (*Satan in Society*, 243).

Physicians, sexual researchers, and anthropologists came running in to prove that ever since the world began and wherever humanity had settled, woman's natural pleasure was to suffer. Talmey, citing as evidence for his scientific conclusions writings by Schiller, Goethe, and Krafft-Ebing, stated: "A moderate degree of submission to the wishes and the will of the man she loves is . . . characteristic of the feminine nature, and is not abnormal. Even entire sexual bondage is not properly pathological, if it be only the means of obtaining or retaining possession of the ruling man. It is not perversion if fear of losing the companion and the desire to keep him always amiable, content and inclined to love, are the motives for submission" (*Woman*, 128).

No wonder that aspiring young artists like Paul Klee were by 1901 busily trying to design artistic projects centered on "fettered female nudes" (*Diaries*, 53). Indeed, with this sort of scientific knowledge to guide them, it is not at all surprising that the "adventurous" young men of the 1880s and 1890s would dream of the attentions of a sweetheart of the type described by Jules Laforgue in one of his poems. He imagined that such a woman might "one beautiful evening, come, desiring but to drink from my lips, or die," and would be so grateful for that bit of masculine condescension that, like a lapdog, she would return to be his slave, as "an escapee, indeed, half-dead / To curl herself up on the

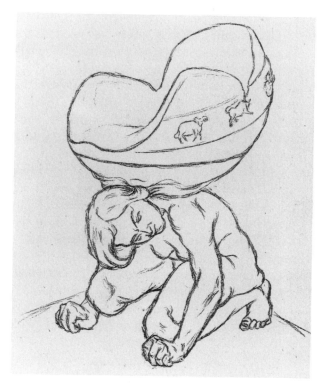

IV, 39. François-Rupert Carabin (1862–1932), "A Seat," drawing (ca. 1893)

mat I will have placed for her before my door. / Thus will she come to me, her eyes alive with folly / And with those eyes she'll follow me everywhere, everywhere!'' (*Oeuvres Complètes,* II, 115–17) In Oscar Wilde's *The Picture of Dorian Gray* (1891) Lord Henry Wotton consoles Dorian for his marginal sense of guilt in having driven the young actress Sibyl Vane to suicide. Remarks Wotton, ''I'm afraid that women appreciate cruelty, downright cruelty, more than anything else. They have wonderfully primitive instincts. We have emancipated them, but they remain slaves looking for their masters, all the same. They love being dominated. I am sure you were splendid'' (116). So much, then, for the need to feel guilty about doing violence to women. The intellectuals had decided that, as Wedekind's character Rodrigo says in *Pandora's Box,* ''beating or love-making, it's all one to a woman'' (122).

Edgar Saltus, an American writer who dabbled ceaselessly in symbolism, in his *Historia Amoris* (1906) enthusiastically described the exploits of the ''real'' Bluebeard, the fifteenth-century French nobleman (and companion to Jeanne d'Arc) named Gilles de Retz. Gilles, he pointed out, ''after marrying Catherine de Thonars, one of the great heiresses of the day, subsequently and successively married six other women. Whether he murdered them all or whether they died of delight is not historically certain.'' The ''key spotted with blood,'' which in the fable Bluebeard gives to his seventh wife, he said, symbolizes ''the eternal curiosity of the eternal Eve concerning that which has been forbidden'' (197).

To keep woman from entering into such forbidden explorations, the late nineteenth-century male mind devised telling contraptions. In the illustrated catalogue of the *Salon des Beaux-Arts* for 1893 one can find—some seventy-five years in advance of similar productions by such arrière-garde purveyors of pop art imagery as Mel Ramos and Allen Jones—two drawings of sculptures in the form of chairs by François-Rupert Carabin. The first of these drawings shows a naked woman, who serves as support for the backing of a chair. The other is even more straightforward: The molding of an armchair has been placed on the back of a young woman who grimaces in pain as she leans forward, pressed in upon herself by the weight of the chair molding she must carry and whose sole, agonized support she is [IV, 39]. This drawing represents in the most literal way imaginable the twisted turn-of-the-century dream of masculine mastery which insisted on seeing woman as merely a piece of household furniture. If woman did not behave according to such expectations, she was declared to be perverse. After all, she ought to know better. History itself was man's witness. Saltus, for one, had quite explicitly delineated the domestication of woman: "In the beginning of things woman was common property. With individual ownership came the necessity of defence. Man defended woman against even herself. He beat her, stoned her, killed her. From the massacre of myriads, constancy resulted. With it came the home" (2). Unfortunately, even with such vivid examples before them of the progress made in the protection of women "even from themselves," many late nineteenth-century women still refused to comply with men's demands that they serve as the seat of all domestic comfort, and the men lashed back at them in angry vengeance. The dualistic pendulum was taking another swing back from the glorification of woman as the sacrificial virgin to her excoriation as a predatory whore. Indeed, as we shall soon discover, even a woman's virginity ultimately came to be seen as evidence of her perverse desires. The "battle of the sexes" was declared an "inexorable law of nature" and an intense forty-year pogrom against women was on, to be suspended only in the face of the more immediate bloodshed of the First World War.

CHAPTER V

Women of Moonlight and Wax; the Mirror of Venus and the Lesbian Glass

The nymph with the broken back was the late nineteenth-century's most graphic casualty of the concept of therapeutic rape. But in the developing war of science and philosophy against the shortcomings of woman, many ordinary housewives as well were falling victim to a form of domestic ambush zealously advocated by some of the period's leading medical authorities. In response to declining birthrates among the middle classes—the very backbone of developing civilization—these authorities called upon husbands everywhere to assert their masculinity behind the drawn curtains of their dignified homes, and to do so by force if necessary. For the newly popular theories of human evolution held that in women signs of evolutionary progress were accompanied by a diminished sexual drive: "The sexual instinct in the civilized woman is, I believe, tending to atrophy," declared Harry Campbell, a prominent London pathologist, in his book *Differences in the Nervous Organization of Man and Woman* (1891).

This supposed withering of the sexual instinct in civilized women made it all the more important that the male take his part in the reproductive function very seriously. For the disappearance of the lower passions in woman imposed on the male the responsibility to ascertain the continuance of the human race. He should therefore realize, Campbell argued, that while in the lower species "the female must lend herself to the sexual act," something an animal could not be expected to do "without sexual instinct," the matter was quite different where the evolved human race was concerned. Using the popular example of "savages and semi-barbarous communities" as the (for an evolutionist) somewhat paradoxical source of guidelines for the modern male, Campbell concluded that, given the ruthless muscle flexing of these "primitive ancestors, sexual instinct in the woman has not been essential to the successful performance of the sexual act." The implication was clearly that the modern husband did not need to be abashed by his wife's reticence in sexual matters, and that a bit of barbarism on the part of the male was essential to the continuance of the human species. Echoing the opinions of many of his colleagues, Campbell told the late nineteenth century male that ''woman need not be a willing agent'' (39), thus sanctioning, indeed, encouraging the practice of domestic rape.

Woman's culturally enforced passivity and ignorance thus became the turn-of-the-century male's excuse to play Ghenghis Khan or some other barbarian conqueror on the home front. In the same year Campbell published his book, Louis Dumur, in the April issue of the *Mercure de France,* had casually remarked, ''As in the case of a fortified camp, there are three ways of taking a woman: assault, trickery, or famine'' (''Little Aphorisms,'' 329). Peaceful coexistence did not figure highly among the cultural values of this period. Physical assault on those women who were showing the very reticence they had been forced to cultivate by their men was counseled by many authorities. Proudhon, in his *On Pornocracy, or Women in Modern Times* (1858), had declared: ''Woman does not at all dislike to be treated a bit violently, or even to be raped'' (267). Auguste Forel, in *The Sexual Question* (1906), remarked confidently: ''It is notorious that many women like to be beaten by their husbands, and are not content unless this is done'' (239). Characteristic of the reliability of the scientific research of these experts is the case of Lombroso, who, in making a similar claim, cited Zola's *Nana* as one of his primary pieces of evidence.

Woman, it had been decided, needed to be educated in matters of this sort even if she resisted her education. Her inherently passive nature required it, for, said Campbell, ''a large proportion of women do not experience the slightest desire before marriage'' (200–201). Once she had been taken by force, she was likely to learn to submit dutifully, for it was part of woman's nature to imitate incessantly. Her passive nature made her incapable of original thought or action, but she had a protean capacity to take on whatever form she was given to imitate. The great Charles Darwin himself, in *The Descent of Man* (1871), had placed the stamp of his all-knowing scientific approval on this tendency in woman: ''It is generally admitted that with woman the powers of intuition, of rapid perception, and perhaps of imitation, are more strongly marked than in man'' (643), he asserted, relying magnanimously on that magical authority, ''general admission,'' so useful to the nineteenth-century scientist.

By the mid-1870s, the idea that woman was inherently an imitator, not an originator, had become one of the most pervasive clichés of Western culture. Because of her propensity for imitation, the stage came to be seen as the place where woman could best express her contribution to the cultural life of civilized society. For what was acting if not a form of imitation? It is by no means an accident that the years around 1900 were years of triumph for actresses everywhere. Women such as Sarah Bernhardt and Eleonora Duse became heroines of mythical proportions. But it was not because they were seen as particularly original that the actresses became celebrated. Instead, they were seen as unusually successful in exploiting the imitative bent of ''woman's nature.''

In the literature of the day actresses were ubiquitous. They served as heroines whose prurient potential was outstanding since, being associated with the stage, they were a self-evident part of the sinful underbelly of turn-of-the-century culture. In addition, casting their heroines as actresses made it easier for writers to show their up-to-date ''scientific knowledge'' of psychology by heavily exploiting the theme of woman's imitative nature. Zola's Nana, with her pitiful histrionics and delectable body, held down the bottom end of the legion of actresses. There was Trilby, of course, with her passive stage capacities only ''activated'' by Svengali's hypnotic spells; and there was Dreiser's sister Carrie, who, being "naturaly imitative" (104) as well as having a "slight gift of observation, and that sixth sense, so rich in every woman, of intuition" (13) was the Darwinian woman personified.

No wonder, then, that Carrie became a success on the stage—although Dreiser's "voice of the Ideal," the intellectual Ames, warns her not to view her acting ability as "talent," for "it so happens," this wise man stresses, "that you have the power to act. That is no credit to you. You might not have had it. It isn't an excuse for either pride or self-glorification. You paid nothing to get it" (485). Dreiser took great pains to make it clear to the reader that acting was for Carrie an intuitive gift, not a talent based on creative ability.

Henry James stressed very much the same thing in giving his reader an account of his heroine Verena Tarrant in *The Bostonians*. James tells us that Verena was "naturally theatrical." In his terms that meant that she had no originality, no personal creative capacity. Like Trilby, she had to be placed in something of a trance, she had to be activated like a robot to become the riveting public speaker she was supposed to be. "Divinely docile," she moved through life very much like a sleepwalker. Like so many fictional women from this period, Carrie included, Verena when she came on stage at first seemed to be "talking in a dream." She proceeded "as if she were listening for the prompter, catching, one by one, certain phrases that were whispered to her a great distance off, behind the scenes of the world" (60–61).

In the eyes of their authors, these fictional women had the characteristic feminine imitative capacity of "reflecting" the world in which they lived without ever really understanding its deeper meaning, its intellectual dimension. Oscar Wilde's Dorian Gray has to come to that terrible discovery of woman's limitations in the realm of creativity as part of his initiation into the "real" world of experience. Watching the young actress Sibyl Vane perform in a local theater, he falls head over heels in love with her, for on the stage she literally becomes the (male-created) women whose roles she enacts. Given her capacity to act, he thinks that she has "genius and intellect." Impetuously he declares his love to her, and now it is Sibyl's turn to be ecstatic: "Dorian, Dorian,' she cried, 'before I knew you, acting was the only reality of my life. It was only in the theatre that I lived. I thought that it was all true. I was Rosalind one night and Portia the other.' " But love is fatal to Sibyl, because her only intellectual existence is in the roles she plays. Reality brings her back into the fold of "common women." Love brings reality. Hence love kills woman's mystery. "I knew nothing but shadows, and I thought them real. You came—oh, my beautiful love!—and you freed my soul from prison. You taught me what reality really is."

But, of course, the real Sibyl is merely a common creature without any poetic qualities of her own. Dorian thus learns that what he had fallen in love with in the first place was woman's capacity to mirror the male mind: Imitation was her only talent, and her descent into reality has now robbed her of this talent. Hence Sibyl has no "value" anymore: "Yes," Dorian cries, his aesthete's soul wounded to the core, "you have killed my love. You used to stir my imagination. Now you don't even stir my curiosity. You simply produce no effect. I loved you because you were marvelous, because you had genius and intellect, because you realized the dreams of great poets and gave shape and substance to the shadows of art. You have thrown it all away. You are shallow and stupid!" (100–101) Dorian, in consequence, throws Sibyl away as if she were no more than a wilted flower. Presumably in a moment of masochistic ecstasy she commits suicide—after all, as we have already seen, Wilde was quite certain that women love to be dominated. Thus, Wilde implies that by dying in a state of dramatic self-negation Sibyl regained some of the mythic status she had foregone by returning to reality.

Woman's nature as a "reflector" could, of course, also be depicted without resorting to her proclivity for acting. Flaubert, in his story "A Simple Heart" (1876), created an elaborate parallel between the self-effacing woman-servant Félicité—the sort of ideal working-class beast of burden every middle-class family dreamed of having—and the parrot which became the object of her personal veneration, itself a classic symbol of imitative behavior in the realm of nature. Woman and parrot: In civilization each lived, at best, a reflected existence. Félicité, as a servant, had rightly spent her life imitating her masters. In worshiping her parrot she instinctively chose to worship the god of imitation, the absurd essence of her meaningless existence.

If, as Harry Campbell had declared confidently, "in imitativeness and lack of originality [woman stood] conspicuously first" (232), it was no wonder that the moon had come to stand for the essence of everything that was truly feminine in the world. The moon, too, after all, existed only as a "reflected entity." It had no light of its own, just as woman, in her proper function, had existence only as the passive reflection of male creativity. The sun was Apollo, the god of light, the moon Diana, his pale echo in the night. As usual, the late nineteenth century adapted elements of classical mythology in a manner which, while shaky in the realm of historical scholarship, perfectly suited the symbolic relationships expressive of its contemporary cultural ideology.

Thus, whether she moved within the most humble sphere of domestic endeavor or within the highest realms of public life, woman was always, first and foremost, reflective and not originative in essence. As an actress she reflected the creations of the male mind, and as a housewife it was her role to do the same: "Woman, given the nature of her consciousness, has been placed between her husband and her children as a living reflector," said Proudhon, "and it is her mission to make concrete, to simplify, and to transmit to the younger brains the thoughts of the father" (87). Indeed, even science had found justification for the link between woman and the moon. Havelock Ellis, in his *Man and Woman* (1894) pointed out that the "curious resemblance" of a woman's menstrual cycle "to the lunar cycle was long ago noticed. More recently Darwin had suggested that the connection between physiological periodicity and the moon was directly formed at a very remote period of zoological evolution, and that the periodicity then impressed upon the organism has survived until the present day" (282). In a sense, then, woman was a natural child of the moon, of Diana.

The seemingly physical bond between woman and the moon was used by writers and artists not only to depict her as the moon goddess—which, certainly, they did in endless variations—but also to explain the natural origins of her pallor: her white skin, her invalid's condition, her "consumptive" passivity. For, as Ellis also pointed out, "the fact that women are thus, as it were, periodically wounded in the most sensitive spot in their organism, and subjected to a monthly loss of blood, is familiar, and has been used . . . to explain numerous phenomena. It has even been suggested that to the weakening influence of this cause we must attribute the early arrest of development of girls in height, muscles, larynx, etc." And although Ellis himself recognized that "there does not seem to be any real ground for this supposition" (282), many of his contemporaries thought otherwise.

For the artists the link between the moon and woman—her weakness, her imitative nature, her passivity, and her emotional waxings and wanings—was a subject with far too many attractive symbolic possibilities to ignore. On its least dramatic level the connection was made through the fetishization of the "beauty" of the pasty, unhealthy coloring of so many of the later nineteenth-century middle-class women who did their best to stay out of the sun. For in their "infirmly delicate, translucent" whiteness, as creatures vividly "incapable of useful effort," it was their function to represent the credit of their husbands and thus be "valuable as evidence of pecuniary strength" (107), as Veblen put it. In 1871 Tennie Claflin spoke derisively of the "marble contour" (23) women of fashion tried to adopt. The grayish, waxy "moonlight" coloring of these women's skin was thus valued highly. When, for instance, Edna Pontellier, in Kate Chopin's *The Awakening* (1899), dares to walk out into the sun without her gloves on, her husband reacts with concern: "You are burnt beyond recognition," he remarks in obvious exaggeration, and Chopin shows him, a characteristic member of the class of seekers after pecuniary decency, "looking at his wife as one looks at a valuable piece of personal property which has suffered some damage" (201).

Where economics teaches a society to value a specific form of beauty, artists are never hesitant to rush in and abet that taste. Thus, the painted beauties of the second half of the nineteenth century virtually all had skins composed of equal parts of moonlight and wax. No painter was more famous for his ability to produce this texture than Bouguereau, who made pasty skins a standard feature of his women. Earl Shinn, whose attitude toward this painter could perhaps best be described as one of appreciative distaste—at one point he called his work "faultily faultless" (*Art Treasures,* III, 76)—was at his best whenever he set out to describe the characteristics of this famous Bouguereau epidermis. He called the skin of the painter's women "parchments scraped down with a razor" (I, 54), and in trying to give a flavor of the artist's "clean and waxen" style (I, 43), he talked about his "polished and marble cutting" textures (III, 67), "as if the canvas had been scraped all over with a piece of glass, and then waxed like a bit of Queen Anne furniture" (II, 11).

The dulled, passive whiteness of moonlight and the sterile glazing of wax were the perfect textures with which to create the ideal women of the later nineteenth century. The woman of moonlight and wax came to exude, through the pasty texture of her skin, the virtuous pliability of the wife as household virgin. After all, the moon goddess Diana (Artemis in her Greek guise) was in mythology a virgin as well as a mother deity, a goddess of fertility, and the guardian of children and childbirth; yet despite all this she remained spotlessly chaste.

In *The Art Treasures of America* Earl Shinn gave an instructive description of Bouguereau's "Charity," a work whose structure and general theme were virtually identical to the painter's "Alma Parens," except that in "Charity" the all-giving mother was seated in a stately architectural alcove rather than out in wild nature. For this work, Shinn wrote, Bouguereau was able to compose "a group so placid, fixed, and elegant, that we seem to be rapt into some ideal state of being, where the accidents of life are unfeared or unknown, and the underlying serenities of existence alone exist in their permanent manifestation." This

> tranquil young "Charity," who sits in her shrine with the tenderness of a peasant-mother and the fixity of an idol, has an eternal look which is reassuring for the permanence of the human virtues; no fear that she will change tomorrow and become uncharitable. The temperament is fixed, her beneficence is everlasting. The babes cling to her with the confidence of instinct;

V, 1. Thomas Wilmer Dewing (1851–1938), "The Garden" (or "Spring Moonlight") (ca. 1891)

one is at her breast, one spells his lesson at her feet, one cowers in the shelter of her robe and looks out upon the world as if its most hostile powers were futile against the asylum built up out of Charity's mantle.

Not "caught up with the storms and accidents of humanity upon her," this Charity is the epitome of womanly compliance, a reflector, a moon of tranquility, showing "the patience of long habits of benevolence, the repose of a native world of goodness that is subject to no change, no chance, no caprice," and "the refinement of goodness that forgets self and spends for others" (II, 30–31).

In its ideal form this moon dream of woman as soundless, emotionless moonbeam continued to populate the paintings of symbolists well into the twentieth century. Artists such as Puvis de Chavannes, Henri Martin, and especially Alphonse Osbert painted women moving effortlessly through, or lying exhausted in, moonlit woods, mysterious creatures of stasis in a turbulent world. Often they were icons of quietude, as in Thomas Wilmer Dewing's "The Garden" (also often called "Spring Moonlight") [V, 1]. Evelyn de Morgan showed woman in both her guises as moon and earth [V, 2], although in contrast to similar depictions by male painters, in her version woman was not a mischievous nymph riding on the moon or stretched out along its edge but rather a real woman imprisoned by the moon and struggling to escape from that bondage—perhaps because she was awakening to the destructive import of the symbolism which had enclosed her. In a similar fashion, de Morgan's "Earth," sleeping and not at all the alert, buxom, all-giving fantasy of such painters as Bouguereau or Otto Greiner [see IV, 3], seems very unhappily caught—almost like a latter day image of Emily Brontë's "Prisoner"—among the harsh, aggressive spear points of a barren, inhospitable mountain range.

Male painters, however, remained oblivious to the merits of such philosophical ambiguities. For them woman simply remained the moon in its various stages: waxing, wan-ing, rising, setting—or, as in Herbert Draper's entry to the Royal Academy exhibition of 1900, a rather forbidding figure reluctant to relinquish her reign, standing at ''The Gates of Dawn'' [V, 3].

The problem with these women of moonlight and wax, as Draper's painting would tend to indicate, was that in their cool, reflected perfection they were too icily self-contained. They often seemed to spend the lonely hours of the night, whose faint light they embodied, not in dutifully beaming back the godlike light of the sun of man but in using it to warm their senses and ''energize'' their world of passive identity.

The roundness of the moon seemed also to symbolize the distant, cool, circular self-enclosure of woman—that isolation in solitary self-enjoyment which had made the female figures in Lord Leighton's ''Flaming June'' and ''Summer Moon'' [see III, 6] collapse into themselves under the soft enticements of the moon. As a matter of fact, Leighton and his high Victorian coadjutor, Sir Lawrence Alma-Tadema, had learned much from Bouguereau's methods of painting feminine skin. Both specialized in what Cosmo Monkhouse, referring to Alma-Tadema in his book *British Contemporary Artists* (1899), called ''the representation of light-reflecting surfaces and textures'' (193). What more perfect light-reflecting surface could there be than the moonlike skin of woman? In the work of both these painters the textures of marble, diaphanous cloth, and feminine skin are coextensive variations on a single theme: opulence, fabulous luxury—in short, conspicuous consumption. In their works, woman and the materials and textures with which they surround her are all reflective of the sun of man, the provider, the conqueror. As a privileged posses-

V, 2. Evelyn (Pickering) de Morgan (1855–1919), ''Sleeping Earth and Wakening Moon'' (ca. 1900)

V, 3. Herbert Draper (1864–1920), "The Gates of Dawn" (ca. 1900)

sion, a mother-of-pearl creature, she was owned and hence unattainable to all but the one man whose reflected light she was.

Arthur Symons, in his poem "Clair de Lune," declares himself hopelessly distant from the woman he yearns for. Even though he has "known" her in the biblical sense, he does not "own" her. He therefore sees "In the moonlit room your face, / Moonlight-coloured, fainting white," yet finds "Lips that are not mine to kiss" and "Eyes that are not mine to keep / In the mirror of mine eyes" (*Selected Poetry and Prose,* 49). If woman was a reflector, a mirror, then the mirror itself was symbolic of woman: a "ghost," a reflection of passion, an elusive image, a self-contained, self-encircling entity. As Symons also said in his poem "Moonrise," in the encounter between the sun and the moon the sun must set, man must tire "from the sorrowful and immense fatigue of love," and helplessly "watch the moon rise over the sea, a ghost / Of burning noontides, pallid with spent desire" (51–52).

The moon represented woman's sterile, self-reflective identity. It was, as Jules Laforgue insisted in his poem "Climate, Fauna, and Flora of the Moon," "the wrath of nothingness," "the circle of Immaculate Conception," a "dead mirror." In another poem called "Clair de Lune" (one of the turn of the century's most overused titles) he spoke of the moon as an "eye as sterile as suicide" (*Oeuvres Complètes,* II, 16). Endlessly fascinated by the resemblance between woman and the moon, Laforgue came to see something lunar in all of woman's attributes. Describing the nymph Syrinx in his *Six Moral Tales,* he pictured her as lifting "her arms toward the pure firmament where Hecate will shine this very evening. As a result of this gesture, her two white breasts rise a little under her diaphanous tunic. Pure and lunar, they efface themselves" (110). Diana, Artemis, and Hecate were the late nineteenth-century's intermeshing trinity of goddesses representing the feminine mentality, linked together without much justifiable authority because they were collectively seen to be symbolic of the same archetypal lunar qualities of woman. The images lingered, were used and reused, their denigratory implications soon becoming cryptic. How many of the more recent readers of Edmund Wilson's *Memoirs of Hecate County* have understood in its title Wilson's subtle promise to his readers of a delineation of the topography of the "imitative, lunar inanities of the feminine mind?"

Woman, as the moon, was characterized by her distance from man. Inevitable disappointment awaited the male who sought to approach her. It was as if the sun, the active principle, must forever struggle with the passive inertia of primal being. The god of solar heat certainly did his best to try and make the floating, circling, whirling women of night, "The Mountain Mists," of Draper's painting of 1912 [V, 4], perish in his ardor, as the couplet accompanying the painting had indicated, but the frustrating thing was that when the sun's energies were spent, and night fell again, the moon was still there, round, immobile, mysterious, inaccessible, self-contained. The primordial battle between night and day thus came to seem symbolic of the battle of the sexes. To conserve the sun's energies for creative purposes of a higher order, the moon should be fed with as little light as possible and be isolated, as it were, from the sun. The late nineteenth-century male had seen to it that such a process of starvation was initiated.

Now, to his amazement, he saw that remarkably little sunlight was needed for the moon's continuation. Woman, it seemed, was entire unto herself, perhaps needing the male to be fructified but otherwise remaining completely self-contained. As part of the undifferentiated primal matter, she lived largely in a state removed from individual consciousness. She was, in the words of Erich Neumann, nature, the primordial, maternal

V, 4. Herbert Draper (1864–1920), "The Mountain Mists" (ca. 1912)

uroboros, the feminine principle which, as "the Great Round, the Great Container, tends to hold fast to everything that springs from it and to surround it like an eternal substance" (*The Great Mother,* 25). Neumann, a close follower of Jungian psychology, was writing about the tradition of the "Great Mother" in the 1950s, but his assumptions and, consequently, his findings were a logical extension of the symbolic structures which operated within the turn-of-the-century mind. Flaubert, in *The Temptation of St. Anthony,* had identified the uroboros as the bisexual primal state of being, about which "Oannes," who appears to Anthony "with the head of a man and the body of a fish," informs the hermit saint:

> I have dwelt in the shapeless world, where slumbered hermaphrodite animals, under the weight of an opaque atmosphere, in the depths of gloomy waves—when the fingers, the fins, and the wings were confounded, and eyes without heads floated like molluscs amongst human-faced bulls and dog-footed serpents.
>
> Over the whole of those beings Omoroca, bent like a hoop, stretched her woman's body. But Belus cut her clean in two halves, made the earth with one, and the heavens with another; and the two worlds alike mutually contemplate each other. I, the first consciousness of Chaos, I have arisen from the abyss to harden matter, to regulate forms; and I have taught men fishing, the sowing of seed, the scripture, and the history of the Gods (230–31).

"Omoroca," the uroboros—the circle of chaos with a woman's body—had been split in two for the sake of evolution. The male half became the creator of mind, the heavens. Woman remained behind as the nurturer of the body, the earth, and needed to be regulated and directed. By herself woman tended to return instinctively to the uroboric conditions, the state of primitive bisexual self-sufficiency, dominated by the feminine urge to physical self-reproduction.

Woman, Otto Weininger had argued in his book *Sex and Character* (1903), "is always living in a condition of fusion with all the human beings she knows, even when she is alone; she is not a 'monad,' for all monads are sharply marked off from other existences. Women have no definite individual limits" (198). Therefore, while woman was the personification of the circle, the symbol of self-containment, of the uroboros, the snake biting its own tail, she was also representative of the static, unindividualized, nondifferentiated being expressive of the unthinking condition of brute nature. Nicholas Cooke, in his *Satan in Society* (1870), had already described a fundamental conflict in the makeup of men and women: "There is far less variety of temperament among women. They seem, in this respect at least, to be cast more in a common mold than men. It would seem that, in the designs of Providence, each man has to follow the paths of a special destiny, and consequently is endowed with special aptitudes. The common destiny of women does not exact those profound and essential differences among them which are remarked among men" (279). Max Nordau—a characteristic turn-of-the-century academic with a talent for saying what everyone else was saying with such an immoderate air of self-confidence that he was considered a true genius—insisted in his book *Paradoxes* (1885) that "woman is as a rule, typical; man, individual. The former has average, the latter exceptional features." Therefore, he continued pleasantly, "there is incomparably less variation between women than between men. If you know one, you know them all, with but few exceptions" (48–49).

The symbol of woman, the self-contained round, the uroboros, began to appear with increasing frequency in turn-of-the-century art. A drawing simply entitled "Woman," which appeared in *Jugend* in 1896, showed a woman suspended in time, caught in a state of suspended animation, in the uroboric circle of her primordial materiality, quite literally represented by the archetypal symbol of a snake biting its own tail. [V, 5]

More frequently the self-containment of woman was symbolized by less obvious representations of her uroboric, generalized being. In the art of the late nineteenth century she was encompassed, decoratively or "organically" or both, by an endless variety of circles which crowded around her in the form of garlands, wreaths, and swirls of cloth, or even in the form of a mundane "tub"—that ubiquitous tool of feminine toiletry which in its convenient circularity made a symbolic statement about the inscrutable self-contained nature of women—in works ranging from the impressionist images of bathing women by Degas [V, 6] to the virtually photographic depictions of the academics. Gaetano Previati, in "The Dance of the Hours" [V, 7], symbolized the struggle between night and day as the conflict between the sun of the world of ideals, beaming into eternity, and the moon of static being, with its bevy of meaninglessly whirling hours, stuck forever in sublunary circularity.

Edwin Howland Blashfield joined the themes of feminine weightlessness and the sleeping woman, whose morality was suspect in the luxury of her repose, and thereby succeeded in linking the idea of a mirror image, or self-reflection, with the uroboric circle to symbolize the always slightly mysterious conjunction of woman, sleep, and poetry [V, 8]. Likewise,

V, 5. Unidentified artist, "Woman," drawing (1896)

V, 6. Edgar Degas (1834–1917), "The Tub," pastel (ca. 1885)

V, 7. Gaetano Previati (1852–1920), "The Dance of the Hours" (1899)

the secret of Rodin's "source" [V, 9] would seem to reside primarily in the uroboric circularity with which, as effectively as a cage, it seems to contain the water sprite within it, thus preventing her from pursuing the males she is poised to pounce upon beyond the boundaries of her appointed realm.

As earth, earth mother, vulval round, moon, and mirror of nature woman was a simple reflection of the world around her. She was the arable soil of the material world. She existed in and for what she mirrored, and unless she mirrored the world of man, she mirrored brute nature, the world of woman, herself. Thus, paradoxically, as long as woman lived among women, she lived alone, completely self-contained. She mirrored other women and other women mirrored her. Her purity was her self-containment, her inviolate sex the mirror of her existence, barren though that existence might be. Yet as long as woman existed apart from man, she existed as Woman, the great undifferentiated, static expression of primal being. Thus, her womanhood was a source of continual fascination to her. To see herself was her only hold on reality: If she was the mirror of nature, then ·water, the natural mirror, was the source of her impersonal, self-contained self-identity. To prevent loss of self she had to reassure herself continually of her existence by looking in that natural mirror—the source of her being, as it were, the water from which, like Venus, she had come and to which, like Ophelia, she was destined to return.

Burne-Jones' "The Mirror of Venus" [V, 10] is a striking representation of this notion. In a landscape as barren as her own self-contained existence, we encounter woman—

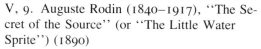

V, 8. Edwin Howland Blashfield (1848–1936), "Sleep and Poetry" (ca. 1886)

V, 9. Auguste Rodin (1840–1917), "The Secret of the Source" (or "The Little Water Sprite") (1890)

V, 10. Sir Edward Coley Burne-Jones (1833–1898), "The Mirror of Venus" (1898)

Venus—in her undifferentiated multiplicity (for the women she is are separated only by the color of their dresses and their hair, not by individualized facial traits or "personality"). She stares at her manifold reflection in a pool of water, "the first mirror," that dangerous "mirror of Venus," the archetypal pool of her identity, circling in profound silence along the interlacing limbs of women's interdependence. The women, who are symbolic of Woman, are perfectly, absolutely reflected in the virginal, unbroken surface of this pool. As they look at themselves they are filled with the wonder of their existence, yet they remain wistfully aware that as soon as the surface of the pond is violated, as soon as the mirror is broken, they will lose their reflection, their interlacing, collective identity, and will, in effect, die.

For, as we have seen, it was part of late nineteenth-century convention that the woman who truly loved a man must die. It was her wish to expire so she might prove her love, but, in fact, she really had no choice. The breaking of the mirror, the transfer of the jewel of her containment, the violation of barren chastity, of woman's absolute purity, whether on an actual, physical level or a symbolic, spiritual one, was, after all, in woman a sign of her total abandonment to the will, the identity of the male. In succumbing to love the true woman was, as it were, forced to break her mirror, to forgo her identity, and instead be spiritually absorbed by her man. If, on the other hand, she recklessly sacrificed her self-containment and broke her mirror, succumbing to worldly desire without being given a secure harbor for her selfhood in the identity of a willing lover, mental disorientation and madness were the inevitable result.

V, 11. Luis Bonnin (b. 1872), "The Water Mirror" (ca. 1901)

V, 12. Ernest Bieler (1863–1948), "The Mysterious Water" (ca. 1912)

At such times woman, crazed and thrown back into her animal state, the uroboric snake uncoiled, would crawl back to her water-mirror, to the source of her precivilized, undifferentiated feminine identity, as, for instance, she is seen doing in a striking drawing by Luis Bonnin [V, 11]. These were the hysterics, the womb-driven women, forever narcissistically in search of their own reflection, hoping to find in the circular mirror of universal undifferentiated womanhood their lost uroboric wholeness. These were the madwomen described by Lombroso and Ferrero in *The Female Offender:* "Their disposition is profoundly egotistical, and their absorbing preoccupation with themselves makes them love scandal and a public sensation" (219). Such women forgot their passive roles and became crazed bacchantes, the sort of women of whom Tennyson's Lady of Shalott, whose "mirror cracked from side to side," was a decorous and still redemptively sacrificial example.

William Holman Hunt's depiction of her madness [see II, 10] was a remarkable visualization of the theme of woman and her mirror, her circular existence and its imprisoning yet stabilizing function in woman's existence. Hunt's lady lives in a world of endlessly recurring circles, but her mirror has cracked, her self-possession is leaving her, chaos is invading her previously well-ordered world, and we are all too well aware of the fact that her rush to death must be just around the corner.

No wonder that, having been given madness and death as their only alternatives to the stabilizing glance in the mirror, women in turn-of-the-century art and literature were forever looking in the mirror, any mirror, but preferably a circular or, even better, an oval mirror. The water-mirror was, of course, best of all, as in Bonnin's drawing or Ernest Bieler's painting "The Mysterious Water" [V, 12], an even further stylized version of Burne-Jones' "Mirror of Venus."

Given the extreme demands made upon women for their sacrificial complicity with male ideals of feminine self-effacement, it is not surprising that the overly optimistic expectations of late nineteenth-century males were most often sorely disappointed by their wives and lovers' perverse unwillingness to "break their mirrors" and abandon their selfhood to become the willing reflections of their men. Before long the theme of the mirror as the symbol of feminine self-sufficiency came to be linked inextricably with its traditional use as a symbol of woman's vanity. The themes of woman's virginal self-containment and her search for selfhood in the mirror of her material being soon blended with man's irritation at woman's failure to sacrifice her ego to his superior being. Thus, a woman's glance in the mirror became representative of her perverse unwillingness to recognize that it was her natural, predestined duty to yield her ego to man's will.

Vanity, self-absorption, the reflected qualities of woman's moonlike existence, her passivity, her imitativeness—all these themes came relentlessly into play in the woman-and-her-mirror theme. Mrs. E. Lynn Linton, judge of mores and the "Dear Abby" of her time, declared in *Modern Women* (1889): "Woman, we suspect, lives always before her glass, and makes a mirror of existence." Women were self-conscious creatures indeed, said Mrs. Linton, but "it is self-consciousness of a very peculiar and feminine sort—a consciousness not of themselves in themselves, but of the reflection of themselves in others" (253–54). No wonder, then, that Swinburne felt called upon to praise her, when she died, as "Kind, wise, and true as truth's own heart" (*Collected Poetical Works*, II, 1217). Cooke thought that, given woman's self-conscious nature, one might "almost believe that a mirror invisible to others, always reflected her to herself." But, he added, "to the 'know thyself,' in its large philosophical sense, she is an entire stranger" (*Satan in Society*, 284). Max Beerbohm, echoing this sentiment sardonically in "The Pervasion of Rouge" (1894), wrote that "loveliness shall sit at the toilet, watching her oval face in the oval mirror" (123). Arthur Symons, thinking of another of his (probably mythical) women in "White Heliotrope," remembered "the mirror that has sucked your face / Into its secret deep of deeps, / And there mysteriously keeps / Forgotten memories of grace" (*Selected Poetry and Prose*, 47). In *The Waste Land*, T. S. Eliot gave a disdainful portrait of a mindless typist, who in her solipsistic universe, terminally self-involved, "turns and looks a moment in the glass, / Hardly aware of her departed lover" (*Complete Poems and Plays*, 44).

Predictably, actresses, who were thought to express most directly the "inherent imitative tendency" of women, were also the ones who, in literature as well as art, were most relentlessly fixated on their mirrors. Once again it was Zola who helped shape the visual representation of this almost criminally self-absorbed woman with his celebrated depiction of Nana's autoerotic interplay with her mirror. "One of Nana's pleasures," Zola tells us,

> consisted of undressing in front of the mirror on her wardrobe door, which reflected her from head to foot. She used to take off all her clothes and then stand stark naked, gazing at her reflection and oblivious of everything else around her. A passion for her body, an ecstatic

admiration of her satin skin and the supple lines of her figure, kept her serious, attentive and absorbed in her love of herself.

When Nana's aristocratic admirer, the Count Muffat, enters at this point, she pays him scarcely any attention and continues to admire herself:

> Nana had grown absorbed in her ecstatic contemplation of herself. She had bent her neck and was gazing attentively in the mirror at a little brown mole just above her right hip. She was touching it with the tip of her finger, and by leaning backwards was making it stand out more than ever; situated where it was, it presumably struck her as both quaint and pretty. Then she studied other parts of her body, amused by what she was doing, and filled once more with the depraved curiosity she had felt as a child. The sight of herself always surprised her, and she looked as astonished and fascinated as a young girl who has just discovered her puberty. Slowly she spread out her arms to set off her figure, the torso of a plump Venus, bending this way and that to examine herself in front and behind, lingering over the side-view of her bosom and the sweeping curves of her thighs. And she ended up by indulging in a strange game which consisted of swinging right and left, with her knees apart, and her body swaying from the waist with the continuous quivering of an almeh performing a belly-dance (220–22).

Nana demonstrates graphically how unimportant, how unnecessary, loving a man is for a woman who has become fascinated with her mirror image as the symbol of her independence from men.

> Nana was hunching her shoulders. A little shiver of emotion seemed to have run through her limbs, and with tears in her eyes she was trying, as it were, to make herself small, as if to become more conscious of her body. She unclasped her hands and slid them down as far as her breasts, which she squeezed in a passionate grasp. Then, holding herself erect, and embracing her whole body in a single caress, she rubbed her cheeks coaxingly first against one shoulder and then against the other. Her greedy mouth breathed desire over her flesh. She put out her lips and pressed a lingering kiss on the skin near her armpit, laughing at the other Nana who was likewise kissing herself in the mirror (223).

Zola's American follower, Theodore Dreiser, also had the heroine of his *Sister Carrie* sit endlessly before the mirror. In his narrative the mirror becomes Carrie's oracle and guide, the stabilizing force of solipsistic introversion which allows her to triumph blindly in a world in which men who ought to see—but cannot see beyond woman—are doomed to perish. "The mirror," Dreiser tells us, convinced Carrie "of a few things which she had long believed. She was pretty, yes indeed. How nice her hat set, and weren't her eyes pretty? She caught her little red lip with her teeth and felt her first thrill of power" (76). Given the fact, Dreiser argued, that most human beings were by nature, "after all, more passive than active, more mirrors than engines" (78), it was logical that woman, the essence of the passive element in being, would find stability before "the glass." Even Carrie's mind is a mirror. "She looked into her glass and saw a prettier Carrie there than she had seen before; she looked into her mind, a mirror prepared of her own and the world's opinions, and saw a worse. Between these two images she wavered, hesitating which to believe." Of course, between conscience, the mind's mirror, and the real mirror, promising mindless pleasure and ease, there could be no real contest, for Carrie possessed only "an average little conscience, a thing which represented the world, her past environment, habit, convention, in a confused, reflected way" (89). Thus the materialistic mirror is woman's reality. "She sees but one object of supreme compliment in this world and that is herself" (100).

Mark Twain said the same thing when he made his Eve look in the water mirror and comment that this ''is where I go when I hunger for companionship, someone to look at, someone to talk to. It is not enough—that lovely white body painted there in the pool— but it is something, and something is better than utter loneliness. It talks when I talk; it is sad when I am sad; it comforts me with its sympathy; it says, 'Do not be downhearted, you poor friendless girl; I will be your friend.' It *is* a good friend to me, and my only one; it is my sister.'' The woman in the water mirror is hence the true object of Eve's ecstasy. Having stirred the water, her own reflection disappears, and Eve is disconsolate until, ''after a little, there she was again, white and shining and beautiful, and I sprang into her arms! That was perfect happiness; I had known happiness before, but it was not like this, which was ecstasy.'' Woman and the primal water which was the symbol of her chameleonic fluidity were thus united. Twain also made sure that his readers would not forget the ''eternal'' closed triangle composed of woman, the mirror, and the moon. ''At night she would not come if it was dark, for she was a timid little thing; but if there was a moon she would come'' (46–47) Lester Ralph, in his superbly executed drawings for Twain's book, played on this aspect of woman's mirror nature; in one drawing he showed Eve dipping her toe into the water and thereby turning its surface into a series of concentric uroboric circles.

Woman's circularity, her disturbing self-sufficiency, made her exist in an environment of perpetual self-reinforcement as long as she gazed into the glass. In 1887 Arthur Symons, in his poem ''Laus Virginitatis,'' described the virgin of his title as she sat before her glass and murmured, ''The mirror of men's eyes delights me less, / O mirror, than the friend I find in thee.'' Symons' virgin was indeed quite pleased with ''the shadow of myself'' she found in the mirror. ''I to myself suffice,'' she declared contentedly, ''myself the limit to my own desire, / I have no desire to roam.'' (*Collected Works*, II, 72) Late in life Symons was still in pursuit of woman and her mirror. In his poem ''Stella Maligna'' from *Lesbia, and Other Poems* (1920), the poet emphasized the circularity of the glance in the mirror. ''She sat before her mirror and she gazed / deep into eyes that gazed at her again'' (39). In the mirror woman's eyes became the eyes of the medusa hypnotizing her, drawing her ever further into herself. But the deeper woman was drawn into herself, the further she drew away from man's civilizing influence and the more dangerous she became. Then her eyes became ''pools where many a drowned hope lies, / They shine above the dead who sleep below'' (36).

The painters loved that gorgonlike aspect of woman's glance into the mirror, pseudomythologizing it, as always, to dignify it as a subject. They especially liked to portray an obscure incident taken from the story of Perseus and Andromeda, described as follows by Dante Gabriel Rossetti in his poem ''Aspecta Medusa,'' itself meant to accompany his drawing of the same subject:

> Andromeda, by Perseus saved and wed,
> Hankered each day to see the Gorgon's head:
> Till o'er a fount he held it, bade her lean,
> And mirrored in the wave was safely seen
> That death she lived by.
>
> Let not thine eyes know
> Any forbidden thing itself, although
> It once should save as well as kill: but be
> Its shadow upon life enough for thee (100).

Rossetti, in effect, tells woman that she should look into the mirror only under man's guidance. Without Perseus to hold the Gorgon's head, without the mirror's reflection of woman's submerged evil nature, woman's glance would become the glance of knowledge, of the most dangerous kind of knowledge: knowledge of forbidden things, self-knowledge. It was, he argued, better to let sleeping feminine egos lie.

Most of the popular fin-de-siècle painters offered analogous visual admonitions to woman not to peek into the mirror of self without the tempering supervision of a man to guide her. Only a truly perverse woman—a lamia, she who was the very incarnation of the temptress, the snake of forbidden knowledge—could dare to do so. The painters were also well aware that a prominent emphasis on the sinful nudity of one or more of these criminally self-concerned creatures was not likely to hurt the commercial viability of their work. Paul Gervais' "The Mirror" [V, 13] is a characteristic example of the sort of production in which the artist's sense of moral concern was decidedly less of a factor than his admiration for the corporeal pulchritude of his self-absorbed sinners. Gervais also showed himself to be a characteristic devotee of the period's reduplication principle in art by painting not only two women in rapt admiration of their reflection but by managing to bring several versions of the theme of circularity and of the water mirror into his image as well.

Other painters showed a greater devotion to the principle of moral seriousness in their explorations of the theme. Giovanni Segantini, for one, hastened to chastise the perversity of woman once again by showing her naked and unashamed, vanity personified, staring deeply into "the source of evil," the water womb of the earth, in which knowledge writhed about among its gigantic uroboric snake rings, destroying woman's modesty, dependence, and self-forgetfulness, and inducing her instead to grow into a flower of evil. [V, 14].

V, 13. Paul Gervais (1859–1934), "The Mirror" (ca. 1907)

V, 14. Giovanni Segantini (1858–1899), "The Source of Evil" (or "Vanity") (1897)

It was clear that a good woman would not care to look into the glass. That so many women spent so much time before their mirrors anyway was indicative of their perverse attraction to the medusa's head. The turn-of-the-century painters did not have to paint explicit analogies of this sort. The woman-and-her-mirror theme and its implications were well enough known to allow them to have their viewers draw their own conclusions. Still, virtually every painter thought it his responsibility to expose the dangerous game woman played when she looked into the mirror. Consequently the painters of the turn of the century did everything they could to produce new variations on the theme of woman and her fascination with her own reflection. There is scarcely a figure painter of the period who did not undertake to paint "woman before the mirror." Many, indeed, returned to the theme over and over again. A few, such as the American expatriate painters Frederick Frieseke and Richard Emil Miller, managed to get a woman with at least a hand-held mirror into virtually every painting they produced.

It is interesting to note that the initial suggestion for Zola's description of Nana before the mirror may actually have come from his friend Manet's interpretation of Zola's heroine, painted in 1877, three years before the writer published his novel [V, 15]. Manet, fascinated by Zola's delineation of the youthful Nana in *L'Assommoir,* painted her in the early stages of her career, as a prostitute, even while Zola was bringing her center stage in his narrative of her postadolescent life.

Manet's familiar image of "Nana" is not nearly as noisily and emphatically a study in feminine self-absorption as Zola's description of the scene was to be, but it is nonetheless fraught with visual intimations of the theme. Standing plumply and prominently sideways in the center of this painting, Manet's Nana has briefly permitted herself to look

V, 15. Edouard Manet (1832–1883), "Nana" (1877)

V, 16. Henri Caro-Delvaille (1876–1928), "Brunette at the Mirror" (1906)

away from the oval mirror in which she has been studying herself, as if to give the viewer a casual and only mildly curious glance. The viewer, indeed, is placed in the role of a gentleman caller, one who is likely to take his own marginal position on the sofa alongside the top-hatted suitor already seated there, to wait impatiently until Nana is through primping and admiring herself and may deign to give them her attention. Manet's message is clear enough: Nana stands self-contained, self-absorbed, largely unmoved by the concerns of the men around her.

Manet, as well as many of the other impressionists who continued to paint figures, notably Renoir, returned to the theme of the woman before her mirror time and again. But the academic painters were anything but backward in their own pursuit of variations on the mirror theme. By the end of the first decade of the twentieth century, as a matter of fact, it had become perhaps the single most prominent excuse for figure paintings at the Parisian "salons de peinture." These paintings usually had simple descriptive titles such as "Vanity," "The Mirror," "Her Reflection," and so on. Henri Caro-Delvaille, in a work he sent to the Venice Biennale in 1909, virtually turned the mirror theme into a version of Twain's self-reflective water mirror, portraying his Eve, like Twain's, ready to jump in as if her image were being reflected in the very waters of the undifferentiated feminine [V, 16].

Usually these painted women were so fascinated by their own images that they sidled up as closely to their mirrors as they could, very much in the manner of Caro-Delvaille's weighty young woman. The odds-on favorite tool of the painters for the depiction of

V, 17. Albert von Keller (1844–1920), "Portrait of a Lady" (ca. 1906)

feminine self-involvement, however, was the hand-held round or oval mirror. With its hint of the moon, its provocative vulval shape, and the versatility it gave the painters when they wished to depict the dramatic gestures of feminine egotism, it was an irresistible prop. Albert von Keller's "Portrait of a Lady," exhibited at the 1906 summer exhibition of the Munich Secession [V, 17], is a typical example of this kind of image. Passionately involved with herself, this young woman becomes the personification of vanity. At the same time, the painting is characteristic of the peculiar, self-denigrating involvement of many society women, especially immediately after the turn of the century, with the painters' antifeminine fascination, an involvement that caused them to permit—even ask—these painters to portray them as creatures of terminal vanity, goddesses of evil, or symbols of bestial passion.

Carl Strathmann's "Vanity" is a less personalized version of the oval mirror theme [V, 18] than von Keller's, but as Strathmann's busy Jugendstil work already indicates, the theme of woman and her mirror was given a considerable boost by the Arts and Crafts vogue of the 1890s, since during this period designers everywhere—in a typical example of the escalating exchange of influences—were turning out massive numbers of elaborately carved oval hand mirrors decorated with wood nymphs, cupids, Ophelias, Venuses, and so on. Mary G. Houston's design is typical [V, 19], combining as it does a fascination with art nouveau's organic line, the vulval ovoid shape, and a relief of an Ophelia-like woman entwined in her own hair, apparently caught just as she was about to float away among the reeds. Advertising, too, was catching on fast. By 1913, for example, *Jugend* carried a full-page ad for a food supplement called Biomalz, in which an enticingly uncovered young woman in tip-top physical shape stares intently into her oval hand mirror while a black servant boy enters with her Biomalz.

Women painters also frequently pursued the theme. Mary Cassatt's "The Child at the Mirror" [V, 20] is one of numerous paintings in which she explored the various ways in which a woman and her mirror were thought to reflect each other. In this painting Cassatt succumbed to the period's appetite for reduplication of symbolic props by placing the mother and child in front of a large stationary mirror, which, one might say, serves as a window into the closed room of their lives. At the same time, the mother coaxes her

V, 18. Carl Strathmann (1866–1939), "Vanity" (ca. 1910)

V, 19. Mary G. Houston (active ca. 1895–1908), design for a mirror (ca. 1899)

V, 20. Mary Cassatt (1844–1926), "The Child at the Mirror" (or "Mother Wearing a Sunflower on Her Dress") (ca. 1905)

V, 21. Robert Reid (1862–1929), "The Violet Kimono" (1911)

playful little daughter to look into a hand mirror she holds in front of the child, who, in turn, has reached for this intriguing new toy. In a manner clearly visible to the viewer, the child's face is reflected in the circular hand mirror, which itself is held suspended before the much larger stationary mirror, in which both mother and child can also be seen reflected.

Cassatt has thus created an extraordinary visual world of hermetically closed circles: first, that formed by the linking pattern of the mother and daughter's arms, symbolizing the unity of their being; second, the circle of the two figures and their reflections in the stationary mirror, symbolizing the interior, domestic containment of their world; and, finally, the child's reflection in the cramped uroboric circle of the hand mirror, in which the daughter is made to see herself as if to prepare her for her inescapable future of self-reflective imprisonment in the echo world of the glass. In addition, the large sunflower pinned to the bodice of the mother's gown is emblematic of the stationary, flowerlike, domestic, reproductive existence both are destined to maintain—an aspect of their lives further reinforced by the manner in which the nude little girl is seated on her mother's lap—as if she were growing, like a giant flower, from her mother's loins.

Clearly, Cassatt was able to bring a considerable element of psychological depth and poignancy into her mirror portraits, thereby adding a critical dimension to her use of the theme. In another of her paintings, "Antoinette at Her Dressing Table" (1909) she placed her subject with her back to a large circular mirror, looking at herself in another much smaller but equally circular mirror held in such a way that while looking into the hand mirror she must inevitably also see the reflection of her own reflection in the larger one. It is as if in the ironic circularity of this play of circular mirrors Cassatt's Antoinette—whose face expresses a subdued wistfulness and perhaps even sadness—has recognized the constricted parameters of her own world. Her comportment would seem to disclose that she knows that in being confined to the world of the looking glass, in being almost literally surrounded by it, she lives not in a state of blissful self-absorption but in a ghetto, a prison house of the boudoir.

A similar elegance of expression and subtlety of characterization distinguish Robert Reid's "The Violet Kimono" [V, 21]. Reid, however, hinted rather less sympathetically than Cassatt at an endless pattern of reduplication, whereby the circular mirror which reflects the woman's features also reflects the presence of a second circular mirror: Not only the lady but her reflection is surrounded by reflections. This woman's world, the artist would seem to say, recedes incessantly further into itself.

"Passion," Rossetti had said in his description of the "True Woman" in Sonnet 57 of *The House of Life,* "in her is / A glass facing his fire, where the bright bliss / Is mirrored, and the heat returned. / Yet move that glass, a stranger's amorous flame to prove, / And it shall turn, by instant contraries, / Ice to the moon" (227). The true woman's only proper glass was her husband—seeing herself reflected in the desiring eye of any other man would make her see the gorgon's head of betrayal and turn her into stone (or, in Rossetti's case, into ice) in response to the observer's intruding glance. Unfortunately, few women were docile enough to respond in such a fashion. Most chose instead to lean forward like Narcissus and use the intruder's eye as the water mirror of vanity.

Indeed, woman came to be seen as Narcissa, the true feminine incarnation of what had once been an image of masculine egotism. As the poet Armand Silvestre remarked in his commentary on a painting by Fichel depicting a woman passionately kissing her own mirror image, "The immortal fable of Narcissus has been brought to life again in this graceful painting. But how much more excusable is this woman than the melancholic hero of the fable, this woman who is shown to us here so madly in love with her own image! Man certainly has better things to do than to contemplate himself in the reflection of his doubtful beauty. Woman's beauty, however, which does not brook blasphemy, is a natural excuse for her who has made of herself a god'' (*Le Nu au Salon des Champs Elysées,* 1891).

Silvestre was a witty, worldly French poet-critic, a friend and sympathetic supporter of the impressionists in their early years. The elegant, libertine commentaries he wrote throughout the 1890s for the long-running series entitled *Le Nu au Salon,* a compendium of the most enticing naked ladies exhibited each year at the spring exhibitions of the Paris salons, made these volumes as popular among intellectuals as they were among more ordinary oglers. Learned and ironic, these commentaries reflect perhaps more perfectly than any other source the attitudes of the comfortable, sexually liberal, ''antibourgeois'' bourgeois male establishment of the turn of the century.

To Silvestre it was clear that ''the Narcissus of fable'' should have been a woman, and that women had been far more inventive in utilizing the mirror of self-love than this clumsy creature of classical mythology had shown himself to be. ''Narcissus,'' Silvestre had remarked in his commentary on an 1889 painting by Carrier-Belleuse of a nude woman regarding herself in a characteristic hand-held mirror, ''was a fool.'' Looking into the water mirror in which his own image had been revealed to him, Narcissus had simply ''sunk into a useless contemplation of himself in the water, to become forever immobile, planted without recourse among other languishing flowers.'' Woman, however, knew better than to be limited by such merely stationary loyalty to self. ''She preferred to carry her source with her. She simply took into her hands a little of the water in which she had studied herself, and so the mirror was invented. From that point on it became her most trusted companion, and she took not a step further in the world without consulting her confidante.''

The mirror, then, came to be regarded as the central symbol of feminine narcissism. The story of Narcissus and the nymph Echo became especially popular because it permitted a convenient conjunction of the themes of woman *as* mirror and woman *in* the mirror. Narcissus, in his self-obsession, was the epitome of effeminacy, of the ephebe, the boy not yet in possession of his masculine ability for self-transcendence, while Echo, the nymph who was capable only of echoing the words of others, was the perfect personification of woman's imitative nature. Together these impossible lovers came to reflect a single image of the feminine identity: imitative, circular, narcissistic.

Turn-of-the-century painters sometimes painted them separately—the nymph roaming through the woods searching for a voice to echo; the boy, resplendent in his youthful effeminacy, transfixed by the water mirror—but mostly they were painted together, as in Solomon J. Solomon's celebrated 1895 version of the theme [V, 22]. Solomon's painting characteristically stressed the almost physical continuity between the two, with Echo leaning inward toward Narcissus and looking longingly up at him, a ''proper woman'' yearning to be ''energized'' by his voice, through which alone she could attempt to express her love for him. The boy, meanwhile, oblivious to Echo's needs, continues to gaze, in a state of destructive, autoerotic self-involvement, at his image in the water mirror below. The painting's design reinforces the idea of the uroboric circularity of self-love by means of a compositional pattern in which clothing and the leaves of the forest circle in upon the inward curve of the imploding but never to be conjoined pair of would-be lovers.

V, 22. Solomon J. Solomon (1860–1927),
"Echo and Narcissus" (1895)

It was inevitable that the unyieldingly dualistic requirement of the mid-century—which held that woman must be incessantly all-giving, and that she must be the embodiment of altruism to balance the economic necessity of the egotistic ambitions of the male—should lead to its opposite: the myth of the completely self-sufficient and hence completely egotistical woman, whose only wish was to gaze into the mirror and spend herself in auto-erotic self-contemplation.

Havelock Ellis, who credited himself with the invention of the term *narcissism,* unwittingly pointed to the manner in which environmental influences upon human behavior tended, in the years around 1900, to be construed as evidence of patterns of motivation which were thought to be inherent in human nature. In his monumental *Studies in the Psychology of Sex,* he discussed the supposedly abnormal nature of the narcissistic impulse by juxtaposing it to the normal feminine instinct for altruistic behavior. With the pride of an explorer finding a long-lost artifact whose existence proves the unbroken continuity of a native art form, Ellis quoted a paper of his entitled "Auto-Erotism, a Psychological Study," first published in 1898, in which he discussed

> that tendency which is sometimes found, more especially perhaps in women, for the sexual emotions to be absorbed, and often entirely lost in self-admiration. This narcissus-like tendency, of which the normal germ in women is symbolized by the mirror, is found in minor degree in some feminine-minded men, but seems to be very rarely found in men apart from sexual attraction for other persons, to which attraction it is, of course, normally subservient. But occasionally in women it appears to exist by itself, to the exclusion of any attraction for other persons.

What is most striking in this passage is that Ellis, seeking to be the objective scientist, thought that he had identified a pathological condition characteristic of woman which, in fact, was a condition whose "symptoms" and characteristics had been developed and categorized progressively over a period of several decades by artists and writers who were themselves responding to the "betrayal" by woman of ideals of behavior imposed on her during an earlier historical period. Both the emblem of the mirror and the glance of Narcissus had been impressed upon Ellis' mind, as the structure of his remarks makes abundantly clear, not by the direct study of feminine behavior but by the art and literature produced by his contemporaries. Like Freud, Ellis was extremely well versed in the culture of his own time. Indeed, as early as 1879 Earl Shinn, in describing a work of the French painter Charles Chaplin, had pointed to the prevalence of the newly fashionable image of woman as egotistical self-admirer. The painting in question, he pointed out, showed "a modish type of the female Narcissus, a nude girl standing on the edge of a lake, in which her swan-colored limbs swim double, swan and shadow" (*Art Treasures,* III, 35).

In a world which stresses the value of individualism above all else, it is a primary requirement for the "self-confident" mind, to remain blind to the logical conjunction of personal ideas and the assumptions held by the "mass" of one's contemporaries. The ideas of "individual" thinkers, more often than not, are largely constructed from contemporary clichés. These clichés have merely been stripped of their baser trappings, of their rhetorical conventionality, in accordance with whatever happen to be the prevailing guidelines for the "individualistic" ego. In philosophy, as in the social sciences, literature, and art, imputations of originality are usually based on matters related to descriptive and declarative style and not on originality of thought.

In line with this pattern, Ellis was content to credit himself with naming, if not quite discovering, the feminine autoerotic tendency of narcissism. Describing the trajectory of the word's entry into psychoanalytic terminology, he pointed to its eager adoption, following the publication of his article, by Paul Näcke and other German psychoanalysts of the turn of the century. "Thus I seem responsible for the first generalized description of this psychological attitude, and for the invocation of Narcissus; the 'ism' was appended by Näcke." Ellis concluded by pointing to the term's subsequent adoption by Freud and his followers. By 1905, the date of the first edition of *Three Contributions to the theory of Sex,* Freud was certainly "acquainted with the conception in its earliest form, for he there adopted the term of 'auto-erotism' with which in my writings it was associated. But in the second edition (1910) there was a reference to Narcissism." Finally, in 1911 Otto Rank wrote "the earliest study of Narcissism on strictly Freudian lines," in which Rank remarked that "apart from one or two very interesting casuistic and literary indications, especially by Ellis, nothing has become known as to the origin and deeper significance of this singular phenomenon." Characteristically, both this remark by Rank and Ellis' subsequent discussion of one of Rank's case studies are prime examples of the manner in which the turn-of-the-century psychosexologists fed on each other's narrow practical evidence, as well as of the literary and artistic inspiration for their theories. In Rank's case a young, "narcissistic" woman "would sometimes feel sexual excitement when seated before a mirror doing her hair, and Rank refers, though only passingly, to 'the apparently very intimate connection between Narcissism and masturbation.' Rank's study," Ellis concludes ingenuously, "full of interest and suggestion, was marked, as his work has always been, by its wide knowledge of the earlier scientific and literary suggestions of the subject in hand" (vol. II, pt. 2, 355–57).

It is thus clear that the "discovery" of narcissism and the autoerotic mentality by the psychoanalytic community trailed behind the vogue for the same subject on the part of artists and writers by quite a bit. Yet the sexologists found in the very cultural creations which had suggested to them the development of this autoerotic pathology of the feminine mind no more than a confirmation of the existence of the syndrome. In addition, Ellis became conveniently convinced by recent suggestions on the part of classical scholars steeped in the same turn-of-the-century art as Ellis "that there really was present in the Greek mind the idea of Narcissus as embodying an attitude of mind which would now be termed auto-erotic." In a footnote Ellis needed only to add that "it is interesting to note that it is Echo who brings us to the Greek explanation of the origin of masturbation. Pan was in love with Echo but could never succeed in laying hands on her, and his father Hermes, out of pity for his unsatisfied desire, mercifully taught him the secret of masturbation hitherto unknown" (vol. II, pt. 2, 348). Thus the psychoanalytic pathology of the autoerotic woman as representing an egotistic inversion of woman's "inherent altruism" was born, and given a distinguished ancestry in virtually a single gesture.

Yet the syndrome's true origins as a "feminine perversion" about which psychoanalysts felt they needed to be concerned are clearly to be found in the dualistic mentality of nineteenth-century male culture, a facet of its disappointment with woman's limited capacity for self-sacrifice. Thus, long before Ellis' "discovery," artists—generally taking their cue from such unbiased students of the feminine mind as Zola and basing themselves on such "clinical" examples as his description of Nana's actions before her mirror—had begun to show woman as not only staring intently, narcissistically, into the mirror but as having fallen head-over-heels in love with her own image, often to the extent of being driven by a desperate urge to kiss and caress herself.

Antoine Magaud's painting of a young woman's autoerotic "Kiss in the Glass" [V, 23] is a typical example of this variation on the mirror theme. Paintings of this sort, which abounded during the twenty or so years surrounding the turn of the century, clearly had their origin in the rediscovery of women's sexuality—and especially in the sensational disclosures about women's autoerotic tendencies by sexologists of the 1870s and 1880s such as Moll, Féré, and Krafft-Ebing, as well as such manic moralists as Cooke. The voyeuristic fascination of both scientists and artists with this subject became the basis for passages of "significant psychological insight" concerning the feminine mind by the most stylish writers of the turn of the century, from Remy de Gourmont and Pierre Louÿs to Gabriele D'Annunzio and Juan Valera y Alcalá Galiano. The last named, in a passage from his novel *Génio y Figura* (1897) quoted by Ellis, has his heroine, Rafaela, admit that her maid's habitual admiration for her beauty, when she dries her after her bath, makes her do "a childish thing which whether it is innocent or vicious I hardly know." Rafaela insists that "what I do is not out of gross sensuality but aesthetic platonism. I imitate Narcissus; and to the cold surface of the mirror I apply my lips and kiss my own image. This is the love of beauty for beauty's sake; the expression of affection in a kiss toward what God has made manifest in that disembodied reflection." Ellis found this passage especially "true to life" (vol. II, pt. 2, 352).

The image of woman kissing herself in the mirror marked a drastic change from the mid-century conception of woman as household nun. The altruist had become an egoist; and as the men who glorified the aggressive mentality in the world of affairs knew all too well, egoists tended to destroy others in search of their own fulfillment. The self-reflective women, the Rafaelas, the beautiful women pleased with themselves because they were

V, 23. Antoine Magaud (1817–1899),
"A Kiss in the Glass" (ca. 1885)

beautiful became a focus for vengeful attacks by turn-of-the century intellectuals. The Lady of Shalott, Mariana, Ophelia—the heroines of mid-century had been so popular because they were the embodiment of altruistic feminine passion. The very sight of virile flesh had made them slide into self-forgetful eagerness to die for love. Now woman had become different. No longer the moon of reflected light, she had become the moon of circularity, of uroboric self-sufficiency. Egotistical, self-involved, she no longer cared a hoot for men; all she cared about was herself. And in ceasing to be self-sacrificial, she clearly became destructive of the masculine ego. The dualistic mind does not tolerate compromises.

In a story called "The Woman of Marble," published in *La Revue de Paris* in 1900, Henri de Régnier, guided like virtually all "progressive" intellectuals of his time by Baudelaire's fundamental misunderstanding of Poe, used the themes of feminine narcissism, self-reflective circularity, the mirror, the moon, and various related symbols to spell the doom of two bewildered males. Supposedly the narrative recorded an event of the sixteenth century, but the story and its characters were fin-de-siècle to the core. A sculptor is fascinated by the physical beauty of Giulietta del Rocco (her surname serving as a not-too-subtle indication of her stony indifference to the artistic subtleties which fired the heart of man). He is a man after the nineteenth-century mold: "I made love and war. The crossing of swords and the joining of lips fascinated me equally." For him to sculpt is to conquer. He sculpts women. He sets out to sculpt Giulietta—nude, of course. Not interested, other than for aesthetic reasons, in the perfection of the living woman, he creates her perfect effigy, "captures" her in marble, and in a sense thereby conquers the demands of the flesh. But Giulietta, perverse as only woman can be, reclaims her effigy in a gesture affirming her narcissistic self-involvement: "Giulietta approached the statue slowly. She tenderly embraced the marble, which seemed to return her caress, and she pressed her ephemeral lips on the eternal lips. Their two smiles touched" (233). Score a victory for the twisted narcissism of woman, for by means of this gesture Giulietta impressed her living being upon the statue. The statue, thus "animated," evades the sculptor's effort to capture woman as an "ideal." Giulietta gets dressed.

Two cousins, friends of the sculptor, enter. They see Giulietta. The tentacles of passion engulf them. Gorgeous young men, the best of comrades who have until now shared all things—including women—they represent the ideal and the real. Alberto is "violent and sensual," while Conrado is "gentle and a dreamer." Since women, as we all know, love to be beaten, the women Conrado made love to always "quickly forgot that he had made love to them, while those of Alberto always remembered their amorous encounters for a very long time" (232). Inevitably love drives the two young men apart. Alberto, since he represents the Real, of course ends up with the real Giulietta. Conrado, representing the Ideal, gets the statue, the "eternal" woman of marble: the artist's creation, perverted by Giulietta's egotistical kiss. Since she has narcissistically "implanted" herself in the marble of her own statue and has, in a sense, become eternal herself, Giulietta dies. That is naturally when the real trouble starts. Marble beauty may be eternal, but it is even more "inscrutable" than woman's living flesh. The spirit of Giulietta's self-containment thus lives, more powerful than ever, in the woman of marble. The statue now becomes the untouchable woman who does not go mad in the name of love but who drives men mad. Woman taking revenge on man through her self-containment is the woman of marble.

As was to be expected, the two handsome young men, those former fast friends, kill each other at the foot of the statue. In creating the ideal woman the sculptor has created the ideal destroyer. He must destroy her—love is war. He hacks at her with his tools. The woman of marble "repulsed my efforts with all her living solidity. It was less a one-sided destruction than a combat. A sharp chip hit my brow. I bled. A kind of fury had seized me, and soon it changed into a rage of vicious intensity. One moment, I felt a sense of shame, as if I were beating a woman. The next I seemed to be defending myself against an enemy." Breasts, arms, roundness of body—it all has to go. Finally "her head, still untouched, rolled to my feet. I picked it up; it was still intact and it was heavy. I wrapped it in my mantle and left the town." On a mountain, in the moonlight, the sculptor finally buries the introverted essence of the eternal woman among the rocks: "I placed the marble head there, but not until I had kissed that fatal, deadly beauty on the lips. It is there that she still reposes, among the roseate tree trunks whose resin drips down as if it were a rain of embalmed, transparent tears" (241). One may assume that, having once gotten rid of self-reflective womanhood in both love and war, the narrator went on to live a productive male individualistic existence.

The turn-of-the-century male's fascination for, horror of, and hostility toward woman, culminating in an often uncontrollable urge to destroy her, to do violence to that perverse, un-Platonic reflection of the Platonic ideal of perfect beauty he was so eager to pursue— these emotions speak all too loudly through this story by one of the major figures of the French symbolist movement. It is a story that could not have been written without the example of the numerous images of narcissistic women, admiring themselves in the mirror or kissing their own image, industriously produced by the painters who preceded Régnier.

In fact, it is striking how much of the work of now-forgotten artists echoes through the writings of the still popular poets and novelists of this period. The exchange of imagery between painters and writers was intense. The painters raided literature, while the writers were endlessly describing and transliterating the paintings they had seen. As a result of our bias in favor of the impressionist image over the sharply delineated three-dimensional modulations of the academic painters, we tend to see as significant only such links as the one between Manet and Zola. But for Zola and the other late nineteenth-century writers who are still read, the realm of significant art was much wider. They may have praised the impressionists and have recognized the value of their innovative approach, but their heads were still full of the images created by the academic painters. The realistic mode of visual representation is brutally unambiguous. The mind likes to play with apparent ambiguities; hence realism in art is easily dismissed as "unartistic" in its often almost ludicrously direct positing of relationships. Language, on the other hand, is the realm of the abstracted sign. That is why we like to hide our own imaginative poverty behind language. The writer—in appropriating, in recreating in words a current cultural prejudice he has seen in its starkest, most simplistic expression in the form of a "realistic" painting—can easily use language to obfuscate, to smudge the outlines of the image and make its import seem deep. Inevitably, however, his more subtle verbally encoded visual image will be retranslated by his less subtle, more conformist readers into a mental gallery of iconic images, a gallery virtually always hung and rehung in accordance with the prevailing cultural commonplaces about human nature and truth. Thus, by retracing the visual origins of the "great" literary works of turn-of-the-century writers we tend to find ourselves uncomfortably back in a world of intellectual clichés which, in their written form, were elegantly obscured by the apparent profundities embedded in the necessary ambiguities of language.

It is this advantage of literature over painting that has made the simplistic dualistic antifeminine content of late nineteenth-century thought survive with much less criticism in the realm of literature than it has in the realm of visual imagery. But we cannot begin to assess the continuing influence of the late nineteenth-century mentality until we realize that Zola's writing links up with the work of Bouguereau and Cabanel as much as with Manet—and that Manet himself had his intellectual origins in those painters, not only through his stylistic negation of their work but also through his unquestioning continuation of much more of their basic mentality concerning male-female relationships than has generally been allowed. The symbolist narratives of writers such as de Régnier, like the symbolist images of turn-of-the-century painters, are weak attempts to turn simplistic cultural prejudices into narrative ambiguities. It is, however, important to remember that the virtually monolithic fascination of turn-of-the-century culture with issues of power and submission, conquest and defeat, evolution and degeneration, and the perpetual state of "war between the sexes" finds a largely uncontested pattern of reinforcement in the apparent narrative complexities of symbolist art, as well as in the visually liberated but intellectually imprisoned images of the impressionists.

Woman's desire to embrace her own reflection, her "kiss in the glass," became the turn of the century's emblem of her enmity toward man, the iconic sign of her obstructive perversity, her greatest weapon in her reactionary war against the progressive male. Since, as we have seen, woman was thought to have much less of a capacity for individuation than man—in Cooke's words, she was "cast more in a common mold than men"—she

V, 24. Fernand Khnopff (1858–1921),
''The Kiss,'' crayon (ca. 1887)

inevitably had difficulty in trying to differentiate men from each other and herself from other women.

Frank Wedekind's Lulu, for instance, is the very personification of the narcissistic woman. We are told that she is ''intoxicated with her own beauty—seems to be idolatrously in love with it'' (69)—so much so that she remarks at one point, ''When I looked at myself in the mirror I wished I were a man . . . my own husband!'' It is clearly Lulu's autoerotic impulse that makes her say so. But her autoerotic focus is coupled with her inability to differentiate, to individualize those around her. Lulu cannot love because all men are merely tools which she uses to love herself more intensely; hence any man will do. Because the men she sleeps with believe, in their ignorance of feminine nature, that in simple exchange for her fidelity to them she will respect their efforts to make her part of their ''individualized'' existences, they fall victim to her, for as *Erdgeist,* as the essence of undifferentiated primal nature, she cares only for the mindless reduplication of her own pleasures. Because she is the generalized spirit of the earth, she is every woman. Wedekind informs us that Lulu is known under a wide variety of names. One man calls her Mignon, another Nelli, a third, perhaps most appropriately, Eve.

V, 25. Dante Gabriel
Rossetti (1828–1882),
"Rosa Triplex" (ca.
1874)

Woman as Eve is every woman. Hence the turn-of-the-century philosophers argued that when a woman kissed another woman it was indeed as if she were kissing her own image in the glass. In the art of the period, the image of woman kissing herself in the mirror was therefore often supplemented by images of a woman kissing another woman, as if she were kissing herself. Fernand Khnopff's "The Kiss" [V, 24] is a typical example of this genre. The origins of Khnopff's image in the theme of woman's narcissistic self-concern are made explicit in a series of sketches by this painter which, in structure and theme, are directly related to this one, except that in them the artist has drawn into the image the outline of a circular mirror, making it clear that the amorous woman's lips are pressed against the glass of feminine self-reflection. A similar exploration of the notion that what women see in other women is what they see when they look in the mirror can be found in Rossetti's "Rosa Triplex" [V, 25], in which the artist painted a portrait in triplicate of the young May Morris as if she were three different women possessed of a single, coextensive state of heavy-lidded languor. These three roses, which are one, join to form a tripartite mirror image of the Eternal Feminine.

Thus, while the sexologists were rediscovering female sexuality in terms of woman's supposed autoerotic fixation during the 1860s and 1870s, artists and writers were discovering the existence of lesbianism as a sort of extension of this supposed autoerotic fixation. But even as late as 1885 Max Nordau—who, as we have already seen, was especially adept at pronouncing popular clichés in a manner which gave them the ring of newly discovered truth—still felt able to say with confidence that "the sight or the idea of a person of her own sex has no power to excite [woman's] sex centre to any form of activity, and hence man must be her ideal of beauty" (*Paradoxes*, 279).

This widespread disbelief in woman's ability to love woman was, in a certain sense, one of the most striking indications of the dislike men such as Nordau had for women in general. For just as it is hard for us to imagine that if we truly love another person others might not, it is equally difficult to contemplate the existence of attachments among those of whom we are disdainful. Except for a relatively rare explicit early treatment of lesbianism such as could be found in Simeon Solomon's gouache of "Sappho and Erinna" (1864), in Courbet's "Sleep" (1862), or in the writings of Baudelaire, his follower Swinburne, and in the sensational—and sensationally popular—novel *Mademoiselle Giraud, My Wife* (1870) by Adolphe Belot (supposedly written because it was the duty of "the writer to signalize and stigmatize certain corruptions" [254]), it was not until around 1900 that the depiction of lesbianism became a popular and, among intellectuals, a popularly accepted theme. This acceptance was largely due to the fact that the assumption that women were undifferentiated beings had become a cultural commonplace by this time.

Just as woman's glance in the mirror, while expressing her autoerotic inaccessibility to the "individualistic" male, had nonetheless still kept her within comfortable voyeuristic distance, so lesbian contact between two women often came to be seen as a simple extension of their autoerotic tendency, and hence open to the same sort of nonthreatening voyeuristic involvement on the part of the male. In 1904 Bernard Talmey succinctly expressed the connection habitually made between autoerotic practices and lesbianism: "The female masturbator becomes excessively prudish, despises and hates the opposite sex, and forms passionate attachments for other women" (*Woman*, 123). It was only after this link was made as a factor of the later nineteenth-century's fascinated discovery of the existence of female autoerotic practices that the recognition of women's capacity to be lovers as well as friends gained a more general currency. Indeed, Talmey emphasized that

while homosexuality has been made a crime in men, it has been considered as no offense in women. The reason for this defect in our criminal laws may be ascribed to the ignorance of the law-making power of the existence of this anomaly. The layman generally does not even surmise its existence. A woman is by nature not aggressive, and the inverted sexual intercourse among women is not as easily detected as in men. Women's attachments are considered mere friendships by outsiders. We are accustomed to a much greater familiarity and intimacy among women than among men. We are, therefore, less apt to suspect the existence of abnormal passions among women. On the contrary, such friendships are often fostered by parents and guardians, such attachments are praised and commended. They are not in the least degree suspected of being of a homosexual origin (147).

V, 26. Georges Callot (1857–1903), "Sleep" (1895)

V, 27. Pierre-Georges Jeanniot (1848–1925), ''After the Bath'' (1908)

V, 28. Louis de Schryver (1862/63–
1942), ''Fond Confessions'' (1905)

The turn-of-the-century painters, however, were eager to show their worldly knowledge about these matters. Starting in the 1880s, they exhibited at the yearly salons paintings in which the more traditional representation of two close friends arm in arm or shoulder to shoulder was replaced by a much more explicitly sexual imagery. At the same time, by giving these women strikingly similar features, artists continued to emphasize the mirror theme, thereby habitually reinforcing the notion that in her generalized, nonindividuated state a woman loving another woman was little more than an extension of woman's inherent autoerotic concerns—as had been the case in such mirror images as that of Fernand Khnopff. Painters also often took their cue from Courbet's "Sleep," in which the interlacing of the women's limbs suggested their more than merely physical resemblance and hinted at the absence of individuation in their identities. Georges Callot's salon entry of 1895 [V, 26], given the same ubiquitous title as Courbet's painting—a title used by late nineteenth-century painters to cover their documentary explorations of a wide range of suspected feminine erotic practices with a *pro forma* cloak of decorum—is a typical example, although it is rather more gentle in its portrayal of the relationship between the two women than most of the paintings dealing with this subject, which often emphasized with protoexpressionist fervor the "bestial" nature of such connections between women.

Pierre-Georges Jeanniot, in his painting "After the Bath" of 1908 [V, 27], came closer to the norm in this respect. Jeanniot (whose inspiration for this work may have derived from another familiar painting by Courbet—his "Girls Resting Near the Seine" of 1856) showed a trio of women in high spirits, presumably after a refreshing dip in the river. The one in the middle has turned to her companion with a livid glint of interest in her eyes and has playfully bared one of her breasts. It is clearly an amorous message to her friend well understood by the third woman present, who is looking in upon what is happening with a knowledgeable smile of anticipation.

Even if the turn-of-the-century painters did not follow Courbet's lead in structuring their images, there were plenty of ways for them to emphasize the coextensive qualities which linked women to each other. Another French painter, Louis de Schryver, in his 1905 salon entry entitled "Fond Confessions" [V, 28], showed the tender embrace of two

V, 29. Eliseu Visconti (1866–1944), "In Summertime" (1891)

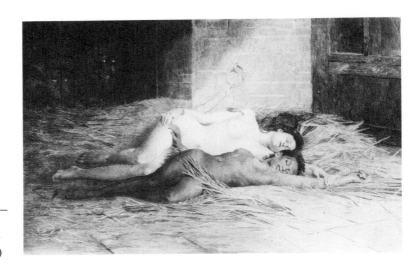

V, 30. St. George Hare (1857–1933), "The Victory of Faith" (ca. 1891)

women who quite obviously had a more than comradely interest in each other. At the same time, he was careful to show their features to be virtual mirror images of each other. The Brazilian painter Eliseu Visconti, in his "In Summertime" [V, 29], painted in Paris in 1891, showed a pair of young women who, it would seem, had already had quite a night of it. He made it appear as if, even though they had separate bodies, they were in fact little more than a single organism. Surely it was not just the debilitating summer heat that gave these youngsters their air of contented exhaustion. The Irishman St. George Hare, in a painting entitled "The Victory of Faith" [V, 30], combined the fashionable woman-as-slave theme with spurious religious content to double his viewers' pleasure. In this knowing late version of the Courbet-style enlacement pattern, the brown and white skins of the two young "martyrs" contrast with and yet echo each other. The Austrian Egon Schiele, in a 1913 sketch showed two women caught in the act, as it were; he made their clothes and bodies blend together to form a nearly abstract pattern suggestive of a single hothouse flower [V, 31].

The German Ida Teichmann, a very fine artist who specialized in brilliant, realistic pencil studies of young women, combined in the structure of a drawing again entitled "Sleep" [V, 32] direct suggestions of the mirror theme with a supremely sensuous depiction of erotic lassitude in the two pairs of young women she portrayed as being caught up in a tender mutual embrace rather than on their way to dreamland in the more decorous arms of Morpheus. The Spaniard Pablo Picasso, who was, at least in his perception of women, a typical child of his time, returned to the " woman echoes woman" theme, showing two "friends" whose facial expression and even body type—indeed, whose every facet—showed that as beings they were interchangeable [V, 33].

The Frenchman Edmond Aman-Jean intimated that in such places as a theater box, where fashionable women could contemplate the imitative talents of their sisters upon the stage, they might also engage in mirror minuets as if they were flowers growing from a single stem [V, 34]. "With a poet like this even sensuality is sentimental, purely intellectual, slightly neurasthenic," Raymond Bouyer remarked in 1907 in an article for *The International Studio* on Aman-Jean's pastels. He added:

> With her strange, feline, rather kittenish expression—gracefully tender, elegantly familiar, enigmatically sweet, clever, voluptuous and roguish—with her air of happy or mischievous

irresponsibility, as the lively pupil of her eye sparkles beneath the arch of her brow, and her mouth opens like a scented flower to show the enamel of her white teeth—the favourite heroine of the pastellist is a sister, or at least a near relative of the sirens or muses . . .

In the work of Aman-Jean, Bouyer decided, "contemporary woman is interpreted by a thinker who can translate soul into form" (290).

The male fantasies concerning woman's autoerotic self-involvement might have remained relatively harmless if the sexologists had not rushed in to declare this "perversion" of woman's "inherently passive nature" an immediate threat to the further evolution of civilization. One popular theory held that by abnormally stimulating her properly sexless physical being, woman tended to grow more and more masculine, tended to take on ever more virile characteristics, thereby destroying the delicate balance created by nature between the active male and the passive female. Talmey sketched the following portrait of the sexual "hyperaesthesia" of the lesbian and the "masculinizing" effect it was supposed to have, which was prototypical of the characteristic twentieth-century stereotype of same-sex-directed women: "She neglects her dress and assumes and affects boyish manners. She is in pursuit of boys' sports. She plays with horses, balls and arms. She gives manifestations of courage and bravado, is noisy and loves vagabondage" (152). This descrip-

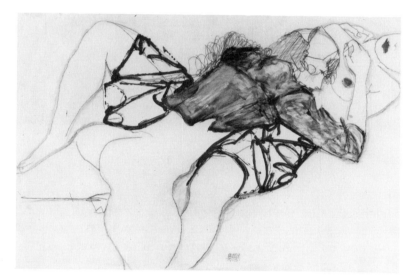

V, 31. Egon Schiele (1890–1918), "The Two Friends" (or "Tenderness") pencil and tempera (1913)

V, 32. Ida Teichmann (b. 1874), "Sleep," pencil (ca. 1905)

V, 33. Pablo Picasso (1881–1973), "The Friends" (1903)

V, 34. Edmond Aman-Jean (1860–1936), "In the Theatre Box," pastel (ca. 1898)

tion places an interesting additional light on the twentieth-century cinema's pervasive theme of the tomboy, who, starting out as a competitor with boys, is ultimately "tamed" and brought to the altar by a victorious male friend, only to settle down to a virtuous life as wife and mother.

These narratives of male triumph did not just symbolize the return of a wayward young girl to her normal feminine duties. They in effect celebrated the "killing" of the sexual instinct in woman as a prerequisite for her return to civilized society. The failure to complete that rite of passage would create a dangerous pack of viragos, of manlike women who, having male characteristics but none of the god-given male capacity for personal individuation and intellectual development, might precipitate society, in Horace Bushnell's words, "down this abyss, and make a final wreck of our public virtue in it," which could not help but be "the end of our new-born, more beneficent civilization" (*Women's Suffrage*, 31).

Lombroso and Ferrero warned that sexuality, once unleashed in woman by whatever means, could only be disastrous because, in effect, the sexual impulse was itself a male impulse. "Normal women," they stressed, "are monotonous; they resemble one another, and cannot be acted upon by suggestion" (204). Active enjoyment of sex awakened an inherent criminal instinct in woman which was nothing short of disastrous. She would become "excessively erotic, weak in maternal feeling, inclined to dissipation, astute and audacious." Such a woman tended to dominate "weaker beings sometimes by suggestion, at others by muscular force; while her love of violent exercise, her vices, and even her dress, increase her resemblance to the sterner sex" (*The Female Offender*, 187).

Lesbian and autoerotic practices in women, moreover, were supposed to be the cause of such abnormal conditions as "hypertrophy of the clitoris," as Talmey asserted (123). This evidence of "viraginity" was exploited by the Belgian artist Félicien Rops to portray the brute, uncontrollable aspects of feminine nature. In such an etching as his "Hermaphroditic Joy," a work that for once seems to have daunted the editors of the magazines of the period, he testified to the characteristically literal bent of his imagination by depicting a woman with a pronounced case of this form of viraginous hypertrophy, her distended clitoris rising from her loins like an erect but (as Rops clearly meant to suggest) laughably slender phallus.

The most common assumption about the effect of autoerotic activity upon woman, however, was even more vicious than the attribution of masculinizing "viraginity" to woman. An exacerbated version of the belief which saw "emasculating debility" to be the effect of autoerotic activity in boys, it joined with notions spawned by the theory of evolution to lash out viciously against all women. According to this popular belief, at least half of the world's population was made up of inherently feebleminded creatures who, brainless and incoherent, were at best decorative flowers but, far more often, blindly strangulating, clinging vines.

CHAPTER VI

Evolution and the Brain: Extinguished Eyes and the Call of the Child; Homosexuality and the Dream of Male Transcendence

The anthropologists, biologists, and sociologists who, in the later nineteenth century, developed "the science of man" are often so extreme on the subject of the relationship between the sexes that reading their remarks is like entering into an insane asylum in which the inmates have written all the rules. On the surface intensely rational and seemingly meticulous in their pursuit of scientific evidence, these scholars tended to pronounce notions about the origin and goals of civilization so catastrophically antihumanist and heartless that it is an immense relief to come across Freud's belated recognition, in 1930, that "if the development of civilization has such a far-reaching similarity to the development of the individual and if it employs the same methods, may we not be justified in reaching the diagnosis that, under the influence of cultural urges, some civilizations, or some epochs of civilization—possibly the whole of mankind—have become 'neurotic?'" (*Civilization and Its Discontents*, 91) Freud reached this understanding at a time when the evils of the Nazi epoch were already apparent to everyone, when attitudes and assumptions he himself had helped to classify and "normalize" as part of human development had already lashed back at him, too.

The truly psychotic, rather than merely neurotic, idealization of a supremely evolved white male and the concomitant assumption that somehow all others were "degenerate" had, as Freud was writing these words, begun to reap its most evil harvest, the genocidal anti-Semitism of the earlier twentieth century. Even the most casual reader of the theoretical disquisitions of the later nineteenth-century exponents of the science of man must at once perceive the intimate correlation between their evolutionist conclusions and the scientific justification of patterns of "inherent" superiority and inferiority in the relations between the sexes, various races, and the different classes in society.

The theory of evolution became a catchall phrase used to justify any and all forms of personal and social aggression upon the constitutionally weak, culturally disadvantaged, and economically oppressed. The nineteenth century had been seeking an image that might replace the much too static earlier conception of a "chain of being," in which each of God's creatures, including humanity, had its proper position as an interlocking link in the mighty delineation of rank in the order of creation, from the lowest organism to the angels surrounding God's throne. Many successful men felt that in a restless world of economic agitation, which had seen members of the nobility become paupers and mere men revered like gods in the realm of economic achievement, to be identified as a simple link in a chain was an affront. The ladder of success was a much better image—and what more perfect ladder than that which had been constructed by evolutionary science?

The men who, in the eighteenth and early nineteenth centuries, had enlisted women to be the keepers of their souls had thereby still expressed an old-fashioned worry about the moral legitimacy of their exploits in the marketplace, even though for quite a while—from Defoe through Adam Smith, Malthus, and Ricardo—the major authorities in the field of economics had tried to reassure them that the depredation of the weak by the strong was a factor of God's will; that "the struggle for existence," the necessary "competition for life," was a calculated feature of God's design. When, at the outset of the nineteenth century, the Chevalier de Lamarck had begun to place emphasis on a kinetic impulse in nature, an impulse to change which was inherent in animate being, it had not been difficult to turn the "chain" of being into a "ladder" toward perfection, on which the economic activities of each generation were like cumulative steps in the progress of humanity toward its ultimate destiny among the angels at God's feet.

When Darwin published his *The Origin of Species* in 1859, aspects of what was to become known as the theory of evolution were, in one form or another, already widely understood elements of scientific theory. It was the simple metaphoric significance of Darwin's outline of species development, as much as the sheer weight of scientific evidence he presented, which made his work so popular. The justification of predatory behavior had already been effectively and popularly argued by Herbert Spencer in his *Social Statics* (1850), and Darwin's conclusions fit quite exquisitely within the ambitious dreams of human progress expressed in Auguste Comte's *System of Positive Polity* (1851–54). The notion that through natural selection the survival of the fittest could be achieved, making progress itself into a logical by-product of the inherent struggle for life, was blissfully easy to understand in a world of intense industrial expansion, widening social inequality, and dizzying contrasts between the rapidly increasing wealth of a few and the abject poverty of many. Spencer had already postulated morality as no more than the convenient but always temporary fixative of an ever-changing configuration of developmental imperatives—he described evil as simply the "non-adaptation of constitution to conditions" (*Social Statics*, 57). The notions of Comte and Spencer now joined with Darwin's theory of natural selection to form the basis for a "science of society" which, in turn, became the "technical" justification for the wholesale subjection of the powerless to the powerful and for ever-increasing levels of social inequality. The casual, convenient juxtaposition of partial truths and arrogant suppositions characteristic of this period was to find its ultimate expression in the institutionalization of genocide as a means of social control.

There is little reason to question the enormous importance of the developments in scientific knowledge which resulted from Darwin's *Origin of Species* and the apotheosis of evolutionist ideas, but historians tend to sidestep the unpleasant aspects of the triumph of Darwinian theory. Histories of the second half of the nineteenth century often delineate romantic images of embattled progressively minded scientists struggling mightily against reactionary prejudice to discover "truth." In consequence, they have tended to overstate the opposition encountered by the evolutionists. Artists, scientists, and researchers can only be said to be truly fighting against prejudice when they are working in isolation and against the prevailing winds. If their work is being followed and accepted with eagerness by a large segment of the intellectual community, as was the case with the evolutionists as early as 1870, and if the mass marketplace of intellectual ideas represented by such excellent general cultural magazines as *The Atlantic Monthly* or *The Nineteenth Century* embraces their cause, then they can no longer be considered lonely geniuses bravely battling prejudice. Moreover, especially when those with money and power lend their support to the endeavors of artists, poets, or scientists, it is likely that these searchers after truth have been enlisted to win the hearts and minds of the public over to the position of the ruling class. By definition, those who do research for the ruling class are part of the establishment.

Given the opportune conjunction between the aggression-and-struggle theory underlying the thinking of the evolutionists and the aggressive, incorporative ambitions of the late nineteenth-century ruling class, there simply can be no doubt that the contentions of the evolutionists spoke loudly to the established forces among the intellectual community of the late nineteenth century. Indeed, Tolstoy was obviously far from wrong when he pointed to the popularity of the new "science of man" and its close link to the forces of the predatory market society:

> The positivist philosophy of Comte and the doctrine deduced from it that humanity is an organism, and Darwin's doctrine of a law of the struggle for existence that is supposed to govern life, with the differentiation of various breeds of people which follows from it, and the anthropology, biology, and sociology of which people are now so fond—all have the same aim. These have all become favourite sciences because they serve to justify the way in which people free themselves from the human obligation to labour, while consuming the fruits of other people's labour (*What Then Must We Do?* 213).

Although she did not have any of Tolstoy's scruples about living off the labor of others, Edith Wharton certainly concurred with him in identifying the later part of the nineteenth century as a time in which "science" had become a rather suspect form of popular concern, a fad. She gave her own ironic comment on her contemporaries' fascination with the theory of evolution in the story "The Descent of Man," published in 1904, where she remarked that in the last quarter century science had ceased to be the concern of just a few professors. "Everyone now read scientific books and expressed an opinion on them." And she continued, "Daily life was regulated on scientific principles; the daily papers had their 'Scientific Jottings.'" Even the spurious notion that science was a realm of daring individualists helped its popularity. "The very fact that scientific investigation still had, to some minds, the flavour of heterodoxy, gave it a perennial interest. The mob had broken down the walls of tradition to batten in the orchard of forbidden knowledge" (8).

In light of this, it should be stressed that very late incidents such as the Scopes trial of 1925—which have often been used as evidence of the continuing trials of the evolutionists—in no way represented the brave struggles of a beleaguered band of scientific pioneers. Instead, the Scopes trial drew so much publicity because it represented the *arrière-garde* grandstanding of cultural middlebrows in the intellectual backwaters of Tennessee. Clarence Darrow's oratory represented the safely ensconced opinion of the big-city middle classes reveling in the self-righteous indignation of those who knew that "science" was being persecuted by superstition. Even the triumph of the "benighted" fundamentalists over the "light" of science, in this respect helped the majority feel as if they were part of a great intellectual struggle whose outcome was still in doubt. Such safe, romantic causes give monolithic societies a taste of the thrill of pluralism.

In his essay "On the Natural Inequality of Men," published in the January 1890 issue of *The Nineteenth Century,* T. H. Huxley had snootily remarked (largely with reference to the ideas of Rousseau) that "missionaries, whether of philosophy or of religion, rarely make rapid way, unless their preachings fall in with the prepossessions of the multitude of shallow thinkers" (1). He was right, of course, in saying that the public at large tends to embrace those who give coherent formulation to ideas which reflect the popular prejudices of the day. But making this declaration ought to have caused Huxley to stop and think. Why should the decent, middlebrow magazines of his day have been so eager to publish his articles—why should Darwin and Spencer have been on everyone's tongue—if the notions they were propounding were indeed as unpopular as Huxley kept insisting? Of course, in an age which had virtually turned the idea of the individual into the manifestation of God-made-flesh, to be a lone crusader was a sign of election to the court of the Almighty. No decent intellectual wanted to be accused of conveying popular ideas or of elaborating commonplaces. In consequence, the hordes of social Darwinists who at the turn of the century loudly spoke their minds, propounding Western civilization's most cherished cultural platitudes, continued to speak with the air of those who were marching to a different drummer even while, in truth, they were waltzing along in unison to the well-regimented strains of the brass band of cultural conformity.

In his *Lectures on Man* (1864), Carl Vogt, "professor of natural history in the University of Geneva, foreign associate of the anthropological society of Paris, and honorary fellow of the anthropological society of London," as the British edition proudly proclaimed, pitted himself against the fundamentalist opposition he still encountered, yet he was already speaking with the voice of a man secure in his authority, cradled by the support of his scientific audience, and out to have a little fun at the expense of his ignorant opposition. It was clear that the days of Michelet's rhetoric of woman's natural sainthood were passing. In describing the fundamentalist conception of creation propounded "more than twenty years ago" by Frédéric de Rougemont, Vogt remarked with infinite sarcasm, "I . . . heard him at a public lecture explain the creation of Eve from Adam's rib, and why God, in his infinite wisdom, had selected the rib in particular, and no other part of Adam's body. 'He took no piece of the head—woman would then have had too much intelligence; he took no piece of the legs—woman would have been too much on the move; he took a piece near the heart, that woman should be all love!'" (3)

But when we consider what "improvements" evolutionists such as Darwin and Vogt made in our conceptions of the role of woman in society, we discover that where woman was concerned the theory of evolution represented a baroquely inscribed license to denigrate and destroy. In fact, Vogt's beliefs were not at all at variance with de Rougemont's claim that women did not have much intelligence. Where de Rougemont used a metaphor and relied on faith to justify this notion, Vogt did much more damage by offering "irrefutable scientific evidence" to support this point of view; and where de Rougemont tried to place women in the gilded cage of intuitive sainthood, Vogt threw them into the dark, ominous dungeon of near-bestial insignificance. In fact, as we shall see, the later nineteenth-century "naturalists"—the biologists, sociologists, and anthropologists who focused on the sex roles of the human species—these early "sexologists," who might actually best be given the generic title of "bio-sexists," were of crucial importance in building a pseudoscientific foundation for the antifeminine attitudes prevalent around 1900. Among these bio-sexists, Vogt must certainly be accorded a position of considerable eminence. His tool for the denigration of woman was the supposed science of craniology, of brain measurement.

Craniology and evolutionary theory came into scientific favor side by side. This is not surprising. As long as, in the pre-evolutionary era, humanity had been seen as a separate creation of God, entire unto itself, it would have been tantamount to blasphemy to regard the various forms and varieties of humanity as either more or otherwise rather less "successful" specimens of God's creation, since that would impugn the parameters and the developmental intentions of God's capacity as a creator. As Carl Vogt was to point out, "the prevailing opinion, hitherto, was, that species are fixed normal types, which may undergo changes within a very limited sphere; that they were the expression of a definite realized idea; that they were the separate unchangeable materials from which, according to a creative plan, the structure of the organic world had been erected" (447). In other words, even if not every human being had developed as far as he or she could have—even if education, ambition, industry, and environment had created differences in actual achievement, God, being fair and equitable, had endowed all human beings, male as well as female, Christians as well as heathens, black as well as white, with the *potential*, the capacity, for achieving grace. Each human being, in other words, carried the God-given tools necessary for transcendence.

But the theory of evolution changed all that, for now a long historical process of natural selection was thought responsible for fundamental qualitative differences among groups of human beings. De facto discrimination among races, on the basis of claims of perverse ignorance and the wilful subversion of God's design, had obviously been around for a long time and had conveniently served as the basis for arguments in favor of slavery; after all, the benighted might properly be coerced into the service of the true God. But now, for the first time, race and sex discrimination could be given a legitimate "scientific" basis rather than the always somewhat tricky justification that it was "God's will."

The lay public was informed of this convenient development with astonishing rapidity. To combat the already diminishing influence of the orthodox religious conceptions which had formed a solid basis for antidiscriminatory activity in the fields of sex and race at mid-century, evolutionary theory had arrived in the nick of time, a resplendent white knight in the service of discrimination. As Charles J. Sprague said in his explication of "The Darwinian Theory," published in *The Atlantic Monthly* in October 1866, this theory showed that, as a result of the process of natural selection, "the largest and strongest get the best food or the most attractive females, and then transmit their strength or their peculiarities to their progeny" (420). Thus, "in the struggle for life, the strongest live; or, in other words, those best fitted to live in the environment endure." This explained the "vast difference between the highest and lowest species of the genus *homo*. Were the race confined to those lowest species, we imagine that European and American pride of nature would go before a grievous fall." Such troubling questions as the Christian's responsibility to other races, which had until now tormented religiously motivated bleeding hearts, became moot virtually overnight, for Darwinian science showed why "the Negro, the Malay, the Mongolian, are almost precisely what they were five thousand years ago."

It was now becoming clear that "some forms may be less plastic than others and give way less readily to the incident forces. These may remain unchanged for a far longer period than subsequent varieties, and be coexistent with them'' (422–23). If nature had not been able to change these ''archaic'' forms, it was foolish to assume that humanity could do anything useful in the matter, and it was certainly impossible to change overnight such characteristics as had required the passing of thousands of years for their development in the white male. Thus Darwinism provided the scientific argument for ever-intensifying racist and sexist attitudes, which became steadily more virulent and apparently unassailable because they appeared to have the full support of ''nature.'' The phrases ''natural selection,'' ''struggle for life,'' ''survival of the fittest,'' and ''evolution'' became the cultural buzzwords of the late nineteenth century. All that was still needed was to show that evolution was not a collective process—not a process involving groups but an individual affair—for the idea that one might have to consider others in the conduct of one's activities was clearly an unattractive impediment to individual incentive.

Herbert Spencer, poised for the chance to prove the merits of predatory individualism as the true evolutionary force in society, came to the rescue almost instantaneously. In 1862, just a few years after the publication of Darwin's *Origin of Species,* he published his *First Principles,* in which he argued two seemingly contradictory points. First, he stated that ''evolution, under its primary aspect, is a change from a less coherent form to a more coherent form.'' This was the sort of process which we can observe in ''societies and the products of social life'' (299), and thus justified calling society an organism in the process of evolution. But it was also true that in the universe as well as in individual organisms, increasing ''density,'' the physical evidence of greater coherence, was the sign of a more complex integration of the parts of an organism. Hence, he argued, it is clear that ''there has been an evolution of the simple into the complex alike in individual forms and in the aggregate of forms.'' Obviously what is more complex is more complex because it contains more ''variety.'' ''Advance,'' Spencer argued [and here we see how easily ''science'' becomes metaphor] ''from the homogeneous to the heterogeneous is clearly displayed in the progress of the latest and most heterogeneous creature—Man. While the peopling of the Earth has been going on, the human organism has grown more heterogeneous among the civilized divisions of the species; and the species as a whole, has been made more heterogeneous by the multiplication of races and the differentiation of them from one another'' (312–13). Thus the ''internal heterogeneity'' of the organism of human society needed to increase if society as such were to evolve properly into a more ''densely structured'' organism.

In consequence, such signs of a greater social complexity as ''supremacy and subordination must establish themselves.'' And, inevitably, ''so long as men are constituted to act on one another, either by physical force, or by force of character, the struggles for supremacy must finally be decided in favour of some class or someone; and the difference once commenced must tend to become ever more marked. Its unstable equilibrium being destroyed, the uniform must gravitate with increasing rapidity into the multiform.'' In other words, the greater the inequalities in a society, and the greater the individual differences among men, the more ''advanced'' that society showed itself to be as an ''organism.'' Inevitably the natural kingpins in such a society were those who were most ''individual,'' most intellectually advanced, and most visibly at the top of the economic ladder.

Thus, the writings of Spencer and Darwinism were combined to justify social inequality and ''economic individualism.'' But just as Darwin had proved the theory of evolution through scientific research, it was clearly necessary to prove the validity of Spencer's social Darwinism through similar research. Spencer had pointed to the manner in which the development of the person reflected the levels of development of the various races.

> In the infant European we see sundry resemblances to the lower human races; as in the flatness of the alae of the nose, the depression of its bridge, the divergence and forward opening of the nostrils, the form of the lips, the absence of a frontal sinus, the width between the eyes, the smallness of the legs. Now as the developmental process by which these traits are turned into those of the adult European, is a continuation of that change from the homogeneous to the heterogeneous displayed during the previous evolution of the embryo; it follows that the parallel developmental process by which the like traits of the barbarous races have been turned into those of the civilized races, has also been a continuation of the change from the homogeneous to the heterogeneous (313–14).

Obviously, a scientific classification of the various levels of development reached by the different races was in order: An organized world needed an organized pecking order. That is where craniology, the science of brain measurement, came in.

Spencer was using the newest scientific information when he noted that in the higher species "evolution is marked by an increasing heterogeneity in the vertebral column, and especially in the components of the skull: the higher forms being distinguished by the relatively larger size of the bones which cover the brain, and the relatively smaller size of those which form the jaws, etc. Now this trait, which is stronger in Man than in any other creature, is stronger in the European than in the savage" (313). Spencer here echoed "findings" of Carl Vogt and other craniologists. Thus, the already flourishing science of brain measurement was linked with the theory of evolution to establish the "scientific" basis of the pecking order not only among the races but also among the various members of the "higher" races. In establishing that order, researchers quite conveniently also found an entirely new and potent argument to justify the natural inferiority of woman.

Carl Vogt was probably the best known and most highly respected of the craniologists who busily measured skull sizes and comparative brain weight of "superior" and "inferior" races, and of men and women. Eager to turn a foregone conclusion into a scientific discovery, he and his colleagues used extremely peculiar standards of measurement. Characteristic, for instance, of the "scientific" procedure of Vogt was the arrogant mixing of convenient assumption and incomplete "fact." This method, combined with a form of metaphoric description which tended to lean heavily on value-charged implied comparisons, could make the conclusion that "savages" and women had "scientifically" been shown to be inferior to the Caucasian male into a perfectly self-evident fact of nature.

Vogt demonstrated the convenience of his procedure in his *Lectures on Man*. After stating that "when looking at a characteristic Hottentot or Negro skull" one could clearly observe that "the face projects like a muzzle," he needed merely to add that "it has been observed generally"—by whom or why was left undiscussed—that the "development of the jaws is in direct relation to the intellectual capacity of a people" (51–52), to reach the (to him) obvious conclusion that a wide variety of non-Germanic peoples all over the globe belonged, using the European as standard, "to the lowest races of man" (52). Quoting the findings of other craniologists, which had showed that "the female skull is smaller" than that of the male and that "the skulls of man and woman are to be separated as if they belonged to two different species," he concluded that "we may, therefore, say that the type of the female skull approaches, in many respects, that of the infant, and in a still greater degree that of the lower races; and with this is connected the remarkable circumstance, that the difference between the sexes, as regards the cranial cavity, increases with the development of the race, so that the male European excels much more the female than the Negro the Negress'' (81).

In a barrage of similar ''scientific'' findings Vogt went on to show a series of ''inherent'' differences between his standard of measurement, the German male, and the ''simious'' Negro. Then, descending to what he clearly considered the bottom of the evolutionary scale, he pointed to even greater '' simian'' qualities in ''the Negress, in whom the arm is absolutely longer than in the male, whilst the upper arm is absolutely shorter.'' Without further ado, he thereupon pronounced the following categorical decision toward which he had been building all along: ''We may be sure that, whenever we perceive an approach to the animal type, the female is nearer to it than the male, hence we should discover a greater simious resemblance [in studies of the ''missing link'' between humans and animals] if we were to take the female as our standard'' (180).

From this conclusion it was but a short step to the decision that everything in the evolutionary process pointed to the fact that the development of woman in general tended to parallel that of the ''inferior'' races rather than the evolving white male. Once that decision had been reached, the race-and-sex equation which was to dominate turn-of-the-century thinking became fully operational. ''The Negro-child is not, as regards the intellectual capacities, behind the white child, Vogt remarked, but ''no sooner do they reach the fatal period of puberty, than, with the closure of the sutures and the projection of the jaws, the same process takes place as in the ape. The intellectual faculties remain stationary, and the individual—as well as the race—incapable of any further progress.'' Thus, concluded Vogt, every indication of evolutionary science led to the conclusion that ''the grown-up Negro partakes, as regard his intellectual faculties, of the nature of the child, the female, and the senile white'' (191–92). The popular turn-of-the-century juxtaposition of women, children, and the ''inferior'' races, had consequently been given a ''scientific'' basis which served as a justification for the imposition of horrendous misery and destruction on untold millions of human beings.

The equation of the intellectual capacities of women and children was, of course, a simple continuation of the fetishized idealization of woman—equivalent, in her innocence, to the child—which had been perpetrated by such mid-century ideologues as Michelet and Comte. Their high-toned rhetoric had already been turned into a formula for denigration by such avowed and widely influential mid-century misogynists as Proudhon and Schopenhauer. The latter, in his essay ''Of Women'' (1851), had unhesitatingly declared that ''women are directly fitted for acting as the nurses and teachers of our early childhood by the fact that they are themselves childish, frivolous and short-sighted; in a word, they are big children all their life long—a kind of intermediate stage between the child and the full-grown man, who is man in the strict sense of the word'' (296). Michelet had said, ''Mother and child are one in that living ray, which restores their primitive and natural unity'' (57). Now Vogt made his appearance to explain the simple evolutionary reason why this was true. Woman was stunted in her evolutionary growth; she remained, as Schopenhauer had said already, forever the child—the ''savage.''

Darwin himself was impressed by Vogt's evidence. Instead of questioning the anthropologist's scientific proof, he quoted him in *The Descent of Man* (1871), coming to the not surprising decision that, indeed, ''the female somewhat resembles her young offspring throughout life'' (702). Citing Vogt directly, Darwin noted that male and female children resemble each other closely, like the young of so many other animals in which the adult sexes differ widely; they likewise resemble the mature female much more closely than the mature male. The female, however, ultimately assumes certain distinctive characters, and, in the formation of her skull, is said to be intermediate between the child and the man. Again, as the young of closely allied though distinct species do not differ nearly so much from each other as do the adults, so it is with the children of the different races of man. Some have even maintained that race-differences cannot be detected in the infantine skull (635).

This is a relatively short passage in a very long book, but the fact that it constituted, as it were, the ultimate approval of the master of evolution for the theory that woman's evolutionary growth had been stunted, guaranteed that Darwin's remarks would have a disproportionately large influence on late nineteenth-century thinking. His words became law and were repeated everywhere as if they were based on the most absolute of scientific truths. From this point on, virtually everyone who wished to say anything about the ''inherent'' differences between the sexes quoted Darwin directly, or indirectly, with the air of adducing incontrovertible proof. Yet nothing is more characteristic of the ''degenerative effects of inbreeding'' among late nineteenth-century intellectuals (to turn the tendentious language of these researchers back upon them) than the manner in which Vogt justified his findings on the basis of an appeal to Darwin's theory of evolution, and the way Darwin put his seal of approval on Vogt's findings by adopting them without further question.

How important Darwin's voice was in solidifying the prejudices of late nineteenth-century thinkers against women can be demonstrated by an examination of the development of Spencer's ideas about women's rights. Spencer had all along been the sort of mid-nineteenth-century ''progressive'' who thought that for the masses the economic necessity of living in filthy slums was a good, character-building experience. (''Whoso thinks that government can supply sanitary advantages for nothing, or at the cost of more taxes only, is woefully mistaken. They must be paid for with character as well as with taxes'' [348]). However, driven by a fervent zeal for the individual's freedom from government intervention, Spencer, writing in 1850, sounded suspiciously like a man not unwilling to give woman a chance to compete equally with the male in the arena of predatory self-amelioration. ''The law of equal freedom,'' he had pointed out, ''manifestly applies to the whole race—female as well as male.'' He added that it remained to be shown ''why the differences of bodily organization and those trifling mental variations which distinguish female from male should exclude one half of the race from the benefits of this ordination'' (*Social Statics*, 138). In consequence, he began to sound dangerously like a precursor of John Stuart Mill. He even defended the notion that women should be given the right to exercise political power—although he admitted that such an idea was ''repugnant to our sense of propriety—conflicts with our ideas of the feminine character—[and] is altogether condemned by our feelings'' (151).

As this remark already indicates, not everything Spencer said in *Social Statics* showed him to be a true champion of the women's cause. His arguments on behalf of women were more likely an attempt to maintain the logical continuity of his thesis in favor of the rights of the individual than a reflection of any special concern for the position of women in society. Thus, when Darwin published his *Origin of Species*—giving Spencer a chance to adjust his own ideas concerning the predatory requirements of cultural amelioration in accordance with the latest findings of evolutionary science—he was only too eager to disparage the position he had taken earlier concerning the capacities, if not the rights, of women. As he pointed out in a new preface to the American edition of *Social Statics* (1864), he would now "make qualifications which, while they left the arguments much as they are, would alter somewhat their logical aspects."

By the time Spencer published *The Study of Sociology* (1873), the influence of Darwin's low opinion of women in the evolutionary scale echoed very directly in his own remarks. No longer were the mental variations between men and women "trifling." Instead, Spencer now declared, "That men and women are mentally alike, is as untrue as that they are alike bodily. Just as certainly as they have physical differences which are related to the respective parts they play in the maintenance of the race, so certainly have they psychical differences, similarly related to their respective shares in the rearing and protection of offspring." Spencer now also insisted that "a somewhat earlier arrest of individual evolution in women than in men" was a natural circumstance, logically "necessitated by the reservation of vital power to meet the cost of reproduction. Whereas, in man, individual evolution continues until the physiological cost of self-maintenance very nearly balances what nutrition supplies, in woman, an arrest of individual development takes place while there is yet a considerable margin of nutrition: otherwise there could be no offspring." This "arrest of individual development" in women resulted in "a perceptible falling-short in those two faculties, intellectual and emotional, which are the latest products of human evolution—the power of abstract reasoning and that most abstract of the emotions, the sentiment of justice—the sentiment which regulates conduct irrespective of personal attachments . . ." (340–42).

From the moment scientists decided that woman's evolution had not progressed beyond the condition of childhood, it became a foregone conclusion that a link was to be discovered between woman's "stunted" evolution and her reproductive responsibility. Obviously, a plausible reason for this phenomenon would have to be found to make it a part of objective science. What better reason, then, what more perfectly metaphorically harmonious motive than to discover that woman was mentally a child because she needed all her "vital energy" to have children. Brain work required much vital energy—and hence brain work was properly the realm of the male. To think was to "spend" vital energy just as much as it took to give birth to a child. Hence men created in the intellectual realm, while women needed to conserve energy to create in the physical realm. Indeed, all of woman's energies were properly required for this process of physical creation. That was why nature had made her incapable of serious thought, for serious thought would compete dangerously with woman's reproductive capacity for the finite amount of vital energy available in any human being. That was also why woman was, in evolutionary terms, no more than "an undeveloped man"—the title of a long chapter in Harry Campbell's disquisition on *Differences in the Nervous Organization of Man and Woman*. And, certainly, it was why "the mind of the average woman appears to be absolutely deficient in the power of coherent, impersonal thought," as George Fleming insisted in 1888 in *The Universal Review* (403).

Consequently, any unnecessary brain activity should be avoided by women, since, as Nicholas Cooke insisted in *Satan in Society*, "in man the substance of the brain has more consistency, more density; in woman it is softer and less voluminous" (279). To the late nineteenth century the idea of a "brain drain" had a disturbingly literal ring. In both men and women the expenditure of physical energy involved the loss of "vital fluids," but certainly no physical activity involved so much of a squandering of these vital fluids as sexual activity. Indeed, since the blood carried the energy needed for the proper functioning of the brain in the male and for the reproductive function in woman, vital energy was deflected every time the body was stimulated sexually. Consequently, as Bernard Talmey warned, "the frequent exercise of the act of copulation leads directly to anemia, malnutrition, asthemia of the muscles and nerves, and mental exhaustion. Immoderate persons are pale and have long, flabby or sometimes tense features. They are melancholic and not fit for any difficult and continued corporeal or mental work" (*Woman*, 170).

If normal sexual contact between the sexes was already so damaging (Cooke had suggested a maximum indulgence in marital pleasures of no more than twice a month to maintain general physical and mental health), it was easy to imagine the effects of "breeding" and childbirth on the mental and physical capacities of women. If the normal maintenance of the "seat of reproduction" already consumed all the vital energy which might otherwise have fed the brain, the extent to which pregnancy and childbirth ravaged woman's mental capacity was truly immense.

August Strindberg, a faithful reader of the anthropologists' disquisitions on woman's natural inferiority to man, expressed his own opinion on the matter in an article for *La Revue Blanche* of January 1895. After having presented the Darwinist and craniological evidence that "between the child, woman, and the inferior races there exists a not negligible analogy," he remarked, rather disingenuously, "How our white woman has become inferior to man is impossible to determine with certainty." But Strindberg certainly wasn't the man to avoid trying. He was much too well versed in contemporary medical science not to know what ravages were wrought upon her brain by woman's servitude to the reproductive process. Indeed, even the effect of *not* having children was enough to explain woman's mental debility. "How could it be otherwise, since even if she is celibate, a quarter of the time of her life, from her twelfth or fourteenth year to her forty-fifth, is taken up by the labor of procreation? Every fourth week she finds herself caught in her menstrual period, which is accompanied by a loss of blood which can surpass two hundred grams." To Strindberg it was clear that "a human body cannot develop in a normal fashion if it is deprived in this way of such a significant quantity of nutritive fluid." In consequence, he thought it "permissible to posit that these periodical bleedings are in part to blame for the arrest in growth and development of woman, and that indeed this anemia of necessity serves to atrophy the brain." That is why it was appropriate to designate the normal physical and mental condition of the grown woman as that of a "sick child" (13–14).

To this line of thought a convinced Darwinian naturalist such as George Romanes needed only add—as he did in a May 1887 article on "Mental Differences between Men and Women" for *The Nineteenth Century*—that "not only is the grey matter, or cortex, of the female brain shallower than that of the male but [it] also receives less than a proportional supply of blood" (657). An image of woman was now created that showed her to be a perpetually "brain-drained" creature whose mental faculties atrophied progressively as she engaged in her reproductive duties. What is more, since the natural function of women was indeed reproductive and not creative, it could be argued that woman's brain was a superfluous anachronistic holdover from more primitive times, when the evolutionary levels of men and women were still relatively equivalent.

For, as Darwin had pointed out in *The Descent of Man*, and as others had speculated before him, "some remote progenitor of the whole vertebrate kingdom appears to have been hermaphrodite or androgynous" (183). This meant that the separation of men and women into two different sexes was the result of a tremendous, and still ongoing, process involving the progressive separation of individual characteristics, in which nature, following the principles of the division of labor, had divided the original androgynous progenitors into men, whose responsibility was to engage in productive, creative, "evolutionary" activities, and women, whose responsibility was reproductive, imitative, and "conservative." All that was still needed, at this point, was proof that the inherent natural necessity of the ongoing mental enfeeblement of the female, in comparison to the intellectual evolution of the male, was a logical aspect of the inexorable forces of species evolution. Spencer's theories of social determinism in the service of progress were perfect to establish just such a connection.

There is a clear correlation, for instance, between the notion that with the progress of evolution men and women had become more *un*like and Herbert Spencer's famous dictum, in *First Principles,* that "evolution is definable as a change from an incoherent homogeneity to a coherent heterogeneity" (332). The increasingly greater separation of the sexes and their functions was, in other words, a necessary by-product of the evolution of man, of the division of labor on a universal scale. Man, creative and intellectual, was destined to soar to ever higher levels of mental achievement, while woman, incapable of the higher forms of evolution, was doomed to remain a simple tool of nature, a domestic animal, one might say, whose sole responsibility was the reproduction of the race. It was, of course, understandable that women tended to resist this instance of the development of a "coherent heterogeneity" in male-female relationships, but to the many followers of Darwin and Spencer the very fact that the battle of the sexes was palpably heating up was an indication that the processes of evolution were working smoothly. For in the struggle for life and its accompanying requirements, the male, being quite obviously the fittest, was pulling away more and more from his originally relatively equal partner, woman. The progressive functional separation of the sexes was, in a direct sense, analogous to what happens in the evolution of organisms in general. Spencer had pointed out that "incipient organisms, setting out from relatively homogeneous arrangements, forthwith begin to fall into relatively heterogeneous ones" (*First Principles,* 387). Inevitably, in the course of this process there was a progressive loss of "functional equality" among organisms.

Unavoidably then, an ever greater division between men and women—and between the "superior" and "inferior" races—would have to establish itself as human society evolved. The evolutionists liked to point to the clear evidence of history to show how much progress had been made in this area. The activities of "primitive" people still showed a lamentable equality of tasks between men and women. Vogt remarked disdainfully:

> The lower the state of culture, the more similar are the occupations of the sexes. Among the Australians, the Bushmen, and other low races, possessing no fixed habitations, the wife partakes of all her husband's toils, and has, in addition, the care of the progeny. The sphere of occupation is the same for both sexes; whilst among the civilised nations, there is a division both in physical and mental labour. If it be true that every organ is strengthened by exercise, increasing in size and weight, it must equally apply to the brain, which must become more developed by proper mental exercise.

In the division of labor established in the course of evolution, the absence of any need to exercise her brain had made woman disdainful of change, which was, after all, a factor of the exercise of the mental faculty. Therefore, "just as, in respect of morals, woman is the conservator of old customs and usages, of traditions, legends, and religion; so in the material world she preserves primitive forms, which but slowly yield to the influences of civilization.'' Even in her physical constitution woman was retentive of more primitive forms: she preserved, Vogt insisted, ''in the formation of the head, the earlier stage from which the race or tribe has been developed, or into which it has relapsed. Hence, then, is partly explained the fact, that the inequality of the sexes increases with the progress of civilization'' (*Lectures on Man,* 82).

Darwin, feeding off Vogt and Spencer, capped the argument in *The Descent of Man* by confirming that ''with woman the powers of intuition, of rapid perception, and perhaps of imitation, are more strongly marked than in man; but some, at least, of these faculties are characteristic of the lower races, and therefore, of a past and lower state of civilization.'' Given man's natural role as predator, it was inevitable that he would become the superior creature, for to survive he had to develop his imagination and his reason. These faculties had been developed in man ''partly through sexual selection—that is through the contest of rival males, and partly through natural selection—that is from success in the general struggle for life; and as in both cases the struggle will have been during maturity, the characters gained will have been transmitted more fully to the male than to the female offspring.'' Thus, Darwin concluded, ''man has ultimately become superior to woman''— indeed, he argued that all indications pointed to the fact that, just as Vogt had contended, the ongoing evolutionary pressures would ''tend to keep up or even increase [the male's] mental powers, and as a consequence, the present inequality between the sexes'' (643–45).

Thus, both nature *and* nurture conspired to drain woman's brain—even to the point of atrophy. She was, in effect, ''naturally debilitated.'' Paul Möbius, an extremely influential German pathologist who concerned himself with the analysis of neurasthenia and hysteria—and who was prominently quoted by Breuer and Freud in their *Studies in Hysteria* (1895)—wrote a much-reprinted essay on the subject bluntly entitled *On the Physiological Debility of Woman* (1898). He continued to update this opus, compiling new evidence of the natural feeblemindedness of all women, right up to the time of his death in 1907. Women, he argued, were, in comparison with men, by their very nature feebleminded. One did not even have to be a Darwinist to recognize that simple fact. The adage ''long of hair, short of wit'' said it all. Using the evidence of craniology and the opinions of the biologists, as well as the studies of Lombroso and Ferrero, Möbius argued that woman was ''spiritually sterile'' (10) and, of course, ''child-like'' in her incapacity for reasoned thought. But, said he, all this was not something to worry about, for woman's mental debility was indeed a necessary aspect of her biological function. ''Excessive use of the brain does not just confuse woman, it makes her ill,'' he declared. The remark that all woman needed to be was ''healthy and dumb'' seemed to him a very reasonable assessment of her natural function as a ''breeding animal.'' According to Möbius, ''If we wish to have women who fulfill their responsibilities as mothers, we cannot expect them to have a masculine brain. If it were possible for the feminine abilities to develop in a parallel fashion to those of a male, the organs of motherhood would shrivel, and we would have a hateful and useless hybrid creature on our hands'' (14). We should therefore be very grateful, the distinguished professor concluded, that nature had made certain that woman could never be anything but a breeding machine, a feebleminded, ''animal-like, dependent'' creature, who, ''just like the animals, since time immemorial, has done nothing but ceaselessly repeat herself'' (8).

Thus the ''brainless'' existence of woman as a creature of simple instincts, the fertile soil of humanity, the ripe fruit of nature, whose instincts should be cultivated to feed the primal needs of the developing male, came to be seen by many turn-of-the-century males as a scientifically proven fact. Maternity should be the sole realm of feminine activity. Intelligent men, warned Möbius, should make certain that their wives would be "healthy women, not brain-ladies." Let's tear down the new-fangled high schools for women, he argued. They have little effect on girls, except to make them "nervous and weak"(30). Instead, let them develop their usable household skills. "Let's protect woman against intellectual activity" (31).

Darwin's conclusion that "the female somewhat resembles her young offspring throughout life" (*The Descent of Man,* 702), with its negative implications concerning woman's intellectual capacities, fits in perfectly with Michelet's mid-century dictum that a woman and her child were one in "that living ray, which restores their primitive and natural unity." Consequently, it is not surprising that many of the more conventional women of the years around 1900 should have found justification for their femininity in being associated with or being like their children.

Late nineteenth-century art witnessed a development of the conventional mother-and-child image which directly reflected the new evolutionist dictum that a mother and her children were essentially coextensive and formed a "primitive and natural unity." These paintings tended to portray a mother and her child, or children, as virtually glued together—jammed into the space of the image as if they were bound together by an almost visible mental and even physical coherence. Sometimes such paintings showed the mother sitting stoically in the center of the image while her brood crawled all over her, sat in her lap, hung on her neck, leaned on her shoulders, and so on, in the fashion of Bouguereau's "Alma Parens." But often, especially in the work of women painters who specialized (as most did) in painting women, this "spiritual conjunction" between woman and child was depicted as if it were a sign of a riveting "intuitive" bond. Many of Mary Cassatt's works reflect this specific affectation. For example, her painting "Caress" [VI, 1] shows a mother and her two children, their heads jammed together in such a way that those of the two children and that of their mother seem to have an almost physical bond—as if we have here an entirely new form of Siamese linkage. John Singer Sargent also made a striking contribution to this theme when he painted "Mrs. Fiske-Warren and Her Daughter" [VI, 2] in such a manner that it appeared as if the daughter's head had been glued to her mother's neck, thus giving the impression that the girl's head had sprouted like a mushroom from the same stem as her mother's instead of belonging to a person with a separate body and identity.

Indeed, it took the cultural lights of the later nineteenth century very little time to recognize the "truths" inherent in the contentions of the evolutionists and psychologists concerning the natural debility of woman. Artists and poets like to be in the intellectual avant-garde—and what greater proof of one's cultural vanguardism than to demonstrate one's awareness of the latest scientific discoveries? The now aged Tennyson, who had more right than anyone to adopt the evolutionist point of view—in effect, he had been writing what amounted to evolutionist verse decades before the publication of the *Origin of Species*—wrote stirring hymns to the evolving soul of man, while apostrophizing the cultural "sleep" of woman with a new, even more condescending sense of pity. "Oh thou," he declaimed in his poem "To a Lady Sleeping," "through whose dim brain the winged dreams are borne . . . Thou all unwittingly prolongest night" (840)—drawing an opportune analogy between a woman's actual sleep and the sleep of her spirit.

Mark Twain, who always liked to have his fun with the natural foolishness of women—in *A Connecticut Yankee in King Arthur's Court* Hank's Sandy is a brainless chatterbox

VI, 1. Mary Cassatt (1844–1926), "Caress" (or "Caresse Enfantine") (ca. 1902)

VI, 2. John Singer Sargent (1856–1925), "Mrs. Fiske-Warren and Her Daughter" (1903)

with a heart of gold—showed Eve to be absolutely incapable of theoretical reasoning. As she remarks contentedly,

> I have learned a number of things, and am educated now, but I wasn't at first. I was ignorant at first. At first it used to vex me because, with all my watching, I was never smart enough to be around when the water was running uphill; but now I do not mind it. I have experimented and experimented until now I know it never does run up-hill, except in the dark, because the pool never goes dry; which it would, of course, if the water didn't come back in the night. It is best to prove things by actual experiment; then you *know;* whereas if you depend on guessing and supposing and conjecturing, you will never get educated" (*Eve's Diary,* 85).

It would be nice to be able to think that Twain was making fun of the late nineteenth-century male scientists who used just this sort of reasoning to come to their conclusions concerning the natural weak-mindedness of woman, but unfortunately all indications are that Twain's victim, too, was woman rather than the Darwinians.

Wedekind's Lulu has a remarkable effect on the men around her. As one of her suitors remarks, when she speaks logic and sense depart. "I no longer know how or what I am thinking; when I listen to you I cease to think at all" (78). Women everywhere seemed to have this draining effect on men in the literature of the turn of the century. Indeed, this was the period which saw the standardization of the "dumb blonde" image of woman which still dominates Western culture in its monotonous, unvarying array of stereotyped representations, virtually all of which found their origin in the novels and dramas of the later nineteenth century. To catalogue them all would be as pointless and repetitious as the mentality of the men (and occasionally the women) who created these terminally brainless women, whose very names—Nana, Lulu, Fifi, Carrie Meeber, Trina Sieppe, Undine Spragg, Effi Briest—were meant to reflect their mental inanition as well as their lowly position on both the social and evolutionary scales.

In art, the proliferation of the image of woman's self-involved glance in the mirror and an ever greater emphasis on her spiritual improverishment and mental inanition went hand in hand. Arthur Hacker's painting "The Drone" [VI, 3], exhibited at the Royal Academy in 1899, expresses this view of feminine debility perfectly. A richly attired woman—clearly the wife of a wealthy man of the world—stands in a garden, a flower among flowers, and contemplates vacantly, weakly, and a little sadly the activities of a drone. The symbolism of the painting was crystal clear to one contemporary commentator, whose words ruthlessly identified this woman's role in the productive world:

> All the beauty of the foxglove in its many colours is thrown into relief by a depth of dark leafage, and into this congregation of colours comes many a honey-loving and honey-gathering bee, the worker and the non-worker, the industrious and the idle. A large drone, earning nothing for all its loud humming, has caught the notice of the stately lady, herself an idler,

VI, 3. Arthur Hacker (1858–1919), "The Drone" (1899)

VI, 4. Frederic, Lord Leighton (1830–1896), ''Bianca'' (ca. 1881)

VI, 5. Edmond Aman-Jean (1860–1936), ''The Mirror in the Vase'' (1904)

who wanders amid this paradise of blooms, and seems at the moment to be in contemplation of her own life, as being possibly no less indolent and useless than that of the drone (*Great Pictures in Private Galleries*, 26).

The woman Hacker has painted, like the drone, has no sting and gathers no honey. A useless, nonthreatening parasite, she stands—sickly and stoop-shouldered, with the drone of the insect she contemplates (and is) filling the beautiful but empty cavity of her head— among the surging, phallic foxglove and dandelions gone to seed. Her existence is meaningless in a fertile and productive environment. In this painting Hacker has presented a disdainful commentary on the ideal of the Victorian middle-class wife as the personification of the extremes of passive virtue—what we have called the consumptive sublime.

The link between the uroboric circularity of woman, her self-reflective glance in the mirror, and the subsequent progressive atrophying of her intellect was too self-evident not to have been exploited by the painters. Karl Kraus, the widely influential Viennese journalist and aphorist, had expressed most men's expectations when he quipped that ''the eyes of a woman should mirror my thoughts, not hers'' (*Selected Aphorisms*, 101). And, indeed, only when woman's eyes mirrored man's thoughts could she be expected to have any expression at all. Hence woman's glance in the mirror, representative only of herself and "the green circle of her glances"—to quote Arthur Symons (*Silhouettes,* 12)—became the endlessly receding echo of her own empty mind.

Lord Leighton's "Bianca" [VI, 4] is another striking example of a woman whose eyes have been extinguished in the virtuous inanity of the contemplation of her own mindless existence, and Aman-Jean's "The Mirror in the Vase" of 1904 [VI, 5] is an example of the manner in which the water mirror theme could be used to comment on the inherent mental debility of woman. The painter did so by juxtaposing two aspects of the ideology of feminine self-containment. First we are shown the self-absorbed woman staring with singleminded concentration into that mirror of water which, in its circularity, represented the primordial, static realm of material reality as well as the volatile, evanescent qualities of ordinary being. Then, beside this self-absorbed woman we are given, at the base of the vase, the "abandoned" woman, the feminine soul as it showed itself when woman was not looking in the mirror: listless, staring vacantly into nothingness, the personification of mindlessness and uselessness.

The apparent sense of mystery which gives the eyes of Rossetti's "Beatrice" [VI, 6] their hypnotic, snakelike quality is an aspect of this woman's saintly vacancy, of this Beatrice's mirrorlike capacity, to use Kraus' words, to reflect the creative male's imperial thoughts. In a similar fashion, Fernand Khnopff, who had learned much from Rossetti's ideal heads, set out to explore the bottomless—and mindless—mystery and circularity of woman's true nature in numerous images of women whose eyes were as gray and impenetrable as a dense fog drifting over a peatbog in Flanders. Such titles as "I Lock My Door Upon Myself" and "Who Shall Deliver Me?", echo sentiments first expressed by Christina Rossetti. The British painter Frederic Cayley Robinson, meanwhile, made a specialty

VI, 6. Dante Gabriel Rossetti (1828–1882), "Beatrice," pastel (ca. 1880)

of weak-shouldered, dead-eyed women who stood or sat around listlessly in vaguely medieval settings [VI, 7]. These works gave him the enthusiastic approval of critics.

There is something deeply troubling in the paralysis which spreads through the extinguished eyes of these women. As Edvard Munch's 1901 lithograph "Sin" [VI, 8] makes perfectly clear by its very title, much of this ocular inanition was due to woman's perverse insistence on actually hastening the desperate process of her brain drain by continuing not only to study herself in the mirror but also to pursue her autoerotic self-involvement. Instead of heeding Doctor Harry Campbell's call to be civilized and let her sexual instinct atrophy, she was letting her brain atrophy by continuing her practice of physical self-stimulation. Most late nineteenth-century intellectuals were perfectly aware of the consequences of that sort of practice, which in Krafft-Ebing's opinion could only result in something barely short of brain death: "Nothing is so prone to contaminate—under certain circumstances, even to exhaust—the source of all noble and ideal sentiments, which arise of themselves from a normally developing sexual instinct, as the practice of masturbation in early years" (*Psychopathia Sexualis*, 188–89). Hence for such followers of Krafft-Ebing as Bernard Talmey, "the greatest and most important problem of sexual hygiene is the prevention of masturbation in children and young women" (*Woman*, 187). In the 1890s, especially in the United States, where, then as now, a ton of prevention was habitually thought to be worth an ounce of cure, this obsession with the "abnormal" sexual interests of women led to a veritable epidemic of the barbaric form of physical assault upon women called clitoridectomy, the surgical removal of the clitoris.

The brain, as we have seen, was the place from which all man's creative energies, including his sexual potency, issued. Ezra Pound was a late spokesman for this conception. In *Pavanes and Divagations* in a postscript to his translation of Gourmont's *The Natural Philosophy of Love,* Pound identified "the brain itself" with the male and hypothesized that it therefore was "more than likely—only a sort of great clot of genital fluid held in suspense or reserve." This likelihood, he speculated, "would explain the enormous content of the brain as a maker or presenter of images. Species would have developed in accordance with, or their development would have been affected by, the relative discharge and retention of the fluid" (203). One could suppose, if one were to continue Pound's logic, that the very creation of images should be considered a form of autoerotic indulgence. Had physicians been equally worried about the incidence of such indulgence among males, consistency might have required the forcible decapitation of budding poets and painters. Such a procedure would at least have spared the world a great deal of that suffering which stems from the perusal of bad art.

This may seem a poor joke (of the sort Twain liked so much) about a serious subject, but it illustrates the callous disregard for the physical rights of women which was characteristic of what Mary Livermore, a women's-suffrage worker of the period, had called "the unclean army of 'gynecologists.' " Certainly the late nineteenth-century medical establishment showed—judging by the frequency with which it resorted to this surgical operation to "cure" what was considered "hyperaesthesia" in middle-class white women—that it fully agreed with Carl Vogt that these women were "intellectually equivalent" to "the inferior races," to "criminals," and "the senile white," for these were the only male victims of the period ever seriously threatened by such practices as "therapeutic castration."

The frequency with which late nineteenth-century painters depicted women with extinguished eyes demonstrates that they, too, were firmly convinced of the general mental

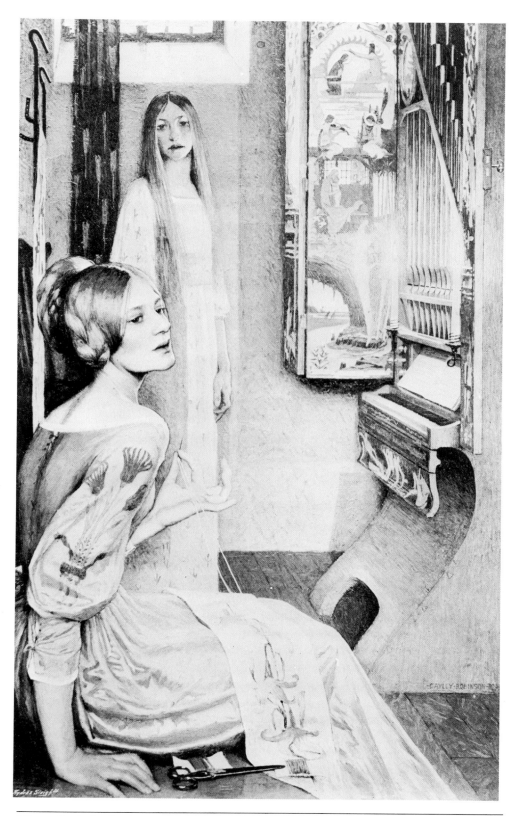

VI, 7. Frederic Cayley Robinson (1862–1927), "Souvenir of a Past Age" (1893)

VI, 8. Edvard Munch (1863–1944), "Sin," lithograph (1901)

VI, 9. Edouard Manet (1832–1883), "Nude Study of a Blonde" (ca. 1875)

debility of women. To recognize the universality of this sentiment one need only compare the many representations of males—whose eyes sparkle with penetration, intense conceptual and perceptual insight, and evolutionary fervor—to these same painters' portraits of women, whose eyes are almost always full of ashes. The expressions of the women in these portraits tend to be fashionably wan, as if to reflect their ebbing intellect, and often their very being expresses the sort of beauty Flaubert saw as tempting when he and his Saint Anthony confronted the devil's most beautiful prostitute: "Her fleshy lips have a look of blood, and her somewhat heavy eyelashes are so much bathed in languor that one would imagine she was blind" (*The Temptation of Saint Anthony*, 327).

The impressionists were no less persistent in the portrayal of women with vacant eyes than their stylistically more traditional colleagues. The joyous colors of their paintings cannot obscure the clearly visible evidence of their disdainful conception of the feminine intellect. Degas frequently emphasized the cavernous darkness he seemed to consider a characteristic feature of women's eyes, and Manet, in such a work as his "Nude Study of a Blonde" [VI, 9], painted a devastating image of a woman whose heavy breasts have become emblematic of her imprisonment in the material world. Indicative of the narrow circle of her being, this woman's eyes lie like dead cinders in their sockets—burnt-out relics of the spirit. The painting is so powerful as an image, so brilliant in execution, so convincing in its characterization, and so apparently realistic that it is tempting to see this work as not just another study of a brainless Nana but a moving psychological portrait of the state of mind of a nineteenth-century woman driven to existential despair by the mental and physical assaults upon her integrity and her self-identity as a person by the men around her. Unfortunately, all indications are that Manet's motive in painting this portrait was no different from that which made Arthur Hacker paint his image of woman as a colorful but "indolent and useless" drone.

In the past, students of impressionism have emphasized the realism of paintings such as this, the revolutionary significance of their concentration upon everyday subjects rather than on subjects taken from mythology and history. But by uncritically participating in the fashionable representation of women as vacuous, empty-headed creatures, the impressionists contributed in substantial measure to the cultural ascendancy of misogynist attitudes.

In fact, the impressionists' special proclivity for the negative representation of women was by no means lost on their contemporaries. In his perambulation of the Salon of 1890, for instance, Maurice Hamel summed up Degas' attitude toward women with remarkable acuity. In Degas' work, he said, woman "is transformed into an illogical creature, almost a dragonfly, almost a doll." Degas was able to do so, Hamel pointed out, "by means of a transliteration of human feelings which remains somehow beyond what is real as a result of the fixed stupor of expression" given by the artist to his subjects. As a result, Hamel concluded astutely, "Degas has pushed to its limits an artistic formula which, in the last analysis, depends upon sentiments of hatred, of aristocratic distaste, and perhaps on a kind of tenderness mixed with revulsion" (*Salon de 1890*, 46–47).

Once mainstream culture rushed in to deify the impressionists, Hamel's brilliant critical analysis of Degas' perhaps only semiconscious motivations in his representation of women could easily be dismissed as an example of the supposedly wholesale and artistically ignorant rejection of impressionism by establishment critics. Those who dabble in culture love myths which posit the artist as an embattled hero. Few historians of the movement therefore care to mention the ruthless accuracy of many of the critical commentaries of the impressionists' contemporaries. Lacking the self-consciousness of the later twentieth-century intellectual concerning matters of sex or race, Albert Aurier, the highly respected critic and symbolist poet, could praise a painter such as Renoir quite bluntly for his ability to denigrate women. In the August 1891 issue of the *Mercure de France* he noted Renoir's propensity to depict women whose extinguished eyes seem to float like empty thimbles on lakes of plump, fruity flesh [VI, 10]. Aurier described these women enthusiastically as "playthings" with the "beautiful, deep, azure, enameled eyes of dolls, of adorable dolls, with flesh molded of roseate porcelain." Renoir had presented an "original and perhaps very wise conception of the famous 'eternal feminine.' " He had succeeded in making his work transcend the trivial, for though "the artist has suppressed virtually completely any elements of intellect his models might have possessed, he has compensated for this by including in his work a lavish display of his own intellect." This sort of aggressive display of male incorporative egotism struck Albert Aurier as perfectly appropriate. Why should Renoir depict woman as beautiful, "since it suffices to show that she is pretty. Why show her as intelligent, or even as stupid, why as false or as disagreeable? She is pretty! . . . Why should she have a heart, a brain, a soul? She is pretty! And that suffices for Renoir, and that should suffice for us. Does she even have a sex? Yes, but one suspects it to be sterile and only useful to our most puerile amusements." As a matter of fact, the critic continued, one could argue that by effacing the identity of his feminine subjects, Renoir had taught us something absolutely crucial about woman:

She does not live. She certainly does not think. We men, though we are all more or less psychologists, are really more foolish dreamers than psychologists; stupidly we insist on attributing to woman our own sentiments, our own emotions, the dreams of beings who live. We choose to give her a complicated heart, a twisted intelligence. We like to declare her to be angel or demon; we enjoy finding her sublime or ignoble, Machiavellian, serpentine, feline! Poor fools that you are, the painter seems to tell us. As if a cat or a viper did not have a thousand times more soul than a woman! Be reasonable, like me, and don't make such a fuss over the pseudovitality of that marvellous and oh so adorably pretty little automaton which the good Lord gave you to amuse yourselves with. Play with your doll; have some fun and attribute to her sentiments she couldn't possibly entertain; breathe life into her with your imaginations, but be careful not to take all of that too seriously, for that would make you as silly as children who, with tears in their eyes and clenched fists, pour invective over an irresponsible plaything! (103–5)

The notion that woman was basically a mindless, vacant creature whose only reason for being resided in her beauty and her reproductive function, having been given visual prominence in the representations of impressionists and academic painters alike, had become a veritable cultural commonplace in the years around 1900. Male ''scientific'' research had shown the general absence of reasoning ability among women. A well-informed writer such as Octave Mirbeau could therefore blithely assert, commenting in the *Mercure de France* on the significance of Remy de Gourmont's *Lilith,* that

the symbolic genesis of woman, as interpreted by Remy de Gourmont, corresponds exactly with the conclusions of anthropological science. Woman does not have a brain; she is simply a sexual organ. And that is the beauty of it. She has but one role in the universe, but that is a grandiose one: to make love, that is to say, to perpetuate the species. According to the irrefutable laws of nature, of which we feel rather than perceive the implacable and dolorous harmony, woman is not fit for anything that does not involve love or maternity. Certain women, very rare exceptions, have been able to give, either in art or in literature, an illusion of creative energy. But those are either abnormal creatures, in revolt against nature, or simply the reflections of males, of whom, through sexual disfunction, they have been able to maintain certain characteristics.

The work of painters, poets, and critics constantly demonstrates how during the last thirty years of the nineteenth century misogyny, the wonders of science, and the theory of evolution had joined to form a holy trinity of saintly masculinity against the regressive entity called woman. The widespread familiarity of most educated members of middle-class culture with this scientific consensus concerning the intellectual and moral debility of women served to infuse many images of this period with a very specific, male-focused cultural content whose meaning has since been obscured.

The present-day viewer, for instance, is not likely to see any special significance in the often-repeated representation in turn-of-the-century art of a domesticated woman looking studiously at a goldfish bowl. Robert Reid's 1911 painting of such a theme [VI, 11] is a beautifully executed example of the genre. A viewer's initial impression at the time when the work was first exhibited (as it still is today, of course) would have been of a superbly painterly, exquisitely elegant image of a young woman, dressed in a kimono bespeaking the leisurely repose of a lady of means, surrounded by art objects—such as the Japanese screen behind her—representative of the height of educated good taste. All that, certainly, is there—and that, fortunately, is primarily what still remains in the image for the present-day viewer.

VI, 10. Pierre-Auguste Renoir (1841–1919), "Torso of a Young Woman" (or "Etude") (ca. 1876)
VI, 11. Robert Reid (1862–1929), "Girl in a Blue Kimono" (1911)

But the educated male viewer of 1911 would have been invited by this image to make inevitable analogies which would have gone far beyond this reading. He would have observed the subtle blending together of the colors of the young woman's kimono and the elegant Japanese screen behind her. He would have seen her not as the proprietor of the objects around her, but as one of the objects herself. He would have identified the woman as part of the "cultural" holdings of a well-to-do member of the leisure class. Recognizing her as property and not as proprietor, he would thereupon have been able to appreciate the appropriate allusion, supplied by the painter, to the theme of the mirror. For the goldfish bowl into which the woman looks is placed upon a shiny, circular table which looks very much like a circular mirror placed horizontally.

Once our viewer would have made those connections, the deeper significance of the goldfish—and the bowl itself—would have become clear. For was not this woman, this mirror of her husband's thoughts, as she glanced into the water in the circular bowl with the circular opening placed on the circular table, seeing something more than just the goldfish swimming aimlessly in a circle? Was not she also seeing the reflection of her own face? Wasn't she a goldfish herself, and wasn't her environment, to a large extent, the goldfish bowl of her own "useless existence?" No wonder, then, that there is such a sad, melancholy, contemplative quality in the woman's expression.

Indeed, there is in this work of Reid's, as in many of his other works—which are now often regarded, as a result of critical misunderstanding of turn-of-the-century usage of the term, as merely decorative—a level of psychological insight into the ambiguities inherent in the decorative centrality, yet intellectual marginalization, of the women of his

VI, 12. Hermine Heller-Ostersetzer (1874–1909), "Goldfish" (ca. 1905)

VI, 13. Gustav Klimt (1862–1918), "Water Snakes" (or "Girlfriends") (ca. 1901)

day which lifts his work into a category of subtle sociocritical statement rarely matched by his contemporaries. Reid's work displays a sensitivity to the point of view of his female contemporaries which is today virtually unrecognized but which made the emotive content of his work not unlike that of Cassatt. The Austrian painter Hermine Heller-Ostersetzer also seems to have wished to emphasize the closure of possibility faced by young women growing up at the turn of the century. In her painting "Goldfish" [VI, 12] the goldfish of the title clearly are the two young girls depicted by the artist with bittersweet affection as they wash and dress in their cramped room. These two, the artist would seem to imply, are really no different from the two small goldfish which can be seen swimming about aimlessly in the tiny, self-contained world of their glass bowl on the dresser.

But subtlety of message was little to the taste of most turn-of-the-century artists. They preferred more obviously symbolic analogies between woman and her goldfish bowl. One could, of course, go no further, when playing upon the analogy between women and goldfish, than the immediacy of representation achieved by Gustav Klimt, whose women, rather than merely looking at goldfish, became aquatic creatures themselves [VI, 13]. Klimt liked to portray the heavy materiality of these fishlike women as they swirled upward, no longer entirely passive but instead vaguely dangerous. These "Girlfriends" (*Die Kunst*'s title) were much like actual "Water Snakes" (the painting's current title). In Klimt's work the theme of woman's erotic self-sufficiency and circularity met up with the water mirror to hint at strange, threatening lesbian passions, at emotional lives lived without men, without recourse to the temperate control of the intellect.

The goldfish bowl was, of course, not the only object which symbolized woman's narrowly circumscribed domestic environment. In a painting reproduced in color in the October 1911 issue of *The Century*, Childe Hassam chose to show that a woman with a vacant expression contemplating a bowl of nasturtiums would do quite as well. Frederick C. Frieseke, in a beautifully composed and executed work published a few months later as a color supplement in the same magazine, showed two gracefully posed young women, their faces barely emerging from an all-encompassing sea of highly decorative carpets, clothes, and upholstery, studiously examining a bird cage. Exhibiting the period's characteristic enthusiasm for disparaging analogies between women and other caged creatures, the painter chose to call this work "The Parakeets" [VI, 14].

In the fantasy world of turn-of-the-century painters women could be found whiling away their idle hours staring at a variety of bowls and cages containing a remarkable selection of flowers and lowly creatures from the animal world, in whose captive state they were supposed to find significant analogies to their own predestined fate in civilized society. If they did not occupy themselves with women's immediate domestic environment, these painters often transported them into "wild" nature to make a similar point. In Paul Chabas' "First Bath" [VI, 15], for instance, a lake becomes the goldfish bowl for a group of little girls. Chabas, one of the most popular painters of his time, had made the portrayal of nude, prepubescent, nearly pubescent, and adolescent girls frolicking in misty lakes a specialty. In a painting of 1914 which he titled "The Moonfishers," he combined a number of mirror themes. Two nude young women are shown in a fruitless attempt to scoop reflected moonlight out of the water in which they are standing into the palms of their hands. In other words, they are in pursuit of a fluid mirror image of moonlight, itself the reflected "feminine" light of a satellite. Thus, in what might be designated as the ultimate symbolic representation of the intellectual marginalization of woman, Chabas has his women, whose cold, fishlike bodies are like the moon, reflect their meaningless existence in a frivolous attempt to capture the reflection of the reflection of the light of Apollo—all this within an environment which in its roundness hints at the goldfish bowl of their lives. For, to emphasize even further what he had already overemphasized, Chabas put his painting into a circular, and hence moonlike, mirrorlike, bowllike frame.

Chabas' prepubescent girls represent another characteristic aspect of the extremely literal mentality of the late nineteenth century: its tendency to turn from women—who, after all, were mentally children—to the children themselves. To escape the frighteningly physical and emotional demands grown women tended to place on them, many men who had expected to find in the women of their own age group the same meek, nunlike qualities they had seen in their fathers' wives, began to yearn for the purity of the child they could not find in women. These men seem to have come to the conclusion that if woman's grown body soiled the passive purity of her childlike mind, it was better to seek all the positive qualities, all the passive, compliant qualities of woman, in the child itself. The very purity of the child seemed to preclude the threat of a sexual challenge. Hence the world of the

VI, 14. Frederick C. Frieseke (1874–1939), ''The Parrakeets'' [*sic*] (ca. 1908)

VI, 15. Paul Chabas (1869–1937), "First Bath" (ca. 1907)

child came to hold a special fascination for the late nineteenth century. In its most familiar form this fascination expressed itself in the relentless exploration of the fantasy world of the child—the world of fairy tales, monsters and dragons, and, most of all, of good old-fashioned chivalry—that pure world of Tennyson's, that wonderful world of brave men and fainting women, of black and white contrasts, of simple solutions.

Indeed, as many argued, the coming of the new Darwinian creed had not at all shown the uselessness of chivalric sentiment among men. On the contrary, it had made its pursuit all the more urgent. For if the differences between men and women were fundamental, if woman was indeed intellectually inferior to man (that much, at least, seemed proven to progressive thinkers), there was still clearly a way of fitting the old pedestal into the new evolutionary chain of being. Chivalry, some decided, need not disappear, even if women were inferior creatures. Indeed, it was argued that they now needed man's condescension more than ever.

Hence, along with the proliferation of images representing the battle of the sexes, there arose also a new emphasis on the Tennysonian notion of man's chivalric duty to woman. Man was to show the quality of his evolution through the great sign of true civilization: his freely chosen "subjection" to woman in everything—such as flower arrangements and the furnishing of rooms—that was petty and unimportant, in everything that did not have to do with being successful in those realms where it really mattered. George Romanes, in his article for *The Nineteenth Century* of May 1887, claimed that "the highest type of manhood can only then be reached when the heart and mind have been so far purified from the dross of a brutal ancestry as genuinely to appreciate, to admire, and to reverence the greatness, the beauty, and the strength which have been made perfect in the weakness of womanhood.'' Romanes admired in women virtues such as ''self-denial, and even self-abasement,'' which he was pleased to identify as belonging ''by nature to the female mind'' (659).

Given the ''marked inferiority of intellectual power'' (655) in women, it thus clearly became man's duty to prevent woman from getting befuddled by thought. That was where chivalry had its function. For, considering woman's ''comparative childishness'' and the fact that her emotions were ''almost always less under control of the will—more apt to break away, as it were, from the restraint of reason, and to overwhelm the mental chariot in disaster'' (657), it behooved the chivalrous male to make woman proud of the high cultural value of her natural inferiority to man. Quoting Darwin, and noting that ''natural selection has been most busy in the evolution of intelligence,'' Romanes concluded that ''we can only regard it as a fortunate accident of inheritance that there is not now a greater difference between the intelligence of men and women than we actually find'' (662). To make sure that woman, that ''psychological plant of tender growth,'' could continue ''to be protected from the ruder blasts of social life in the conservatories of civilization,'' man should be ''more lenient in judging the frailties of the opposite sex,'' which were, after all, a necessary complement to the characteristics of masculine strength. And, Romanes went on, ''it is a practical recognition of these things which leads to chivalry; and even those artificial courtesies which wear the mark of chilvalry are of value, as showing what may be termed a conventional acquiescence in the truth that underlies them'' (661).

It is this yearning to return to attitudes of mythic simplicity which underlies the determination of the Pre-Raphaelites and such figures as William Morris to pursue the representation of chivalric themes in art. The endless recurrence of fantasies of medieval encounters between strong, fair-hearted males and cringing women who needed to be rescued represented the symbolic continuance of the mid-century cult of the household nun. The child-mother, the saintly virgin, needed the strong arm of her newly grown son. Painters such as Edwin Austin Abbey and Byam Shaw, who made the chivalric genre their specialty, thus became a conduit for the transmission of ideas deliberately designed to retard the advance of the women's cause, since in the works of these artists woman was either a wicked witch or a helpless child whose tender arms reached out not in desire but in sweet gratitude for the brave, pylonlike chivalric constancy of her magnificent knight. Not all women were swept off their feet by the lure of chivalry. As Rosa Mayreder, a brilliant turn-of-the-century analyst of the psychology of men's fear of women, pointed out in *A Survey of the Woman Problem,* ''Gallantry, that frivolous and hypocritical attitude, bestows upon woman the mere semblance of pre-eminence in order really to push her back into that place among children and minors which masculine lordship is determined she should occupy'' (133).

Major developments occurred during this period in the establishment of a literature and art which was supposedly specifically directed toward children but which was actually populated with fantasies about childhood fabricated by the grownups who wrote the narratives and the artists who drew the pictures. These fantasies were directly linked to the chivalric dream and a general longing for the world of the child as a reliable substitute for the perversely untrustworthy dependency of woman on man. Inevitably the adults' fantasies about the purity of the child's mind also found expression in similar fantasies about the purity and ideal harmony of the child's body.

Emile Zola, whose novel *The Sin of Father Mouret* was an elaborate and, as we have already seen, a decidedly prurient exploration of the temptation of innocence, presented in his portrayal of Mouret a very good analysis of both the substance and the dangers inherent in the adult fascination with the ideal of childhood innocence. As he struggles with his worldly urges, Father Mouret makes the characteristic late nineteenth-century direct link between the Holy Virgin, the saintly mother, and the blessed innocence of the child. Apostrophizing the effigy of the Virgin in his room, Mouret exclaims, "Make me a child again, kind Virgin, powerful Virgin. Make me five years old. Take my senses, take my virility. Let a miracle destroy the man which has grown in me." To be a child again is for Mouret to be rescued from the dangers of desire:

> I want never to be anything but a child walking in the shadow of your dress. When I was very small, I folded my hands to say the name of Mary. My cradle was white, my body was white, all my thoughts were white. I saw you distinctly, I heard you call me, I went to you in a smile of rose petals. And nothing else. I did not feel, I did not think, I lived barely enough to be a flower at your feet. Men should not grow up. Only blond heads should surround you, only a race of children who love you, their hands pure, their lips healthy, their limbs tender, without dirt, as if they were slipping out of a bath of milk. You kiss a child's soul on his cheek. Only a child can say your name without making it dirty (96).

Zola's portrayal of the male's desperate longing for the purity of childhood and the parental embrace of his virginal mother is, as with virtually everything in this writer's oeuvre, overdone, but it is especially significant because it also places emphasis on the Virgin Mother as child. Mouret has a special affection—passion would be a better word—for this image of the Virgin Mother, because in the feminine fleshiness of her maternity the not-so-immaculate mother of the real world could become an object of desire. "She was not yet a mother; her arms did not hold Jesus out to him; her body did not have the rounded lines of fertility." Moreover, the Virgin Mother had never called on him to come to terms with the masochistic temptations woman held out to the marginalized late nineteenth-century middle-class male. "This one had never been awesome, had never talked to him with the sternness of an all-powerful mistress whose sight alone makes men bury their heads in the dust. He dared look at her, to love her without fear of responding to the soft curve of her light-brown hair" (94–95).

The Reverend Charles Dodgson, better known as Lewis Carroll, was clearly possessed of the same sort of fascination—and suppressed desire—as Zola's Father Mouret. His yearning to be part of the innocent world of the child was equal to his desire to escape from the improper implications of passion introduced into his world by the adult women around him. An early photography buff, he began to take pictures of female children, among them Alice Liddell, the real Alice whose fictional adventures in Wonderland he was to chronicle. Since Lewis Carroll was a friend of such artists as Henry Holiday, it seemed only natural that they would team up to "design" photographs the way one might compose a painting. The adolescent daughters of friends obligingly posed for versions of adult themes, such as that of the "Sleeping Woman." Holiday would draw sketches of adorable little girls in the full innocence of their nudity to serve as compositional guidelines for Carroll to use in exploring the artistic potential of photographing children in the nude. "Naked children," Carroll wrote to Harry Furniss, "are so perfectly pure and lovely." (*Confessions of a Caricaturist*, I, 106) He therefore proceeded to photograph prepubescent girls in the nude (boys did not appeal to him in the same manner), as if he were "painting" standard versions of adult nymphs. Most of the photographs he took he destroyed before he died, but one which was preserved shows a little girl in the standard pose of the nymph with the broken back, so popular among the painters of the time.

VI, 16. Thomas C. Gotch (1854–1931), "The Child Enthroned" (ca. 1894)

VI, 17. Paul Chabas (1869–1937), "Seaweed" (ca. 1909)

It all seemed very pure, this exploration of the soft vulnerability of childhood, very "ideal"—but it is obvious that these men were playing with the fire that turns innocence into sin. From the wife as household nun to the image of "The Child Enthroned," as Thomas Gotch inevitably came to see the ideal of childhood purity [VI, 16], was not a great step. But, as we have seen in the case of Henry Holiday and Lewis Carroll, one pose logically led to another. Consequently the search for the lineaments of the mother's lost innocence in the features of the child could easily take the form of a rediscovery of the enticements of woman in the physical body of the creature whose mental equal she supposedly already was. As a result, the portrayal of the child in its naked innocence often came to echo the representation of woman in art.

Balzac, in *The Wild Ass's Skin* (1831), had already given a remarkable indication of the manner in which men's yearning for signs of childhood purity in women and parallel fantasies concerning the ambiguous innocence of sleep blended to form a hothouse dream, of woman as child, as an unthreatening creature conjoining compliance and amorous abandonment. As his tormented hero Raphael awakens after an all too brief interlude of playful happiness with his infinitely adoring, infinitely innocent near–child bride Pauline—who had earlier told him fervently that her only wish in life was "to become your servant again, now and forever" (215)—he has a fleeting glimpse of paradise:

> He gazed at his sleeping sweetheart who had her arms round his neck, thus expressing even during sleep the tender solicitude of love. Lying gracefully like a child, with her face turned

to him, Pauline seemed still to be looking at him and offering him her lovely mouth, with lips parted by her pure and regular breathing. . . . Her attitude of graceful self-abandon, so trustful, blended with the charm of love the adorable attractions of sleeping childhood. Even the most ingenuous of women still comply, during their waking hours, with certain social conventions which place limits on the free expression of their souls; but sleep seems to restore in them the spontaneity that adds so much beauty to childhood. Like one of those sweet creatures straight from heaven whom the faculty of reason has not yet taught to curb their gestures, nor the faculty of thought to veil their gaze with secret bashfulness, Pauline just now was incapable of blushing at anything (241).

The woman-child, then, suggested to the nineteenth-century male not only innocence but also the absence of any resistance to the particulars of masculine desire.

No artist was more assiduous in exploiting the prurient possibilities of the woman-child than Paul Chabas. In countless paintings with themes analogous to his "First Bath" or his often-reproduced "September Morn," he titillated his audience with suggestive images of adolescent girls in poses of extreme vulnerability, usually standing in water, doing nothing in particular, or at most gathering "Seaweed" [VI, 17] to signify their intimate link with primal nature. His images knowingly exploited the prurient implications of these supposedly idealized images. In productions of this sort the conjunction of suggestions of extreme feminine vulnerability and the wild, uncontrollable realm of brute nature once again served to suggest to turn-of-the-century male viewers the notion of "therapeutic" rape. Zola's infinitely pure lovers in *The Sin of Father Mouret,* for instance, are surrounded by signs of the "life force" which manifestly require their ultimate compliance with the rituals of fertility: "Lilies offered them a guileless asylum after their lovers' walk amid the ardent concern of sweet honeysuckle, musk-scented violets, verbenas exhaling the cool smell of a kiss, tuberoses releasing the swoon of fatal seduction. With their extended stems, the lilies put them in a white pavilion under the snowy roof of their calices, brightened only by the subtle golden drop of their pistils" (134).

Thus, even the symbol of purity itself, the lily, became suggestive of the end of innocence. To the late nineteenth-century intellectual, one condition of being ceaselessly suggested its opposite. Seeing purity, the late nineteenth-century mind was titillated—or disturbed—by thoughts of sin. In his *Le Nu au Salon: Champ de Mars* of 1892, Armand Silvestre, mulling over a painting by Armand Point entitled "Puberty," talked about how in Point's young girl "the disturbing unfolding of woman's essence," would awaken "thoughts until now mysteriously slumbering in deceptive serenity." Thus the appeal of puberty was the prurient certainty of carnal knowledge, the shadow of evil which, for instance, hovers threateningly behind the young girl in Edvard Munch's familiar painting also called "Puberty." Images such as these represent the marketing of an aggression that dare not speak its name. They represent rape, a promise of unopposed carnal knowledge offered to the viewer—who became the sole proprietor of the innocence of the awakening woman—a fantasy of "love without fear."

To feed this prurient desire, the exhibitions of the period featured hundreds of paintings and sculptures of naked adolescent girls, usually seated, eyes shut, and waiting—as in a sculpture by Antonio Ugo entitled "Pubescit," which was exhibited at the Venice Biennale in 1901—or perhaps standing resignedly at the side of a brook. Sometimes groups of such virginally pure children could be found waiting in line to take a peek at motherhood, as in a line drawing by Ida Teichmann published in *Jugend* in 1918, or they might celebrate midsummer night while rowing in expectant nudity toward adulthood, guided by

VI, 18. Susanne Daynes-Grassot (b. 1844), "Child Before the Mirror" (1912)

VI, 19. William Sergeant Kendall (1869–1938), "Reflection" (1909)

VI, 20. Andrea Carlo Lucchesi (1860–1925), "The Bud and the Bloom" (ca. 1906)

an obviously harmless old man, as in a painting Axel Gallen-Kallela exhibited at the Salon des Beaux-Arts in 1909.

But pubescent girls were not the only ones showing disturbing signs of impending adult desires. As the late nineteenth century moved further and further back into the world of childhood, artists kept turning up new connections between these children and the mysterious realm of the adult—and hence far too knowledgeable—woman. Dubiously focused child versions of the woman-and-her-mirror theme were painted in the early years of the twentieth century by numerous artists, including women painters such as Susanne Daynes-Grassot, whose "Child Before the Mirror" [VI, 18] attributes both precocious sexuality and intense self-absorption to a child who cannot be much more than eight years old. William Sergeant Kendall, an American painter who spent considerable time in Europe before settling down to a teaching career at Yale University, was a contemporary of—and male counterpart to—Mary Cassatt as a mother-and-child specialist. He painted many of these knowledgeable children before the mirror, with titles such as "Narcissa," "Crosslights," and "Reflection" [VI, 19]. They were eagerly sought after by cognoscenti and were widely reproduced in popular magazines. Paintings such as these, which often emphasized analogies between the actions of nude little girls and the familiar poses of vanity or physical arousal given to adult women by turn-of-the-century artists, show how thin the line separating "idealist" art and child pornography had become. The pose of the opisthotonic nymph, for instance, was exploited by Andrea Carlo Lucchesi in his sculpture "The Bud and the Bloom" [VI, 20], exhibited at the Royal Academy in 1906, in which this sculptor, who was famous for the immediacy of his eroticism, ruthlessly sexualized the child, the "bud" in this group, by giving a truly extreme backward curve to her torso as she leans against her full-grown companion's knee.

Carl Larsson, who through his drawings became a mainstay in the world of childhood entertainment much as Lewis Carroll had through his writing, played with the same sort of fire as his British counterpart had in his photographs of nude children, when he chose to straddle the line between sentimentalism and obscenity in his certainly not entirely innocent portrayal of "The Little Girls' Room" [VI, 21]. Not even the world of the smallest toddlers, in fact, could escape the eroticized imagination of the turn-of-the-century

VI, 21. Carl Larsson (1853–1919), "The Little Girls' Room" (ca. 1895)

VI, 22. Bruno Piglhein
(1848–1894), "Christmas
Morning" (ca. 1890)

VI, 23. Léon Perrault
(1832–1908), "Bacchus
as a Child" (1897)

painters, who, having been told by the highest scientific authority that woman remained a child all her life, inevitably came to recognize traces of woman in all stages of childhood.

The Munich-based painter Bruno Piglhein, for instance, found an analogy between a favorite artist's pose of the ravished woman and the pleasurable exhaustion of a toddler on "Christmas Morning" [VI, 22]. Rarely has so much technical ability been lavished on so meretricious an undertaking. In France Léon Perrault made the eroticized depiction of toddlers a specialty. His 1897 painting of "Bacchus as a Child" [VI, 23], in which he placed his infant in the characteristic pose of the nymph with the broken back, is an especially repulsive example of this painter's inability to make a distinction between children and women. In 1873 the Dutch painter Matthew Maris painted a reclining little girl with flowers in her hair, a pair of butterflies in the air above her (emblematic of the child's nature), and an extremely knowing glance in her eyes [VI, 24]. No doubt the image would have driven Lewis Carroll to ecstasies of admiration.

Undoubtedly, not all these artists were consciously responding to prurient motives, and the intentions of some probably remained focused on a search for images representa-

VI, 24. Matthew Maris
(1839–1917), ''Butterflies''
(or ''L'Enfant couchée'')
(1873)

tive of ''ideal'' innocence. But Satan is a nasty character and sets his traps where one would least expect them. And so these innocents, these children, began to take on, in the fantasies of the painters, the outlines of the female temptresses they were trying to escape. Minds filled with the lustful shapes of worldly women soon discovered in children the lineaments of sin.

Thus a genre was born in which crass child pornography disguised itself as a tribute to the ideal of innocence, and even children fell victim to man's fearful retreat from women who knew too much about the sins of the flesh. Afraid to deal with women who were strong and independent and who dared make demands, the late nineteenth-century male molded the child into the image of a woman he could handle. The helplessness, weakness, and passive pliability of ignorance he could no longer find in woman he began to attribute to the child. No wonder, then, that among the nightmare visions Flaubert gave his Saint Anthony was that of ''a beautiful dusky child amid the sands, which revealed itself to me as the spirit of fornication'' (*The Temptation of Saint Anthony,* 21). It is a sad irony that, notwithstanding these excesses, many women encouraged the trend; overlooking the erotic content of these paintings, they chose to see instead in these children portrayed as women an idealized version of the innocence many of them still revered as an ideal.

It is, therefore, important to stress that often, though certainly by no means always, the primary audience for these eroticized infants was not men but women, who, as a result of decades of propaganda, had themselves been swept up into the ideology of ''ideal femininity.'' They saw in these sensuous toddlers and in the equally meretricious confections of voluptuous feminine inanition produced by such painters as Renoir, a ''true'' expression of the ideals of femininity their mothers had been trying to drum into them. Often these were women of the leisure class who, in their heart of hearts, knew that they had not succeeded in becoming quite as vacuous as they ought to be and must have upbraided themselves for their own perversity, for knowing what they should not have known, for having done things they would rather forget. To these women Bouguereau's eroticized toddlers and Renoir's vacuous female flowers became icons: images to be hung in their homes as the centerpieces of leisure-class shrines to the ideal of the virginal mother. These works became the *penates,* household gods dedicated to the idea of woman as nun, to which their daughters went to do penance for their inability to be as demure, unthinking, and passive as they were supposed to be.

Still, the late nineteenth century's fascination with carrying every notion to its extreme is nowhere expressed more strikingly than in these peculiar graphic examples of "high fashion" child pornography. Such paintings represent a visual equivalent of the intellectually stylish "anti-infantine" sentiment which became fashionable along with the —obviously far less esoteric—antifeminine stance. The very same images which made many of the women nostalgic for a lost age of innocence came to suggest to the "scientifically" informed men of the period new outposts of childhood criminality. This response probably had its origins in Lombroso and Ferrero's contentions about the vicious amorality of the child. An unevolved savage, the child was, in their eyes, only prevented from ravaging the earth because it was physically helpless; in this respect, too, the child was discovered to resemble woman. In *The Female Offender* this duo of fearless scientists exclaimed, "What terrific criminals would children be if they had strong passions, muscular strength, and sufficient intelligence; and if, moreover, their evil tendencies were exasperated by a morbid psychical activity! And women are big children; their evil tendencies are more numerous and more varied than men's, but generally remain latent. When they are awakened and excited they produce results proportionately greater" (151).

Statements such as this, made by the most highly respected scientific researchers of the period, obviously offered license to artists and writers to find tendencies of the feminine criminal in the child. "The inability to be affected by ideal states" (157) detected by Harry Campbell in the psychological makeup of the child obviously led it to behave like a woman in situations of emotional stress. Paul Adam, in a very nasty article "On Children" published in *La Revue Blanche* in 1895, saw the perverse erotic characteristics of woman as magnified in the behavior of the female child. He contended that these female children, aged between eight and thirteen, "found a perverse pleasure in watching sedentary middle-aged men expose themselves to them for a few pennies." Only the hypocrisy of popular songs and sentimental disquisitions prevented the public from realizing that female children had an inherent tendency toward prostitution. "It has been reported," wrote Adam lugubriously, that the elder of the two daughters of a shepherdess in the Touraine, who had been "lent out habitually to a farmer," actually "burst into tears when, with the arrival of puberty, she discovered that she 'could no longer have fun with the boys without fear of pregnancy.'" Adam warned that "virtually all vices fester in the mind of the child." What we should realize is that "evil in adults is a sign of their not having grown up. In the taverns, in the places of debauchery, in the prisons, it is the mental tone of the child which animates and motivates" (350–53).

Jules Laforgue's Andromeda, in his *Six Moral Tales,* is a thoroughly depraved little vixen. "This savage little adolescent is all steel," he wrote. Thus the search for innocence in the child as a factor of man's horror of woman ultimately backfired in the same fashion in which man's quest for the virginal woman had unearthed the carnal animal. The quest had come full circle, for, in the words of Lombroso and Ferrero, "the offences we have detailed prove the likeness between women and children, since they might be described as offences committed by big children of developed intelligence and passions" (267).

In consequence, if turn-of-the-century painters indulged their search for sentiments of piety by going forth among children, they often found a more ambiguous haven than they had expected. This led to the images of psychological ambiguity we have already examined, as well as to full-fledged expressions of paranoiac obsession with the nature of the

VI, 25. Léon Frédéric (1856–1940), ''The Stream'' central panel (1900)

child. How nightmarish painters' dreams of infantile flesh could ultimately become is graphically demonstrated in Léon Frédéric's monumental triptych ''The Stream'' [VI, 25], in which this artist, ostensibly to illustrate Beethoven's ''Pastoral'' Symphony, created with insane literalness the ultimate representation of the familiar equation between water, women, and the world of the child in a carnal orgy of infant flesh. When images of this sort, of this extreme paranoia, arise in man's imagination, can Buchenwald be far behind?

Certainly such images as Frédéric's ''The Stream,'' whatever the long-range impact may have been of the mentality they represented, show how intensely the generation of men who had been brought up by household nuns was torn by conflicting desires and expectations. While the new evolutionary science had undercut the religious focus of the search for a soul-guardian, the theory of evolution had also implanted a new, urgent sense of responsibility in the young vanguard intellectuals, making them yearn to pursue the ideal and thereby bring humanity closer to the world of disembodied essences, of ''pure mind.'' Indeed, pure mind was to be the new soul, a soul whose creation, whose ''evolution'' was the responsibility of the young artists and intellectuals.

Exhibiting a fearful avoidance of the debilitating effects of ''sexual indulgence,'' these brave youths tried to bring the new world of mind into being in art and thought. The mothers of 1850, in their devoted attention and frequent ignorance of sexual matters—

certainly in their sense of shame about all things related to the pleasures of the body—had encouraged their sons' devotion to the notion of virtuous abstention. Mother—the all-giving, the innocent, the invalid—became their image of true womanhood. She was their preferred companion in the quest for the ideal. Among the women of their own generation, who were often too spirited and too impatient with the roles imposed on them to follow their mothers' example, these young men could not find the passive virtues they had learned to love in their mothers, so they tended to turn to celibacy, to "ideal" friendships with other males, or, as we have seen, to the search for innocence and purity in children, who were, after all, still "sexless."

In her sardonic story "Dionea" (1888) Violet Paget, whose pen name was Vernon Lee, provided an interesting sketch of this type of man of the 1870s and 1880s. Her specimen is a sculptor named Waldemar. An idealist to the tips of his chisels, he has married Gertrude, the girl of his mother's dreams, a veritable tubercular saint, "with her thin white face, a Memling Madonna finished by some Tuscan sculptor." Gertrude's "long, delicate hands" are ever busy, and "her rarely lifted glance" is "more limpid than the sky and deeper than the sea." In true mid-century fashion, she had tamed the animal in Waldemar: "He seems to me, when with her, like some fierce, generous, wild thing from the woods, like the lion of Una, tame and submissive to this saint" (259). As an artist Waldemar seeks to express the ideal in the real: "He seems to feel the divineness of the mere body, the spirituality of a limpid stream of mere physical life. But why among these statues only men and boys, athletes and fauns, why only the bust of that thin delicate-lipped little Madonna wife of his; why no wide-shouldered Amazon or broad-flanked Aphrodite?" The answer is not difficult to guess. "What do I want with the unaesthetic sex, as Schopenhauer calls it" (256), exclaims Waldemar. The female figure, he argues, "is almost inevitably inferior in strength and beauty; woman is not form, but expression, and therefore suits painting, but not sculpture. The point of a woman is not her body, but (and here his eyes rested very tenderly upon the thin white profile of his wife), her soul."

As we are soon to discover, Waldemar has good reason to avoid depicting "earthly" women. Except in the tubercular madonna, the anorexic saint who has sacrificed her sexuality to her "soul-role," the flesh of the adult female spells desire—and desire, Waldemar would claim, along with other late nineteenth-century artists in search of a proper casing for the evolving mind, spells death to the soul. Waldemar thus sought the ideal form in the material volume of men and boys—for him they seemed to preclude temptation—for how could man be sensuously tempted by man? As we have seen, many among his generation went in search of that same ideal in the child—but even the female child had become eroticized in the hothouse minds of their brothers.

Thus, in a world in which arousal spelled doom to the male in search of the ideal, in which the adult woman and even her pubescent daughter all too often had abandoned her responsibility to be a madonna, a state of restful detumescence—of peace, the absence of temptation, of removal from carnal involvement—seemed available only when the artist sought the outlines of the ideal in those real forms which unavoidably, it would seem, still retained the shapes of innocence which had been abandoned by woman, by the temptress Eve. In consequence, some artists enthroned the pubescent boy along with the pubescent girl. After all, Darwin himself, in *The Descent of Man,* had said that "male and female children resemble each other closely, like the young of so many other animals in which the adult sexes differ widely; they likewise resemble the mature female much more closely than the mature male" (635). Thus, it seemed legitimate to seek in the male child those

VI, 26. William Adolphe Bouguereau (1825–1905), "The Wet Cupid" (1891)

VI, 27. Elihu Vedder (1836–1923), "Superest Invictus Amor" (1889)

shapes of the ideal which in the feminine body became so rapidly contaminated by base desires.

Indeed, around 1900 more and more men like Vernon Lee's Waldemar stylishly began to assert the superiority of even man's physical presence over that of woman, encouraged by Schopenhauer's disdainful remark that "it is only the man, whose intellect is clouded by his sexual impulses, that could give the name of *the fair sex* to the undersized, narrow-shouldered, broad-hipped, and short-legged race" ("Of Women," 300). A new admiration developed for the special beauty of the male. Man, it was now argued, with numerous quotations from Plato, had all the "soft," physical attractions of woman, plus the male's exclusive capacity for intellectual transcendence. The ephebe, the sensitive male adolescent, not woman, was the true ideal of aesthetic beauty. Indeed, the young, fully grown male, the "blond god," his lingering freshness enhanced by muscular development and physical strength, seemed to many artists and intellectuals fleeing woman the personification of the magnificent, aggressively evolving mind of man.

But it was soon clear that even in this realm passion could rear its ugly head. Bouguereau, who, as we have already seen, was a genius at turning even holy water into an aphrodisiac, rushed in with such paintings as "The Wet Cupid" [VI, 26], which was exhibited at the salon in 1891. No more sexually enticing adolescent could have been offered by the master, but unlike Oscar Wilde, who was jailed after being prosecuted for general incitement to pederasty, Bouguereau was given official accolades and honors. Other artists also found a sensuous purity in young male bodies which seemed to put the lusher contours of woman to shame. For example, Elihu Vedder, in his version of the cupid theme entitled "Superest Invictus Amor" [VI, 27], created what must surely be one of the most elegant, most subtly erotic versions of the ephebe-equals-woman theme in late nineteenth-century art. Indeed, the manner in which Cupid has grown in this painting is quite astonishing. He has become a young, independent adolescent, no longer the foolish little child who was Venus' natural companion.

The later nineteenth century had generally been content to use Cupid as a tributary symbol for the denigration of woman as child. In most paintings a baby Cupid was shown frolicking with a grown woman, symbolizing the fact that "the passion of love," physical desire for another, was in fact childish—a matter for women and breast-fed putti but not for grown, intellectual men. If the mental level of woman was equal to that of the toddler, let her toddle along with Cupid in the realm of physical dependency. That is why turn-of-the-century artists had such a field day painting little winged amoretti in conjunction with slow-brained, big-hearted, luscious infantile women.

But now, in such a painting as that by Elihu Vedder, love had grown up, become a truly masculine "idea." Soon he was to lose his wings altogether and be placed before the mirror, as in the Canadian Paul Peel's presumably feminine but, like Vedder's "Amor," physically purely androgynous "Venetian Bather" [VI, 28]. He became incarnated as "The Youth of Ulysses" [VI, 29] in a striking painting by Herbert Draper. A veritable inundation of heroic mythological males—from mild, effeminate ephebes to supermasculine muscle men—began taking over what had been considered feminine tasks. The sensitive man had arrived in turn-of-the-century art. In what must be the true apotheosis of kitsch, many years before the advent of Tiny Tim, the adolescent male was made to tiptoe through—if not the tulips—then at least the daisies of the Germany of 1900, as the personification of a truly maudlin "Spring," in Heinrich Friedrich Heyne's painting of that title [VI, 30]. Jean Delville, in "The School of Plato" [VI, 31], appropriated virtually every pose which had previously been given to woman in art—except, of course, that of the nymph with the broken back. Ideal love clearly was a task only to be entrusted to man, since woman had shown herself to be untrustworthy, not the mother of the ideal but the idol of perversity.

Indeed, such an event as the 1895 trial of Oscar Wilde for homosexuality was much more central to the intellectual concerns of the turn of the century than has generally been allowed. It was not just a matter of a favored servant of the ruling class having overstepped the boundaries of what was permitted. Wilde's position also confronted the ruling class with an embarrassing interpretation of commonly accepted ideas. Wilde's platonic idealism was shared by a very large group of poets and artists throughout Western culture. It was an idealism heavily based on Darwinian conceptions, joining a virulent hostility for the petty bourgeoisie with an intense fascination with power—a dream of the Nietzschean superman forging ahead ever more securely toward a new state of being in which the mind of man might transcend its physical prison. These men saw the ideal they were searching for in the supremely powerful musculature of the triumphantly predatory male god of imperialist achievement, the miracle man who could at any time prove his mastery in the struggle for existence.

Since most of the artists and writers who were nourished by the dream of masculine transcendence did not themselves fit this physiological ideal very closely, they tended to idealize those who seemed to personify these characteristics. They became devotees of young blond gods with rippling muscles and steel blue eyes. At the same time, the hatred these artists expressed for the bourgeoisie was fed by what they perceived as the failure of the middle class to cultivate transcendent ambition. Instead of wishing to conquer new worlds, sail toward new horizons of physical and mental achievement, the bourgeoisie continued to find satisfaction in paddling around aimlessly on the pond of mere monetary concerns. Yet, for all the disdain expressed by the antimaterialistic intellectuals, their yearning for the ideal was undercut by an intense fear of competition *from* woman and *for* woman.

Born, for the most part, into middle-class households whose precepts about the nature of women had been taken from Michelet and his ilk, yet growing up in an age which saw

VI, 28. Paul Peel (1861–1892), "A Venetian Bather" (1889)

VI, 29. Herbert Draper (1864–1920), "The Youth of Ulysses" (1895)

women resist their marginalization, assert their sexual presence, and turn the conditions of their marginalization into an aggressive weapon, these men were often overcome by an antifeminine horror born of the disparity they observed between the idea of the perfect, sexless woman and the reality of healthy, assertive female bodies.

Nicholas Francis Cooke, with his self-appointed mission to expose the terrible secret doings of young girls and his fascinated accounts of the dangers of feminine masturbation, represents an early example of that clash between expectations fed on ideals and a rude confrontation with the real. In a scene which would have provided the dramatists of the Restoration with much material for unabashed comedy, he described his own initiation into the world of sin: "At the age of eight the writer was lodged, at a watering place, in the same room with three girls, respectively ten, twelve, and fourteen years. The elder of these little misses succeeded effectually, during the few weeks' association, in inducting her companions into the science of reproduction, while the male member of the quartette was aptly used in illustration of the subject." Cooke confessed his hostility later in life when he saw the perpetrator of this indignity upon his person parade "her young daughters in the most fashionable circles of one of our most fashionable cities." In distress, he inveighed against "the custom of permitting children of different sexes to sleep in the same bed, or in the same room . . . even where the excuse of poverty is wanting" (57–58).

Cooke, in a sense, was the victim of a generation of parents who had come to assume, as a universally accepted truth, that all women were by nature sexless. They regarded any outbreak of sexual curiosity among their daughters as unthinkable, and had thereby created the conditions for that confrontation between the ideal and the real which, in its turn, spawned a generation of men mired in awe, fear, shock, and horror of the perversity of

VI, 30. Heinrich Friedrich Heyne (b. 1869), "Spring" (ca. 1899)

VI, 31. Jean Delville (1867–1953), "The School of Plato" (1897)

woman as she tried to break out of the prison of sainthood to which she had been relegated by the generation of 1850. To that sense of shock on the men's part can be added anger at having been deprived of the rights of their fathers. For the male children of the 1870s and 1880s had at first expected to be given their own chance of enjoying the pleasures of mastery. They had, after all, observed lordly power to have been the birthright of their fathers, as they watched their mothers try to be the all and everything, the very footstool, to their husbands. Instead they encountered resistance from the new generation of women. Somehow man's very right to woman as property was being undermined by her demands for a modicum of equality and independence.

Jules Lemaître, writing on Michelet's *Love* in 1898, forty years after the book's first publication, and treating it as if it were an archeological find worth rediscovering, described the attitude current among his contemporaries very effectively, contrasting it specifically with that common among an earlier generation.

> Virtually all those among our writers who have "lectured" on love have kept primarily to a demonstration meant to indicate that they have not been duped by woman; they have been eager to show that they were armed with the most ferocious sort of experiences, that they were capable of the most subtle and defiant analyses, and that, indeed, they were not so incapable of perversity themselves when it came to that. These men are pessimistic, libertine, a little foppish. And they overestimate woman's complexity in order to make us believe more completely in their own profundity and the extent of their personal researches.

There was in the love for woman professed by these writers, Lemaître pointed out, always an "element of hatred." It was "no more than a displaced, furious version of egoism, an exasperated expression of the instinct of ownership (*La Revue de Paris*, vol. V, no. 5, 732–33).

Lemaître's remarks are extremely instructive, and offer an excellent insight into the prevailing attitudes and motives of his contemporaries. Unfortunately, he was not able to propose anything more constructive than a slightly moderated return to the worship of the child-woman of Michelet. But the element of fascinated hatred, the sense of deprivation, of purity unmasked as perversity which he identified in his contemporaries' responses to "woman" helps to explain the origins of much of the late nineteenth-century male intellectuals' fear of, and gradual withdrawal from, women into celibacy or active homosexuality.

This generation was the first to have grown up in a world in which the rhetoric of evolution was on everybody's tongue, giving new currency, indeed, a contemporary urgency to the meanings underlying the works of classical literature and philosophy which boys had to read as part of their education. It was, in consequence, almost a foregone conclusion that the more imaginative among them would slide into fantasies of power fed by the rebellious, sophisticated evolutionist ideas their conservative teachers still abhorred. The theory of evolution gave an interesting new meaning to the otherwise rather boring Platonic dialogues they had to decipher.

A characteristic example of the flights of evolutionary fancy to which the "grown" intellectuals were led by the fusing of Darwinism and Platonism can be found in the remarks of an anonymous reader, part of a heated debate published in *La Revue Blanche* in 1896 concerning the homosexual tendencies of Richard Wagner. In these remarks, neo-Platonic idealism and evolutionary theory mingle to breed some remarkable notions. "As does his soul," wrote the anonymous reader,

> so the physiology of man has its mysteries. The specialization of function is a rule in the evolution of beings, and that evolution is not yet complete among the higher mammals. May not Nietzsche's *Uebermensch* be an anthropoid creature in whom the functions of love will be entrusted to organs which shall not condescend to any sort of pluralism? Do not protest against this notion in the name of common sense: The range of hypotheses is immense and that of our experience infinitely limited. Renan, in his *Dialogues Philosophiques,* conceived of a chosen race among whom, as a result of the hyperculture of the organs of thought at the expense of all others, humanity would become a luminous brain able to dream, I would imagine, of destinies that would dazzle our imagination (X, 305–6).

It was this evolutionary fantasy, conjoined with the dream of power embodied in the popular image of a young, blond, godlike male with steel blue eyes who was firmly in control of the world's future, which made the writers eager to flex their verbal muscles in literary combat, not only to show that they could denigrate women as well as the next man but also to prove their ability to deride the weak in general. That effort gave rise to such high-flying literary endeavors as Emile Tardieu's catalogue of "The Psychology of the Weakling" in *La Revue Blanche* in 1895, in which he delineated, as characteristic of these unmanly men, such traits as caution, tolerance, respect for others, and disdain for violence. Marginal to the real power sources in a world of growing imperialist conflict, these writers found in verbal belligerence, in the ruthless denigration of the imbecility of the masses, the infantile inferiority of nonwhite races, and the brainless inanity of women their own version of Henri Bergson's philosophy of action. They discovered the glories of the "gratuitous act" as evidence of their power. Because they were—so they thought—brilliantly active in mind, they might team up effectively with the musclemen of physical action so that together they could scale the heights of evolution, unburdened by the weak, the poor, and women.

The links which existed between the late nineteenth-century male's search for an appropriate means of evolutionary transcendence of the "dishonorable temptation" of physical contact between the sexes and his tendency to seek out, in response, the companionship of other men is outlined very directly in the thought of André Gide. This author's mouthpiece, Corydon, in the dialogues advocating homosexuality Gide published under that title, also established an extremely instructive link between "masculine idealism" and the aggressive mentality characteristic of the turn of the century. "I believe," Gide had Corydon say, "that periods of martial exaltation are essentially homosexual periods, in the same way that belligerent peoples are particularly inclined to homosexuality" (143). The reason why Gide sought to find a correlation of this kind undoubtedly has to do with the turn-of-the-century's focus on hero worship and the cult of the blond god. Gide, like many of his contemporaries, wished to dissociate himself specifically from the blanket accusation of degenerate effeminacy habitually bestowed on homosexuals during this period. Instead, these men were eager to stress the masculine, intellectual, and idealist choice expressive of the logical processes of natural selection which, they felt, was more truly representative of their arguments in favor of an atmosphere of "pure male interaction."

In a very direct sense, the elements involved in this glorification of the ideal purity of exclusively male relationships among a number of turn-of-the-century artists represented a reassertion of the ideal of male dominance and female submission held by their parents. Unfortunately, woman was showing a perfidious unwillingness to continue to exist passively as the yielding soil from which the male could grow to achieve transcendent power. In response to this feminine refusal, the dominance-submission pattern which was seen to be necessary for the growth of the superman was recreated by another male, the artist, who, for the sake of the blond god of the future, was willing to assume the yielding function so perversely abandoned by woman. "I say," insisted Gide's Corydon, "that the passionate attachment of another person, or a friend of the same age, is as capable of self-denial as any feminine attachment" (149–50), adding that it was better for the soul of the impressionable boy to mature among males. If "some older person is to fall in love with him, then I believe it is best for him that this person should be a friend of his own sex." Indeed, concluded Corydon, linking his previous remarks directly with the ideal of male transcendence, "I think that this friend will jealously watch and guard him, and himself exalted by this love, will lead him to those marvelous heights, which can never be reached without love. If alternatively he falls into the hands of a woman, it can be disastrous for him" (151).

Thus, one of the most characteristic aspects of the development of male same-sex attachments at the turn of the century was the notion of writers and artists that somehow woman had failed the test of being the proper nurturer of the adolescent blond gods who represented the storm troopers of the soul poised to stage humanity's invasion of the world of disembodied essences. Since, on the other hand, the artists had a much clearer perception of woman's proper function than woman herself, they were willing to provide the example and take woman's place. Consequently much of the strong homosexual impulse in the culture of the period expressed itself in an arrogant siding with the dominant male heterosexual assertion of the inherent superiority of the male in all things. Men, it was decided, were clearly better at doing the woman's job in a personal relationship than women themselves.

Men, then, after having placed women in a ghetto of intellectual inferiority, used this very determination of women's inferiority as a reason to invade this ghetto and become its rulers. Unfortunately, for the most part, women stood by in bewilderment and accepted the rule of a male authority which had pronounced itself as particularly sensitive to the interpretation of the female psyche.

In a humanely structured society, a person's choice of partners in love and pleasure should clearly be entirely his or her own concern. Since all forms of self-identification within the structure of a society are dependent on a complex conjunction of elements of instruction, as well as accidents of environment and individually determined preferences in the realm of physical enjoyment, it seems clear that Gide's contention that homosexual relationships as such are as natural as heterosexual ones is a perfectly reasonable one. But, of course, once a decision about the parameters of "natural" behavior is reached by a committed advocate of a specific type of sexual relationship, the temptation is to go a little further, as Gide did, and contend that, all things considered, the form of physical enjoyment one has personally chosen is the most natural. However, once that unnecessary and exclusionary step has been taken, a world of dominance and submission reasserts itself, a world dependent on exclusive authority and the formulation of repressive moral precepts for the enforcement of one preference over another.

Clearly, the sex of one's partner in love and pleasure has absolutely nothing to do with one's ability to discern right from wrong in the realm of humane social coexistence. That ability is a direct factor of one's distance from the ideological focus of one's own time. It is, in any case, undoubtedly true that, as Gide's Corydon asserted, "in homosexuality, as in heterosexuality, there are all shades and degrees: from Platonic love to lust, from self-denial to sadism, from healthy joy to moroseness, from natural development to all the refinements of vice. And between exclusive homosexuality and exclusive heterosexuality there is every intermediate shade" (23).

The mental attitude which caused most male-focused relationships at the turn of the century to be virulently misogynist was the same as that which made the woman-focused relationships of males virulently misogynist: the assertion of a distinct, natural, and evolutionary superiority of the male in both the mental and the physical realm. It is thus important to remember that the rather significant element of homosexual dissociation from woman which is part of the turn-of-the-century misogynist focus cannot in any way be seen as the "cause" of the wholesale denigration of women during this period. Certainly celibate and homosexual writers and artists joined in, expressing attitudes which were coextensive with, or otherwise parallel to, the misogyny of the heterosexual male, but these attitudes were simply those which predominated among male artists and intellectuals of whatever sexual persuasion. Nothing could be more inaccurate than to assert that the prominence of homosexual idealism around the turn of the century was responsible for the prevalence of misogyny, although it is certainly true that the period's virulent misogyny contributed to the attraction of homosexuality among intellectuals. The apparent increase in homosexuality, which may have been actual or merely the result of a lessening of social circumspection among males inclined to same-sex relationships, quite clearly expressed itself in terms of a suspicion, often developing into an outright fear, of the strength of women's sexual appetite and their eagerness to usurp the arena of male privilege. But these same suspicions were a motivating factor in the ultimate choice of celibacy which led numerous members of the intelligentsia during this same period to embrace the Church, usually the Roman Catholic Church, with a great deal of intensity.

In fact, it would be as inaccurate to say that homosexuality was the "cause" of the misogyny pervasive among turn-of-the-century intellectuals as it would be to blame the women painters of the period for perpetuating male prejudices about feminine attitudes. For while it is true that most of these women painters tended to side with their male colleagues against Woman, the insistence of late nineteenth-century culture on highly structured concepts of individual achievement made it necessary for them—even if only to gain attention—to take on male attitudes which would differentiate them from "the mass of women."

The popularization of the theory of evolution, and the premium it placed on individuality as a sign of a person's "election" to the most advanced echelons of the intellectual community, to the legions of supermen, had a tremendous impact on the evaluation of artistic achievement during the last decades of the nineteenth century. It is by no means accidental that numerous fundamental innovations in style and means of representation in painting developed during this period. As conservative critics of the time never tired of emphasizing, many artists were beginning to pursue what was new virtually for its own sake, to prove that they were original and not imitative.

And here, in a peculiar manner, antifeminine sentiment and the cultural avant-garde linked up once again. As we have seen, every scientist worth his salt pointed to the "innate" characteristic of woman to be incapable of "individual invention." Harry Campbell in his *Differences in the Nervous Organization of Man and Woman,* had insisted that "in imitativeness and lack of originality [woman] stands conspicuously first; indeed, it is essentially in this particular that the masculine intellect shows its superiority over the feminine" (232). The artists, who as a group were increasingly being accused of effeminacy by their more practically minded contemporaries, felt the pressure of the ideal. They, even more than their "noncreative" brethren, shuddered to think that they might be dismissed as unoriginal, as having female—read weak—minds. Said Campbell, "A strong tendency to imitate goes along with the type of weak will we are now considering, and is the mark of an ill-developed mind. Imitativeness, as Herbert Spencer observes, is 'shown least by the highest members of civilized races and most by the lowest savages . . . It shows us a mental action which is, from moment to moment, chiefly determined by outer incidents' " (355).

Thus, to be a stylistic innovator in the arts was to use artistic muscle to push aside any potential accusation of artistic effeminacy. To be original was to be masculine in the best sense. To rebel against the "fathers"—the great satraps of academic art—was to be the toughest kid on the block in the realm of art. The antifeminine mentality and the artistic avant-garde went hand in hand, and so it is not surprising to find that the critics' worst attacks were leveled at the "hordes" of women artists who were showing woman at her "imitative" worst by flooding the world with "unoriginal" paintings expressive only of their ability to copy the styles of their masters.

The influential art critic Thadée Natanson, for instance, ridiculed the "preposterous" feminine attempts at art displayed at the fifteenth salon of women painters and sculptors in a review for *La Revue Blanche* in 1896. He peppered his attack with wholesale accusations of imitativeness. The manner in which these women, he sneered, could only demonstrate their "perpetual preoccupation with attempts to rival their professors through busywork and to use their teachers' most predictable mannerisms as the only significant secrets they managed to cull from those men's famous procedures is insufferable." Since nothing was to be found at the exhibition "which did not reveal, if not ideas that were too undeveloped, then clear evidence of the overdeveloped hair of these meticulous performers, might one," he asked rhetorically, "not have to conclude that the exclusive attribute of man's companion was to copy him clumsily, to make reproductions of his work?" There is, he concluded, "between the decoration of a bosom and the composition of a painting, between the dressing table and the work of art a distance which the brains of these ladies—be it due to an essential inferiority or a lack of education—simply cannot transcend'' (X, 186–87).

Natanson's attitude was typical of the opinions of the more progressive critics. It can be said that while the theory of evolution, which validated the pursuit of the ''new'' at all costs, was undoubtedly the driving force behind the institutionalization of the concept of the avant-garde, the forms of antifeminine thought which had been formulated on the basis of the theory of evolution were also a powerful spur to artists to go in search of new forms of expression. Even the concept of an avant-garde was a powerful tool against the legitimation of women in the visual arts, for if a woman proved her ability to do what the famous male painters of her day were doing, she could be dismissed as a ''typical female imitator,'' whereas if she ventured into uncharted realms of visual expression her productions could be seen as incompetent, while similar experiments of male artists might be regarded as daring innovations. In addition, women who wanted to be painters were habitually told to emphasize ''feminine'' subject matter. For instance, S. C. de Soissons, in discussing the women artists of Boston, remarked,

> As in Europe, so it is true here; there is no lack of *women painters,* but there is an absolute lack of *paintings by women,* paintings which express their peculiar views, which show the feminine spirit. It is doubtless true that this womanly way of looking at things, which is delicious if one takes it for what it is worth, has a right to be translated under an artistic form. One can understand that women have no originality of thought, and that literature and music have no feminine character; but surely women know how to observe, and what they see is quite different from that which men see, and the art which they put in their gestures, in their toilet, in the decoration of their environment is sufficient to give us the idea of an instinctive, of a peculiar genius which resides in each of them. The unfortunate fact is that they do not know their own genius, do not understand it, do not appreciate it and do not cultivate it. Woman should not take care of the intimate rapport of things, she should look at the world as a gracious and moving surface, infinitely shaded; she should leave success to itself, as if the world were a theatre of fairies, an adorable procession of passing impressions.
>
> Strictly speaking woman only has the right to practice the system of the impressionists; she herself can limit her efforts and translate her impressions and recompense the superficial by her incomparable charm, by her fine grace and sweetness.

De Soissons excoriated the successful women painters of his day—such as Rosa Bonheur and Marie Baskirtscheff—for not sticking to the appropriate feminine sentiments he saw as characteristic of impressionism. Instead, he insisted, the majority of these painters had

> a hatred for feminine visions; they make every effort to efface that from their eyes. Many even succeed in assimilating happily our habits of vision; they know marvellously well the secrets of design and of colors, and one could consider them as artists, if it were not for the artificial impression which we receive in regarding their pictures. One feels that it is not natural that they should see the world in the way in which they paint, and that while they execute pictures with clever hands they should see with masculine eyes (*Boston Artists,* 75–78).

Thus women artists found themselves trapped in a state of prejudgment no matter what they tried to do. ''Art has only one gender,'' Proudhon had stated categorically, ''it is masculine'' (*La Pornocratie,* 152)—and that dictum became the credo of the turn-of-the-century avant-garde. In this respect the ostracism of women in the field of art was directly linked to the general male suspicion that to educate women in anything but matters related to housekeeping was to invite trouble.

The men of the nineteenth century in general had, as we have seen, stressed the unimportance, indeed, the dangers of education for women. Such an education would only confuse these feebleminded creatures, bring them in contact with polluting ideas whose misunderstood complexity would contaminate the virginal purity of their minds, and hence interfere with their role as soul-keepers. When women began to struggle free from their enforced bondage to ignorance, men sprung the trap of statistical "science" to prove that women were inferior to them. Indeed, ever since the blossoming of studies based on categorical comparison in the mid-nineteenth century, sexist and racist theorists could justify their contentions on the basis of the objective truth of scientific samples. For by ignoring the historical ramifications of their comparisons, they could come to whatever subjective conclusion they wanted to assert. When you take three women, keep them in total ignorance until they are twenty, then pit them against three men of twenty who have had a first-rate educational environment from birth, it is easy to prove intellectual inferiority. The argument of historical conditioning and its effect can subsequently be dismissed, as late nineteenth-century scientists did constantly, as a feeble effort to contaminate the objective findings of scientific research with subjective conjectures.

However, the late nineteenth-century bio-sexists rarely went so far as to take three women and three men, in other words, to take a scientific sample. They usually took Woman (a strange scientific category indeed) and pitted this generalized entity against three, four, or a dozen carefully selected men. The scientists who did this most blatantly are among those who have come to be most completely hidden from our contemporary awareness, yet they influenced the thinking of twentieth-century social scientists most— not directly, but through the intermediary offices of figures such as Herbert Spencer, William Graham Sumner, and T. H. Huxley, whose taste for all-inclusive pontification made it possible for their twentieth-century followers to downplay these theorists' viciously racist and sexist attitudes and focus on less controversial aspects of their schemes of social development.

A blind trust in anything that had been designated as a scientifically obtained finding, a willingness to see truth in any semirational conjecture which had been presented in the name of objective knowledge, shaped the attitudes of the lay public in the years around 1900. It encouraged the development of the aggressively racist and sexist social structures in Europe and the United States which, in turn, were responsible for the genocidal atrocities of Nazi Germany. For in the atmosphere of quietist collaboration with anti-Semitic attitudes which prevailed among intellectuals everywhere during the early years of this century, as much as in the official policies of the Third Reich, the conditions for genocidal behavior had been established.

As we have already seen, and as we shall have reason to explore further, the link between anti-Semitism, general racist attitudes, and the antifeminine mentality is very direct, since all three stem from identical initial assumptions. It can be said without exaggeration that the psychological "gynecide" advocated by the turn-of-the-century male intellectual avant-garde was a first manifestation of the forces which would make the actual genocidal policies of Nazi Germany not only culturally acceptable to the German populace but a logical historical outcome of the extravagant false science of general turn-of-the-century culture. For it was once again inevitable that the dualistic symmetry of nineteenth-century thought should extrapolate from the legitimate theory of the evolution of species the specious theory of potential devolution, the "degeneration" of society. Once again the male mind fastened on woman as the principal culprit.

CHAPTER VII

Clinging Vines and the Dangers of Degeneration

Reality rarely fits into the dualistic mold. Those who seek their truths at either end of the rainbow of false expectations invariably find pyrite instead of gold. Existence is a nexus of interlocking elements of potential and failure, the staging area for acts which, with every intimation of completion uncover unexpected facets of a new discouragement. In desperation we construct our absolutes, the poles of our dualities, which, once in place, turn the incoherence of being into a fool's paradise of absolute truths. No longer confused, we can stalk the world to find our villains and eliminate the evil which stands in the way of our success. The good man in the white hat can now shoot the man who wears black to make us feel good. When the black-hatted heel seems to be winning, we boo. And, best of all, when a confused creature in gray enters the scene, we can cheer no matter who shoots him.

A dualistic world promises us that the silent movie of our emotional attachments can keep on unwinding forever. But one day, unfailingly, Chaplinesque, we find ourselves up to our necks in celluloid. If we get stuck in the mess, we will choke, and someone will dump us—reels and ratchets and all—into the solid-waste compactor of time passing. If we chance to get out of the mess before choking, we fire the projectionist, hire one with better credentials, and start over again. We may have nothing to show for, but we'll be consoled, for (we have been told) nothing's been lost. The absurd world of dualistic "meaning" is, however, not merely static but wasteful of talent and possibility. While setting up shams of progress and creative development, the pendulum of binary oppositions destroys those who are seeking real change and kicks them unceremoniously into the pit of wasted possibility.

In the nineteenth century, the "century of progress," one of the favorite dualistic ploys to inhibit real change was to trample on the rights of woman and to make her out to be the beast of the Apocalypse. By mid-century the pendulum had already swung furiously in the mind of Proudhon. He saw woman as "courtesan or as housewife; there is no inbetween;'' and he declared "I'd rather see my daughter dead than dishonored!" (*La Pornocratie, 262*) For him it was clear that the idea of the platonic androgyne existed only in the most precisely circumscribed marriage formula. "From the point of view of intelligence and conscience, as well as that of the body, man and woman form one totality, one being in two persons, a single, true organism." Proudhon saw this separation of a single organism into two beings as a useful "first application, by nature itself, of the great principle of the division of labor" (46). He unhesitatingly aligned himself with Comte in viewing woman's role as that of the household nun, of "the incarnation of the Ideal" (11), who leavened the predatory responsibilities of the male with an appropriate modicum of piety. "Woman," he cried enthusiastically, "are you honest, are you good? Well, then, I shall canonize you; indeed I shall do more. I will kneel before you, adore you, love you" (157). Restating an argument of Schopenhauer's, Proudhon remarked that since woman had "less moral energy than man" this made her ideal as a moderating influence, for "man tends more to insist on letting rigid, hard, pitiless justice prevail, while woman tends to reign by means of the heart. However, the qualities of love, charity, pity, and grace, which bring woman's "perfections to their highest point, also testify—from the point of view of pure justice—to her inferiority" (38–39).

It is clear that in Proudhon's mind there were only two possible social developments: evolution or degeneration. The road to progress was masculine aggression, the road to destruction sappy effeminacy. "A nation, after having risen with virile energy, can become *effeminate* and even collapse" (emphasis in original). That was what had happened to the Persians, the Greeks, and the Romans. However, "while a race can become effeminate, it can also make itself more virile through its work, its philosophy, and its institutions" (72). There were, then, only two viable choices in the real world, "either the subordination of women, guaranteed by the modesty of their position in life, or the degradation of men: We must choose" (86). Clearly, to give woman any role beyond that of the household nun was to release the forces of degradation. Women who wanted to usurp part of man's place in creation were going against nature, becoming mock-men themselves, caricatures of masculinity, viragoes. "Lily-livered masculinity makes itself the accomplice of women's audacity, and we see the appearance of those theories of emancipation and promiscuity whose last word is pornocracy. And there is an end to civil society" (74).

Thus, even before the publication of Darwin's *Origin of Species* in 1859, the pattern into which the denigration of woman was to fall had been set. Proudhon, like Sarah Ellis, saw only one function for woman, and that was to be the meek guardian of the "minor morals" of the world. If she moved beyond that role, her function as the "spark" to man's search for the ideal would be negated and the lower forces in her nature would be released, for "it is a law of nature among all animals that the female, pressed by the reproductive instinct, and making a great fuss, searches out the male. Woman can simply not escape that law. Nature has given her a greater penchant for lewdness than man; first of all because she has a weaker ego, and liberty and intelligence therefore struggle less fiercely in her against the animalistic tendencies, and secondly because love is the great, if not only, occupation of her life" (41).

Consequently, when Darwin, in *The Descent of Man*, unhesitatingly accepted the notion of the natural inferiority of woman, and also began to stress the dangers of "reversion" in the development of species, it was easy to put two and two together. The great theorist of evolution, as we have seen, had already expressed the opinion that under normal circumstances the intellectual distance between men and women was bound to increase as a factor of natural selection. Carl Vogt had pointed to woman's tendency to "preserve, in the formation of the head, the earlier stage from which the race or tribe had been developed, or into which it has relapsed" (*Lectures on Man*, 82), and had thereby raised the spectre of reversion. In the world of animals Darwin subsequently observed "two great branches—the one retrograding in development and producing the present class of Ascidians, the other rising to the crown and summit of the animal kingdom by giving birth to the vertebrata" (182). He defined the principle of reversion as a natural phenomenon "in which a long-lost structure is called back into existence," and referred to the fact that the Greeks might have "retrograded" in part "from extreme sensuality; for they did not succumb until 'they were enervated and corrupt to the very core' " (160). Finally, he clearly linked the descent of man to "a hairy, tailed quadruped, probably arboreal in its habit," while he spoke of the homologies man "presents with the lower animals—the rudiments which he retains—and the reversion to which he is liable" (*The Descent of Man*, 696).

For the impressionable late nineteenth-century male, this was all the evidence needed to perceive a massive spectre of degeneration hovering just above the horizon, ready to drag the gloriously evolving species of man back into the jungle. Scientists, intellectuals, artists, and ordinary men everywhere began to worry about the dangers of reversion. Humanity, after all, was just at a point where there seemed to be every indication that real evolutionary developments were taking place. Even Tennie Claflin, the fiery feminist, had admitted that "the minds of all men, everywhere, are being roused with a more comprehensive state of action." For, she said, "they are daily brought into contact with the progressive ideas and thoughts of the world, which continually modify their opinions, views, and even methods of thought" (*Constitutional Equality*, 26). Just at this crucial point, however, the evolving male had to face the horrendous possibility that, in Carl Vogt's words, "man may by arrest of development sink down to the ape" (202). Could there be a more frightening prospect facing the potential superman of creation?

Clearly, any signs of reversion would have to be detected and eradicated to prevent anything of the sort from taking place. Degenerative tendencies were spotted everywhere, and the discovery (in others, of course) of atavistic traits indicative of what Darwin had called reversion became a general pastime among intellectuals. Sexism and racism found new and urgent justification in thoughts about the excruciating dangers inherent in the potential political and social power of the "feebleminded." Max Nordau merely expressed what was on everybody's mind when he published a hugely successful treatise simply but clearly entitled *Degeneration* (1893), in which he outlined the evidence of this dread condition, which he saw as pervading the effeminate concerns of a society in disarray.

Nordau dedicated his book to Cesare Lombroso, the frenetic phrenologist who had, as he pointed out, developed the notion of degeneracy "with so much genius." It was an appropriate tribute. Lombroso and Ferrero, in *The Female Offender*, pointed to the fact that in general extreme specimens of feminine degeneration showed decidedly masculine characteristics. However, these aspects of male tendencies in women criminals were not representative of an evolution of the female brain but rather a clear sign of that reversion Vogt and Darwin had warned about. Since the human male had struggled out of a condition of sexual indeterminacy (those bisexual, hermaphroditic primal origins of the higher organisms) into true masculinity, it was clear that the roles of men and women were meant to be completely separate. It was therefore woman's role in evolution to become more and more feminine and not to take on masculine qualities. For a woman to take on such masculine qualities was actually a sign of reversion, a sinking back into the hermaphroditism of that indeterminate primal state—just as it was a clear sign of analogous degeneracy in the male to show himself to be effeminate. If, in Harry Campbell's words, man had "to exercise mind and body more than the woman," it stood to reason that in the normal order of things these aspects of male identity, "already rendered stronger than in the woman through sexual and natural selection, should by this means become stronger still." On the other hand, it was incumbent upon woman to cultivate her "refinement of nervous organization" (*Differences in the Nervous Organization of Man and Woman*, 82–83). If she did not, if she insisted on cultivating masculine characteristics, that was a clear sign of her degeneracy.

Most evolutionists soon decided that feminism was the clearest example of this form of masculinizing degeneracy. To highlight this perception, Mrs. E. Lynn Linton in 1891 published an extensive three-part denunciation of what she called the "wild women" for *The Nineteenth Century*. These wild women—even her choice of title indicates that Mrs. Linton had read her Vogt and her Darwin—showed, in their "desire to assimilate their lives to those of men," clear signs of "the translation into the cultured classes of certain qualities and practices hitherto confined to the uncultured and—savages." The jig was up: The traits of degeneration clearly showed in the features of the feminist. "In obliterating the finer distinctions of sex she is obliterating the finer traits of civilization," not realizing that "every step made towards identity of habits is a step downwards in refinement and delicacy—wherein lies the essential core of civilization." Thus, the signs of reversion were carved into the virago's being: "Unconsciously she exemplifies how beauty can degenerate into ugliness, and shows how the once fragrant flower run to seed, is good for neither food nor ornament" (596–98).

The beginnings of the feminist movement in the Brontëan prison of the mid-century's idealized mother-woman had a formidable impact on the thinking of the late nineteenth-century male. In the dreams of the evolutionist idealists he found the perfect counterargument to the intellectual aspirations of women. Starting in the late 1860s, a crescendo of chants, which by 1900 had become a thundering Wagnerian chorus of experts, pointed to the degenerative impulse behind feminism. As early as 1870 Nicholas Cooke had warned that "if carried out in actual practice, this matter of 'Woman's Rights' will speedily eventuate in the most prolific source of her wrongs. She will become rapidly unsexed, and degraded from her present exalted position to the level of man, without his advantages; she will cease to be the gentle mother, and become the Amazonian brawler" (*Satan in Society*, 86).

The artists of the turn of the century were fascinated by the notion of the masculinized woman as savage. Whenever they could they showed the warrior woman, the feminist—that naked brawler—in all her dubious masculine power, titillated as well as frightened by the thought of all that feminine brawn. Max Beckmann's "Battle of the Amazons" [VII, 1] demonstrates in the Michelangelesque masculinity of these warrior women's bodies that they were enemies to be contended with. At the same time, in this giant-sized painting, Beckmann reinforced the prevailing notion that women who strove to take on masculine traits were degenerate creatures, turning his women warriors into grotesque, bestial beings whose revolting, pudgy masses of greenish flesh were only barely contained by vast expanses of mottled and distended skin. The soldiers struggling with these mollusclike monsters are battling for dear life on what appears to be an alien planet straight out of a science fiction writer's nightmare.

VII, 1. Max Beckmann (1884–1950), "Battle of the Amazons" (1911)

VII, 2. Wilhelm Trübner (1851–1917), "Battle of the Amazons" (1879)

Beckmann's painting strengthens the impression that, at least thematically, the work of the German expressionists is in many cases no more than the pursuit *ad nauseam* and *ab absurdum* of the conventional antifeminine subject matter of their more academic colleagues. It would be encouraging to think that by doing so they were deliberately attempting to discredit these painters' misogynist point of view, but all indications are that painters such as Beckmann continued to paint *in* the antifeminine tradition rather than *against* it. It is, for instance, perfectly obvious that Beckmann's "Battle of the Amazons" takes its cue, in visual organization as well as in ideological intent, from Wilhelm Trübner's painting of the same title, which was painted more than thirty years earlier [VII, 2]. The principal difference between the two paintings is to be found in Trübner's much smoother, more idealized, and deliberately classical treatment of the amazons' bodies. In this respect the link between Trübner and Beckmann is not at all unlike that between Bouguereau and Renoir. After all, the difference in the treatment of the nude to be found in the work of these two painters depends primarily on the *kind* of sugar coating they applied to their subjects. In a similar fashion, Thomas Couture and Cézanne were brothers under the skin at least in their enthusiastic pursuit of bacchanals. If a concern with painterly method and stylistic innovation is to be seen as the only thing that matters in art, comparisons of this sort are, of course, completely meaningless. It is only if we are willing to grant that content has a function in early modern art not unrelated to that which it had in the more conventional forms of fin-de-siècle art that such conjunctions can have any significance.

In any case, the theme of battling amazons was the turn of the century's favorite vehicle for the graphic delineation of that perpetual battle of the sexes whose principal function was to minimize the damaging inroads of viraginous females into the ever-threatened realm of male supremacy. To men such as Nicholas Francis Cooke, for instance, the virago was "a sort of mistake of Nature." The only men who could possibly be interested in them would have to be freaks themselves, for "these masculine women nearly always ally themselves with blanched males, weak physically and mentally" (280). Clearly, such alliances could only lead to further degeneration. The remarks of the Countess of Jersey, in an article published in the January 1890 issue of *The Nineteenth Century*, are instructive in identifying the practical reasons for men's virulent opposition to any change in woman's position. The countess reacted directly to the evolutionary theories, which held that while man advances intellectually woman either remains the same or tends to deteriorate with the passage of generations. With more than a touch of irony, she pointed out that "it would be strange, indeed, if while the male half of creation is always beating record physically and scientifically (to say morally and artistically would be to introduce argument), the female half were to behave as birds are accused of doing, who build their nests exactly as their predecessors did when they issued from the ark." Citing examples of women who had engaged in feats of athletic daring a century earlier, she questioned the prevailing notion that an active life actually contributed to the degeneration of feminine functions. She also touched upon the masculine fears underlying the movement of women into the public arena. Have women, she asked rhetorically, "actually deteriorated physically and unfitted themselves for maternal duties by the lives which they now lead? And is it quite just to assume that, as is very commonly taken for granted, when a woman attempts anything which is more ordinarily done by men, she therefore desires to emulate or rival man? Is it not conceivable that she may sometimes like the work or sport for its own sake, without any thought of competition with the other sex?" (56–57)

Rosa Mayreder, in *A Survey of the Woman Problem*, also emphasized that the "idea that woman stands neither above nor below man, but beside him in human communities" was "frequently attacked, especially by the defenders of the masterful type of amorist, as a feeble invention of modern feminine thought, or even as a product of deterioration" (237).

It is clear that a fear of competition in the workplace lay behind much of the high-minded arguments against the viraginous tendencies of the New Woman. In writing their diatribes against the unnatural acts of the feminists, the psychologists and physicians—who insisted that ultimately men had nothing to worry about since, in the long run, woman's every attempt at emancipation would of necessity have to collapse as a result of her fundamental and irremediable constitutional weakness—were clearly protesting too much. Economic issues kept entering into the argument. Men's fear of the potential competition of women in the workplace helps to explain why—given the very small percentage of women actually involved in the feminist movement during the later years of the nineteenth century—the amount of frightened attention they received was so intense. Moreover, the feminist issue was perfect fodder for the extreme dualistic imagination of the period. As in Proudhon's case, any sign of a move toward personal independence on the part of a woman was enough to make the later nineteenth-century male think of catastrophic events. This, in turn, would lead to elaborately justified predictions of the demise of "civilization as we know it." Even to think of giving woman the vote was seen as a guarantee that humanity's long slide into the abyss of degeneration had begun. In addition, virtually everybody agreed that all this represented a truly new development.

It was "that wonderful nonsense called 'women's rights'" that had started it all. Tolstoy, for instance, insisted in *What Then Must We do?* that "within my own memory it has spread more and more widely" (353). Indeed, for Tolstoy the matter was quite simple. Women were saints if they chose to bear children and "consciously submit to that eternal, immutable law, knowing that the hardship and labour of that submission is their vocation'' (354). On the other hand, if a woman chose not to have children, she was a vile degenerate: "Every woman, however she may call herself and however refined she may be, who refrains from child-birth without refraining from sexual relations, is a whore" (357).

Tolstoy's remarks are characteristic. The search for woman as the lily, the paragon of virtue, had carried within itself the discovery of Lilith, of woman as snake, the inevitable dualistic opposite of the image of virginal purity. The nun who failed to die in pursuit of virtue risked being suspected of worldly concerns, of a betrayal of her own sacrificial, civilizing mission. There exists no better delineation of the manner in which the poles of this duality were created than the trajectory of Ruskin's search, in *Sesame and Lilies,* for the flower of inspiration. Ruskin's demand that in her guardianship of the home as "vestal temple" woman "must—as far as one can use such terms of a human creature—be incapable of error" (60) represented such a patently impossible task that it booby-trapped woman's household pedestal with nitroglycerin: One good shake and the lady's reign was over. Ruskin's ideal of woman was for her a one-way ticket to hell—and her explosive pedestal was the missile that would get her there. "So far as she rules, all must be right, or nothing is. She must be enduringly, incorruptibly good; instinctively, infallibly wise— wise, not for self-development, but for self-renunciation" (60).

It was this simple dualistic opposition which animated most late nineteenth-century intellectuals and made it easy for them to see in each and every woman who did not conform to the ideal of womanhood—instituted by their fathers and reflected in the virtuous inanition of their mothers—a dangerous backslider in humanity's quest for evolutionary transcendence. Such regressive women were tools of the degenerative forces of nature against which humanity had to do battle in order to evolve. August Forel, who was what one might term a "bio-sexist moderate," was widely respected as an expert on the relations between men and women. His position, clearly delineated in *The Sexual Question* (1906), can be seen as representing the fin-de-siècle's "reasoned norm" on the issue. "The modern tendency of women to become pleasure-seekers, and to take a dislike to maternity, leads to degeneration of society. This is a grave social evil, which rapidly changes the qualities and power of expansion of a race, and which must be cured in time or the race affected by it will be supplanted by others" (137).

Forel's remarks show how closely allied in the minds of the men questions of imperial power and the concept of racial and sexual struggle had become around 1900. What Forel called "the fundamental weakness of the feminine mind" (138), a weakness also to be found in the "inferior races," would become man's undoing if he took his eyes, even for a moment, off his goal of evolutionary transcendence and muscular masculine mastery.

What, exactly, was that pie in the evolutionary sky? Joseph Le Conte—an influential American follower of Darwin, Haeckel, and Agassiz; professor of geology and natural history at the University of California at Berkeley; and mentor of Frank Norris—spelled matters out quite specifically in his book *Evolution: Its Nature, Its Evidences, and Its Relation to Religious Thought* (1888–91), joining fashionable elements of neo-Platonic idealism, pantheism, and evolutionary science. "There can be no doubt," said Le Conte, "that we are now on the eve of a great revolution" (280). This revolution was to take the form of man's transcendence into a new realm of material being, in which he would move from an "outer lower life" to an "inner higher life," not shedding the "necessary work-clothes" of materiality, but letting the "life-force" inherent in all material things and representative of the presence of God in nature distill itself, as it were, into an embodied spirit. This transformation of man into immortal mind could take place at any moment, for evolution had shown "that there is nothing wholly exceptional in such transformation with the sudden appearance of new powers and properties; but, on the contrary, it is in accordance with many analogies in the lower forces, and therefore *a priori* not only credible but probable." Hence "the passage from one plane upward to another is not a gradual passage by sliding scale, but *at one bound*" (emphasis in original) (314–15).

To reach this higher plane man would have to divorce himself, as it were, from nature: "Spirit must break away from physical and material connection with the forces of Nature" much "as the embryo must break away from physical umbilical connection with the mother" (321). Le Conte's images in describing this transcendence of the spirit were, as a matter of fact, all strikingly suggestive of the iconography of the battle of the sexes. "Nature," he said, "may be likened to a level water-surface. This represents unindividuated physical and chemical force." Ultimately the transcendence of man into the realm of the ideal would be his liberation from—would constitute his being "lifted above"—the world of the "earth-mother" into the higher plane of the masculine spirit. Currently man could be likened to a "commencing drop" being pulled by "some individuating force" above the level surface of the water: "In man spirit emerges above the surface into a higher world, looks down on Nature beneath him, around on other emerged spirits about him, and upward to the Father of all spirits above him. Emerged, but not wholly free— head above, but not yet foot-loose."

The present state of man, then, was one of struggle, the struggle between mother/earth and father/mind over the eternal essence of man. Any letup in man's effort to overcome the downward pull of the water represented a total collapse of his chances for transcendence, a reversion to his animal state. "Even though the drop be nearly completed, if we remove the individuating or lifting force, the commencing drop is immediately drawn back by cohesion and refunded into the general watery surface. But, once complete the drop, and there is no longer any tendency to revert, even though the lifting force is removed. This represents the condition of spirit in man" (320).

In practical terms, Le Conte's metaphoric representation was meant to convey that currently man

> is possessed of two natures—a lower, in common with animals, and a higher, peculiar to himself. The whole mission and life-work of man is the progressive and finally the complete dominance, both in the individual and in the race, of the higher over the lower. The whole meaning of sin is the humiliating bondage of the higher to the lower. As the *material* evolution of Nature found its goal, its completion, and its significance in man, so must man enter immediately upon a higher *spiritual* evolution to find its goal and completion and its significance in the ideal man—the Divine man (330).

Few of Le Conte's contemporary readers could have been so obtuse as to overlook the obvious link between animal nature, sin, and unindividuated womanhood implied in his scheme, whereby "spirit, unconscious in the womb of Nature," would awaken and strive "to attain, through a newer birth, unto a higher life." Even if his readers had overlooked it, there were plenty of other writers around the turn of the century who were ready to make the connection even more explicit. Indeed, there is probably no better way to come to an understanding of the extremes of spurious antifeminine theorizing to which the turn-of-the-century mind had become addicted than to examine the then world-famous writings on "sex and character" of the young Viennese theorist Otto Weininger. In 1903 Weininger, scarcely twenty-three, published a book of that title which almost instantly became required reading among the intellectuals of the time.

Weininger's ideas were, of course, by no means original. Everything he asserted had been asserted before. Many of his ideas had already been elevated to the status of commonplaces among the intellectuals of the 1890s. Moreover, his philosophy was a peculiar grab bag of notions taken from Plato, Schopenhauer, Kant, Darwin, Spencer, the social Darwinists, the virulent misogynists of the *Mercure de France* and last but certainly not least, Freud, who had read and admired a version of Weininger's manuscript even before its publication.

What makes *Sex and Character* especially significant is that it took a large group of existing notions and organized them into a system that seemed to be scientific and consequently made a great deal of sense to the author's contemporaries. In other words, Weininger's book flattered the reader's self-esteem by permitting him to say, "I knew it all along! I was right about these matters, and here at last is the scientific truth!" The book became a huge success and was read avidly throughout Europe by laymen and professionals alike, both in its original edition (it was still in print in Germany even after the Second World War) and in numerous subsequent translations. Young intellectuals kept it on their bedside tables, reading a few pages of its wisdom just before they went to sleep, comforted by its assurances that they were the bright hope of the future, and that any pangs of regret they might just then be feeling for having loved and lost a fashionable and charming lady friend were simply foolish and regressive, to be blamed more on the young lady's inability to recognize and appreciate true spiritual excellence than on any faults of their own.

Among the few younger men who did not fall under the book's spell was Ford Madox Ford, who in a long essay entitled "Women and Men," which was published in *The Little Review* in 1919, described the extraordinary effect of Weininger's opus on his acquaintances in London.

> The most important, as it is the most singular, of contributions to modern literature on the sex question is an extraordinary work called in English "Sex and Character". This book is noteworthy because it had an immense international vogue. It was toward the middle of '06 that one began to hear in the men's clubs of England and in the cafés of France and Germany—one began to hear singular mutterings amongst men. Even in the United States where men never talk about women, certain whispers might be heard. The idea was that a new gospel had appeared. I remember sitting with a table full of overbearing intellectuals in that year, and they at once began to talk—about Weininger. It gave me a singular feeling because they all talked under their breaths. I should like to be precise as to the strong impression I then received, because if I could convey that impression exactly I should give a precise idea of what is the attitude of really advanced men toward woman-kind (40–41).

If *Sex and Character,* in Ford's words, "had spread through the serious male society of England as if it had been an epidemic," its success had been equally electrifying throughout the rest of Europe. The book even caused a heated exchange of letters between Wilhelm Fliess and Freud because, as Fliess claimed, Weininger had "stolen" some of his ideas and Freud had been to blame since he had carelessly transmitted Fliess' intellectual "property" to the young theorist.

According to Weininger, long ago, in humanity's brute past, all beings had been bisexual. The beginnings of humanity's move toward sexual differentiation had been its first step on the road to progress. The greater the evolution of mankind, the greater its move toward the attainment of purely male and female entities. What was holding back human progress was that even now "sexual differentiation, in fact, is never complete" (5). In the mathematical equation which created the sexual, that is, reproductive entity in nature, the masculine represented the positive pole, the feminine the negative. Woman might be the locus of the physical—one might almost say, the mechanical—reproduction of humanity, but the male was the locus of the brain, of humanity's capacity for spiritual understanding. Therefore, the more truly male man became, the more spiritual he would be, whereas the more completely feminine woman became, the more materialistic and brainless she would be. "There is," quipped the young philosopher, echoing Möbius, "some historical justification for the saying 'the longer the hair, the smaller the brain'" (68). It should be understood, he stressed, that "women really interested in intellectual matters are sexually intermediate forms" (70). Even so, "there is not a single woman in the history of thought, not even the most man-like, who can be truthfully compared with men of sixth-rate genius" (69).

Thus, for Weininger the more truly feminine a woman was, the more she would be devoid of any spiritual intelligence. As a matter of fact, "in such a being as the absolute female there are no logical and ethical phenomena, and, therefore, the ground for the assumption of a soul is absent" (186). Hence true femininity expressed itself in a vacant stare. The less males associated with these carriers of regressive materialism, the less likely they were to be held up in their progress toward the ideal of true masculine spirituality. In addition, said Weininger, "I shall show reasons in favor of the possibility that homosexuality is a higher form than heterosexuality" (66). Women were doomed to be excluded from the heaven of true spiritual transcendence, for, given the fact that sexual differentiation was proceeding apace in the wake of human progress, they were fated to grow ever more obtuse in their femininity. Moreover, "in spite of all intermediate conditions, human beings are always one of two things, either male or female" (80). Women had, by nature, only a yearning for what was materialistic and immediate. It was true—even if highly unfortunate—that males still had sexual drives (a remnant of the feminine in them), but those were only periodically recurring vestiges of the regressive impulse. "Man possesses sexual organs, her sexual organs possess woman" (92). Hence "the condition of sexual excitement is the supreme moment of a woman's life. The woman is devoted totally to sexual matters, that is to say to the spheres of begetting and reproduction" (88).

The result of all this was that women were, in essence, human parasites. They could not live without men or without each other. In a sense they were interchangeable, undifferentiated beings, for the capacity to differentiate was a characteristic of the intellect, of genius. True genius yearned for true individualism and stood sternly and ruggedly alone. According to Weininger, "The birth of the Kantian ethics, the noblest event in the history of the world, was the moment when for the first time the dazzling awful conception came to him, 'I am responsible only to myself; I must follow none other; I must not forget myself even in my work; I am alone; I am free; I am lord of myself'" (161). The great man "contains the whole universe within himself; genius is the living microcosm" (169). Weininger's absolute male therefore needed no others: "To acquiesce in his loneliness is the splendid supremacy of the Kantian" (162). On the other hand, "the female is soulless, and possesses neither ego nor individuality, personality nor freedom, character nor will" (207), all of which served to show that regressive, materialistic anti-individualistic political philosophies such as communism were basically the weak conceptions of benighted men who, like Karl Marx, were suffering from terminal cases of effeminacy.

Effeminacy, indeed, was the main obstacle against human progress. Men who, against their better nature, were enticed by women—those superficial, materialistic, animalistic, phallus-worshiping sexual automata—into believing that sex could release them from their god-like, lonely responsibility to seek individual perfection and spiritual transcendence, were doomed to sink to woman's level. The enticement to "pairing" was woman's ultimate weapon against man: "Pairing is the supreme good for the woman; she seeks to effect it always and everywhere. Her personal sexuality is only a special case of this universal, generalised, impersonal instinct" (260). Weininger concluded it was time for man to realize that, "as the absolute female has no trace of individuality and will, no sense of worth or of love, she can have no part in the higher, transcendental life. The intelligible, hyperempirical existence of the male transcends matter, space, and time. He is certainly mortal, but he is immortal as well" (284). On the other hand,

> Women have no existence and no essence; they are not, they are nothing. Mankind occurs as male or female, as something or nothing. Woman has no share in ontological reality, no relation to the thing-in-itself, which, in the deepest interpretation, is the absolute, is God. Man, in his highest form, the genius, has such a relation, and for him the absolute is either the conception of the highest worth of existence, in which case he is a philosopher; or it is the wonderful fairyland of dreams, the kingdom of absolute beauty, and then he is an artist. But both views mean the same. Woman has no relation to the idea, she neither affirms nor denies it; she is neither moral nor anti-moral; mathematically speaking she has no sign; she is purposeless, neither good nor bad, neither angel nor devil, never egotistical (and therefore has often been said to be altruistic); she is as non-moral as she is non-logical. But all existence is moral and logical existence. So woman has no existence (286).

If women were at all sincere in wishing to contribute to the spiritual betterment of mankind, they "must really and truly and spontaneously relinquish coitus. That undoubtedly means that woman, as woman, must disappear, and until that has come to pass there is no possibility of establishing the kingdom of God on earth" (343). At the same time, "man must free himself of sex, for in that way, and in that way alone, can he free woman" (345).

It did not disturb Weininger a bit that this would tend to wreak havoc upon the continuation of the human race. After all, he insisted, "the rejection of sexuality is merely the death of the physical life, to put in its place the full development of the spiritual life." Triumphant and resplendent in the warm glow of his own consistency, he added, "That the human race should persist is of no interest whatever to reason" (346). Until such a time as that glorious disappearance might come to pass, it was the male's primary task to eradicate signs of effeminacy wherever they might be discernible.

The characteristic link between women and the "degenerate races," so apparent to Carl Vogt, also caught Weininger's attention. He insisted that Jews, blacks, and orientals had, through inbreeding or the inability to respond to evolutionary impulses, become effeminate and had consequently degenerated. Judaism, especially, Weininger maintained, "is saturated with femininity" (306). Like women, Jews were unable to recognize the link between individual evolution and property, for, said Weininger, "property is indissolubly connected with the self, with individuality. It is in harmony with the foregoing that the Jew is so readily disposed to communism.'' After all, ''the true conception of the state is foreign to the Jew, because he, like the woman, is wanting in personality'' (306–7).

It was, Weininger concluded, the absence of an overriding desire for dualistic absolutes, a disdain for the central importance of binary oppositions, which most separated the woman and the Jew from the glorious world of masculinity. ''Greatness is absent from the nature of the woman and the Jew, the greatness of morality, or the greatness of evil. In the Aryan man, the good and bad principles of Kant's religious philosophy are ever present, ever in strife. In the Jew and the woman, good and evil are not distinct from one another'' (309). Just as a woman could only gain existence through her self-submergence in the male, so, Weininger insisted, the Jew should strive toward self-obliteration in the cause of evolution and Aryan masculinity. ''To defeat Judaism, the Jew must first understand himself and war against himself'' (312). Looking about him, and not liking the effeminacy he saw—even in himself, for he was a Jew—Weininger carried his reasoning to its logical conclusion. A few months after the appearance of his book he committed suicide—presumably a first token sacrificial gesture in a process he hoped might ultimately lead to wholesale ''gynecide.''

True, Weininger was clearly mad. His extremism has made it easy for cultural historians to dismiss his writings as an aberration. Yet, as we have seen, his contentions were simply one more step on the same road of dualistic extremism trod by the majority of turn-of-the-century intellectuals. Perhaps one way to define madness is to say that it is the affliction of those who take the basic premises of the society they live in too literally and pursue the logical consequences of that society's underlying assumptions with too single-minded a determination. Everywhere about him Weininger could have seen—and did see— a confirmation of his theories. The art and the attitudes of the 1890s created Weininger. His contemporaries did not consider him mad. They did anything but dismiss his book; they read it and loved it. Was Weininger mad or was it his environment? Suggestive prefigurations of his theories, moreover, abounded in art around 1900. The notion of woman's mindless materiality and her singleminded absorption in the contemplation of her own sexuality—which led to the endless repetition of the mirror theme in art and to the cultivation of the woman with the vacant stare by the impressionists—also resulted in the depiction of woman as that earthbound creature Weininger passionately despised and saw echoed in the Jew.

Woman had become the personification of the sins of the flesh. Seen as jealous of man's exclusive capacity for spiritual transcendence, she was thought to be intent upon doing everything in her power to drag the male back into her erotic realm. Even if she had no overtly vicious intentions, woman's presumed incapacity for independent thought and her inherently parasitic being led her to interfere with the intellectual development of the male. That was the general consensus. Weininger claimed to have unmasked her chameleonic nature. ''Women are matter, which can assume any shape'' (293–94). In her passion for pairing woman was merely expressing the ''endless striving of nothing to be something.'' But in making these statements Weininger had simply updated and made more fashionably diabolical what had already become a commonplace about the inherently dependent nature of woman.

This commonplace was popularly expressed in the image of woman as a clinging vine. Mark Twain's Eve put the ideal representation of woman's role as follows: ''He is strong, I am weak, I am not so necessary to him as he is to me—life without him would not be life; how could I endure it? . . . I am the first wife; and in the last wife I shall be repeated'' (*Eve's Diary,* 107). This sentimental portrayal of the absolute dependency of woman on man was widely seen as the proper "vinelike" condition for woman. It was also expressed as such by Borg in Strindberg's *By the Open Sea:* "He desired to be loved by a woman who would look up to him as the stronger; he must be the idol, not the worshipper; he must be the trunk on which the weak branch was grafted'' (127).

Estella, that flower of womanhood and the heroine of Ignatius Donnelly's *Caesar's Column* (1890), an extraordinarily popular American utopian romance, asked her hesitant lover,

> What kind of weak heart or a weak head have you, not to know that woman never shrinks from dependence upon the man she loves, any more than the ivy regrets that it is clinging to the oak and cannot stand alone? A true woman must weave the tendrils of her being around some loved object; she cannot stand alone any more than the ivy (197).

Donnelly meant this to be a passage in praise of true womanhood, in praise of the sort of dutiful dependency on the magnanimity of the male that might allow the weaker vessel to share—if only through reflected glory—in the potential for transcendent evolutionary development inherent in the male. He was trying, in other words, to express the sort of sentiment George Frederick Watts aimed for in his painting "Love and Life." According to a note which accompanied it in the Grosvenor Gallery exhibition of 1885,

> Love is represented by the winged figure of a youth and Life by that of a young girl, who, clinging to Love, is being guided by him over the rough places of a rocky precipice which both are ascending together. Love is leading the way and helping Life by his support and tenderness to climb the difficult path—emblematic of the struggling conditions which are the portion of all human existence. The half-extended wings of Love shade the rays of light from beating too fiercely on the delicate figure of Life. Love's footsteps can be traced on the rocky ascent by the daisy flowers which have sprung up in his track. The atmosphere is bathed in the gold of light and the blue of space. As the figures ascend the air becomes more golden. Love, while helping to endure and overcome the struggles of existence, leads upwards into purer and brighter conditions. The truth which the artist has tried to embody in this picture is that Love in its widest, most universal sense—in the sense of charity, sympathy, and unselfishness—raises life upward; that humanity is helped by tender aid on the one hand and by tender trust on the other.

This is certainly an affecting universal symbolism which on the surface would seem to have little to do with the battle of the sexes. But it is no accident that in Watts' painting the figure of godlike Love is that of a young man—with wings, no less—and that the clinging figure of Life is that of a helpless (and wingless) girl.

The juxtaposition of woman's "tender trust" and man's evolutionist struggles found its clearest ideological expression in Owen Wister's *The Virginian* (1902), in which the conquest of the West and the conquest of woman are equivalent features of the "regeneration" of superior Anglo-Saxon "masculine courage and modesty" (251). These qualities, like the Virginian's woman, Molly, had been "watered" by "the pale decadence of New England" (250). In this novel woman's dependence on man, her natural function as a clinging vine becomes a factor of the true male's test of manhood. For a man could only properly become "the trunk on which the weak branch was grafted" when he proved his ability not to be swayed by a woman's weak will. He must meet his challenge, must overcome woman, in order to be able to save her. In a scene which was to become an archetypal part of the cinematic Western of the twentieth century, the Virginian rejects Molly's siren call, which is cast as a plea to him to forego his duel in the sun, the ultimate assertion of his manhood. Molly insists that if he goes through with it, "there can be no tomorrow for you and me." Agonized, but solid as a bull's skull-bone, the Virginian thereupon proves that the evolution of the species is still in the proper hands:

> The blue of the mountains was now become a deep purple. Suddenly his hand closed hard.
>
> "Good-by then," he said.
>
> At that word she was at his feet, clutching him. "For my sake," she begged him. "For my sake."
>
> A tremble passed through his frame. She felt his legs shake as she held them, and, looking up, she saw that his eyes were closed with misery. Then he opened them, and in their steady look she read her answer. He unclasped her hands from holding him, and raised her to her feet.
>
> "I have no right to kiss you anymore," he said. And then, before his desire could break him down from this, he was gone, and she was alone (344).

Thus the Virginian asserts his right to become the oak around which Molly can weave the tendrils of her being by demonstrating that he is indeed the evolved, "full-blooded" man whom Owen Wister adored with such a true lover's fervor. In fact, in writing about his hero Wister repeatedly slipped into passages of wistful, indeed, wishful admiration: "The Virginian looked at her with such a smile, that, had I been a woman, it would have made me his to do what he pleased with on the spot" (180). The master and his willing slave was the ideal: woman as a weakly clinging appendage to the forward-striding male. Unfortunately, the flesh was weak and woman delicious. Her clinging was not of the saint but the sinner. And in man's striving for the sun, "Hecate's boat," as Oscar Wilde put it, tended to cloud his vision.

In his poem "The Garden of Eros" Wilde, in fact, had outlined the task in somewhat more abstract terms than Wister: "My soul / Passes from higher heights of life to a more supreme goal." And in Tennyson's "By an Evolutionist" the poet had droned,

> I have climb'd to the snows of Age, and I gaze at a field in the Past,
> Where I sank with the body at times in the sloughs of a low desire,
> But I hear no yelp of the beast, and the Man is quiet at last
> As he stands on the heights of his life with a glimpse of a height that is higher (811).

It was clear that man would have to make this last move to transcendence alone. In his search for the new evolutionary ideal, man had been able to break the chains of materiality which, according to Plato, had forced him to sit and face the wall of the cave. But even while man was leaving the cave, spurred by the promises of evolution, woman had insisted on remaining there, her static being obsessively hypnotized by the shadows on the wall, just as Flaubert's Félicité, the central character in *A Simple Heart,* had been transfixed by the parrot which she resembled.

As a result of the vogue of evolutionary idealism, a widespread cultural trend developed which viewed woman as an undifferentiated part of nature, a "henid" in Weininger's term, while man was an individual, a "monad." This notion provided a marvelous new excuse for the male, enlarging his inventory of arguments in defense of a double standard which justified his own philandering and condemned his wife—or any woman—for even thinking of doing the same. Since each man was an individual, this argument went, a

woman who decided to sleep with several men showed her inability to recognize the glorious contours of her chosen mate's unique masculine individuality. In her unwillingness
to cleave to one man, she thereby offended the special dignity of every other man's personal identity as well. But the male who roamed from woman to woman was seen as a
poet in search of the ideal. Among the variegated shadows of a single undifferentiated
form, he made a ceaseless, heroic search for the perfect embodiment of his sense of beauty.
Hence a man who strayed was not breaking his trust to his wife, since every woman was,
after all, no more than a simple particle of that great body of nature called Woman.
Maurice Maeterlinck, in *The Treasure of the Humble,* expressed this notion as follows:
''If, like Don Juan, we take a thousand and three to our embraces, still shall we find, on
that evening when arms fall asunder and lips disunite, that it is always the same woman,
good or bad, tender or cruel, loving or faithless, that is standing before us'' (93–94).

Neo-Platonic idealism thus counseled woman to ''stand by her man'' under all circumstances, since her only identity could come from the man whose satellite she was,
while it was equally clear that the man gained nothing from the mass of undifferentiated
being that hung upon his arm in a perpetual state of suspended spiritual animation. Indeed,
woman's perpetual ''sleep of the spirit'' gave added philosophical significance to those
images of somnolent women artists of the period liked to paint. These sleeping creatures
were merely responding to the simple, hedonistic, animalistic requirements of their being.
Since they could not think, they were likely to spend their waking moments doing mischief. When asleep, however, they could do no harm and were hence doubly pleasant to
look upon.

In fact, woman, the ''mental automaton'' of Max Nordau's writings, never experienced, in the manner men did, ''the longing to diverge from the race, and found a new
species, of which she would be the primal type'' (*Paradoxes,* 53). On the other hand, in
evolved males ''the superior development of the judgment and will-centres . . . produces
a truly human genius, which is the highest manifestation of the organic perfection attained
by man up to the present day'' (197). Because of this fundamental divergence, woman, as
Laforgue had said, misunderstood the spiritual motive underlying the emotion of love,
namely, man's ''infinite aspiration toward the Ideal'' (*Six Moral Tales,* 266). Instead, as
even Abba Goold Woolson had to admit, many ''ornamental young ladies'' were taking
their roles as clinging vines too literally. ''Accepting as they do, the oak-and-vine theory
as regards man and woman, they make no attempt to stand erect without support, but
swing wide their graceful tendrils in every breeze, if haply they may find some oak to
cling to'' (*Woman in American Society,* 43).

The problem with the clinging vine was clearly that she tended to smother the main
trunk. An anonymous aphorist writing in *Jugend* in 1904 cautioned, ''Cling to your loved
one, but like a garland of flowers, not like a chain!'' (4) The point was that woman should
cling to the male but should also avoid tempting him beyond his capacity to withstand her
enticements. Still, complete abstention from woman was not generally regarded as wise
either.

Dipping, as was everybody's wont, into the inexhaustible fund of historical examples
favored by turn-of-the-century intellectuals, Bernard Talmey pointed to the overly harsh
requirements of true self-control by citing the case of Origines, who ''found sexual abstinence too difficult and castrated himself. For that reason he never was canonized, for the
spirit should kill the flesh'' (*Woman,* 8). Talmey was, rather sensibly, concerned with the
fashion of idealistic abstinence which had taken hold of his male contemporaries in search

of evolutionary transcendence, and he tried to point out, as Wister had, that one cannot conquer that which one fails to confront. As Joseph Le Conte was to stress, "true virtue consists, not in the extirpation of the lower, but in its subjection to the higher. The stronger the lower is, the better, *if only* it be held in subjection" (emphasis in original). The true conflict between man and woman, the true battle of the sexes, was that between the male evolutionary spirit and the female hunger for the bestial satisfaction of her degenerative material passions. It was essential for the Virginian to leave his woman as she lay sobbing on the floor, and to stalk into the dust to kill his opponent, no matter how much weak-willed undifferentiated femininity tried to seduce him into relinquishing his task by means of the soft contours of her warm body. All was well with the world as long as the male kept this measure of control, and under those conditions man was perfectly justified in having his physical fill of woman. As Le Conte had explained, the "higher is nourished and strengthened by its connection with the more robust lower, and the lower is purified, refined and glorified by its connection with the diviner higher, and by this mutual action the whole plane of being is elevated" (375).

For the painters of the period, the struggle between the "higher" and "lower" was a continuing source of intense fascination. However, while the higher occasionally triumphed in their work, their intimate familiarity with the precise contours of "the more robust lower," personified by bare, buxom womanhood, seems to have left them with a rather more pessimistic forecast of the outcome of the struggle between evolution and degeneration than that offered by Le Conte.

Still, the portrayal of man in all his spiritual nakedness, standing on the promontory of his achievement, was an ever-tempting subject. Characteristic is Ferdinand Hodler's depiction of this scene in his "The Look into Eternity" [VII, 3], reproduced in *Jugend* in 1908. In a 1901 entry in his diary the young Paul Klee, beset with culturally inspired dreams of transcendence, had placed himself squarely on Hodler's promontory when he exclaimed ambitiously to no one woman in particular, "Lean on me and follow me; when the abysses gape below, close your eyes. Trust my step and the high, icy spirit. Then we two shall be like God" (*Diaries,* 54).

The sculptor Ludwig Habich placed a "Praying Youth" [VII, 4] on the idealist pedestal. His miraculously perfect specimen of young manhood—one knee still anchored to the earth, the rest of his body inexorably drawn upward by the force of his intense spiritual concentration—was a veritable incarnation of Weininger's intellectual monad at the very moment of his transfiguration from simple material being into pure masculine spirit. Another German artist, the painter and illustrator Hugo Höppener (better known as Fidus) produced a stream of very popular paintings, watercolors, and drawings during the first years of the twentieth century (which he proceeded to market himself in a series of reproductions), which combined the promontory-poised figure of Hodler with the intense upward reach of Habich's youth. Each of these he called a "Lichtgebet," a prayer to the light, to the masculine sun of transcendence. Fidus—who was a "back to nature" enthusiast of the Leni Riefenstahl variety, as well as, in his advancing years, a staunch adherent to the Aryan cause of Hitler and the Nazis—firmly believed in woman's role as no more than the willing seedbed for future supermales. To suit herself to that role, she was to remain as fertile as nature and as yielding as the plants. If, instead, she took on airs and dared to exhibit creative rather than re-creative energies, it was only right that she be stepped upon, even eradicated, by man in his quest for the great light of wisdom in the sky—as one of his designs for a book indicated all too graphically. Fidus also liked to

VII, 3. Ferdinand Hodler (1853–1918), "The Look into Eternity" (1903)

VII, 4. Ludwig Habich (1872–1949), "Praying Youth," bronze (ca. 1912)

portray the upward-striving male as he reached for the stars, his legs encircled by the clinging arms of a naked seductress, who was clearly intent upon keeping him anchored in earthly lust.

Numerous other painters—usually German, sometimes British—explored the image of the Aryan youth in search of spiritual perfection. Ludwig Fahrenkrog's "The Holy Hour" [VII, 5] is characteristic of such productions. A godlike Aryan youth, the finest specimen evolutionary eugenics had to offer toward the production of a future master race, stands in the open fields among the flowers, arms raised in Fidus' fashion to receive the spiritual benediction of the rising sun. Behind him, to the left and at an appropriate distance, we see the respectfully kneeling figures of his doting parents, huddled together in awe of their son's impending triumph over matter. Also well behind our budding *Uebermensch,* to the right, we find the abject, prostrate figure of another man, most likely his brother, who quite obviously has failed to cultivate his masculine intellectual abilities in the requisite manner and is therefore doomed to crawl upon the earth like a mere animal, a virtual slave to his baser passions. His place is rightfully among the women; the nurturing, protective wife; the demurely kneeling yet proudly honorable vestal figure of his virgin sister; and, nearly hidden behind the others (as she should be), the weeping, terrified figure of a Magdalen, who hides her shame by burying her face in her skirts. Fahrenkrog's opus is one of those expertly painted but monstrously maudlin works of bourgeois kitsch/idealism so popular among the mainstream intellectuals of the fin de siècle, that led directly to the very similar productions of the official artists of the Third Reich. When we are faced with works of this sort, we begin to understand why it took so little time for the modernist aesthetic to take hold among the younger painters. What is disheartening is the realization that most of these younger painters, though eager to escape the straitjacket of Fahrenkrog's academicism, remained content to carry that painter's ideological baggage with them into the brave new world of modernism.

The theme of the striving young male and his burdensome female companion, meanwhile, was pursued in France by figures such as Bouguereau, who added a touch of cheerful Gallic levity to the heavy-handed disquisitions of the Germans when he painted his "Cupid and Psyche" [VII, 6]. In the classical myth Psyche is the personification of the human soul, possessing the large, beautiful wings of a butterfly. Cupid (also known as Amor or Eros) is, in this myth, the more lubricious of the two. But Bouguereau adjusted the story to fit late nineteenth-century expectations by making Cupid more godly than Psyche: He gave him the wings of a seraph. In addition, he showed him flying upward in pursuit of a higher love than that represented by Psyche. The latter, her head voluptuously tilted backward in Cupid's embrace, her body limp and exhausted, was given strange, ineffectual, preposterously small wings resembling those of a moth rather than a butterfly. Bouguereau thus made her incapable of independent flight, forcing her to hang onto our godlike boy and weigh him down. A 1901 woodcut by Charles Ricketts illustrating the same myth linked it even more explicitly with the symbolism of male transcendence and feminine interference. Ricketts' "The Flight of Cupid" showed Psyche to be no more than

VII, 5. Ludwig Fahrenkrog (1867–1952), "The Holy Hour" (ca. 1912)

VII, 6. William Adolphe Bouguereau (1825–1905), ''Cupid and Psyche'' (1899)

an archetypal clinging vine clutching desperately at Cupid's leg as, seraphic wings unfolded, he tried to fly from her embrace. Ricketts, in fact, did not bother to give Psyche any wings at all. For him she was no more than a typical turn-of-the-century woman in an unmade bed.

Clutching women who held men down were a staple of the art of the 1890s. Laurence Koe's magnum opus for the Royal Academy exhibition of 1896, entitled ''Venus and Tannhauser'' [VII, 7], paid voluptuous tribute to the work of Richard Wagner, whose operas unquestionably have the dubious distinction of providing the late nineteenth century with the narrative context for many of the details in its iconography of misogyny. (Weininger, for instance, regarded Wagner as ''the greatest man since Christ's time'' [*Sex and Character,* 344].)

Wagnerian and mythological themes were favorite vehicles for the depiction of the struggle between the higher and lower forces driving humanity, between the masculine and the feminine elements, which, as we have seen, were often thought of as warring with each other even in a single body. ''The Soul's Struggle with Sin'' (the title of an 1895 painting by Sigismund Goetze) was to be fought wherever man might encounter woman. The battle between the sexes was heating up. Painters from Gustave Moreau and Emile

Lévy to George Frederick Watts and Franz von Stuck depicted the lamentable fate of Eurydice, whose beauty made Orpheus yearn to look back at her, thereby causing her to fade once more into the material womb of the earth. As a result she became indirectly responsible for the death of this brave idealist, who had vainly struggled against the temptations of the flesh in order to lead woman into the empyrean of thought. For Sir Edward Burne-Jones the familiar myth of Phyllis and Demophoön took on an entirely new, aggressive significance. Instead of the suicidal nymph of ancient myth, she now became more than just a clinging vine: a serpentine creature, a strange, enticing monster of beauty holding a terrified Demophoön in the double grip of her hypnotic, snakelike eyes and her viselike embrace.

John William Waterhouse, who often elaborated on the visions of the British romantic poets of an earlier turn of the century, showed his version of Keats' ''La Belle Dame Sans Merci'' at the Royal Academy in 1893. He depicted her as entwining her prey with the double enticements of her eyes and her hair, the latter serving as a symbolic lasso. Given the period's cliché that long hair was virtually synonymous with mental debility, poets and painters found woman's tresses to be a particularly apt medium for the symbolic depiction of the dangers of the clinging vine. The manner in which women's hair was fetishized in the late nineteenth century is a perfect example of the processes of ''cultural entrapment.'' The mid-century cult of the superfeminine female had led to an ever greater emphasis on golden tresses, with the result that by the 1870s, as Abba Goold Woolson pointed out, long hair having—by males—been ''declared a glory to woman, she heaps upon her head such a mass of heavy, cumbersome braids, and skewers them on with such a weight of

VII, 7. Laurence Koe (active 1888–1904), ''Venus and Tannhauser'' (1896)

metal hair-pins, that, enslaved by this fashion, she can dream of Heaven only as a place where it will be permitted her to wear short hair'' (*Woman in American Society*, 138–39).

Yet, by the end of the nineteenth century this fetish of femininity forced upon her by current fashion came to be seen as a sign of woman's intellectual weakness and her regressive materiality, causing Oscar Wilde to remark snootily, ''To mesh my soul within a woman's hair / And be mere Fortune's lackeyed groom —I swear / I love not!'' (*Poems*, 184) In ''Atalanta in Calydon'' Swinburne has Althaea caution Meleager that it is dangerous to love a strong, belligerent woman: ''Not fire nor iron and the wide-mouthed wars / Are deadlier than her lips or braided hair. / For of the one comes poison, and a curse / Falls from the other and burns the lives of men'' (*Collected Poetical Works*, II, 265). In this same poet's ''Laus Veneris'' woman's hair has virtually become the gorgon's head: ''Ah, with blind lips I felt for you, and found / About my neck your hands and hair enwound, / The hands that stifle and the hair that stings / I felt them fasten sharply without sound'' (I, 22).

The soul-destroying women of poets such as Swinburne had obviously found a formidable weapon in the snakelike flexibility of their golden tresses. The Scottish painter Thomas Millie Dow painted a kelpie [VII, 8]—one of the many water-based temptresses of fin-de-siècle art—apparently in an effort to warn his viewers that men who might be attracted by this lady's demurely alluring nudity had better observe the threatening configuration of her hair, which he painted as if consisting of a knot of slender vipers sliding

VII, 8. Thomas Millie Dow (1848–1919), ''The Kelpie'' (1895)

VII, 9. Evariste-Vital Luminais (1822–1896), "The Flight of Gradlon: A Breton Legend" (ca. 1884)

down over her back to occupy the rock on which she was sitting. In his poem "Divisions on a Ground," Arthur Symons clearly delineated the dangers which were likely to beset any man who allowed himself to get within reach of such tempting tentacles:

> There is a woman whom I hate and love:
> This is my sorrow: she has bound my neck
> Within the noose of her long hairs, and bound
> My soul within the halter of my dreams,
> And fastened down my heart into one place,
> Like a rat nailed upon a granary door;
> And she has gone a farther way than death (*Collected Works*, II, *Poems*, 58).

Hair was not the only lure with which woman attempted to drag the male soul seekers back down to earth. Every feature of her material being could serve, and every legend and historical incident was combed for new suggestions of woman's perverse ability to unseat the male from his lofty spiritual promontory. As the artists explored these sources, they discovered that not even the nuclear family was safe from infestation by evil feminine hangers-on. The French painter Evariste-Vital Luminais found a striking example of this in a Breton legend and promptly painted the incident [VII, 9]. The homely narrative, with its gruesome content, was eagerly explicated in *One Hundred Crowned Masterpieces*, the coffee-table book in which Luminais' opus was lavishly reproduced.

> Gradlon was an ancient king in Brittany, his capital, d'Is, being situated on what is now known as the Bay of Douarnenez, in the province of Finisterre [*sic*]. According to the legend, the city was once surprised and overwhelmed by a sudden rising of the Ocean. There was time alone

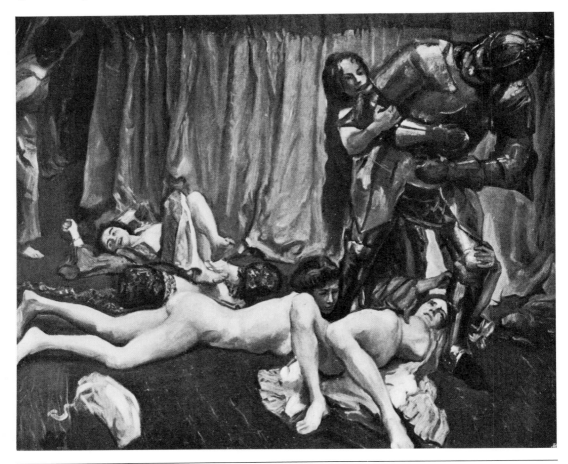

VII, 10. Max Slevogt (1868–1932), ''The Knight'' (1903)

for the swiftest flight. Mounting his horse, with his daughter behind him, and accompanied by Saint Gwenolé, they urged their desperate way through the surging waters. At a critical moment the monk cried: ''Disembarrass thyself of the demon who shares thy saddle, for it is she who by her disorders has drawn down the anger of Heaven.'' The king, recognizing in these words the voice of God, had the courage to abandon his daughter, and was enabled to reach safe ground at a place now called Douarnenez.

Max Slevogt, with his characteristically German excessivism, turned the clinging vine/clutching woman theme into a veritable war zone in his 1903 painting entitled ''The Knight'' [VII, 10], in which an exemplary Wagnerian specimen of medieval chivalric masculine strength and intellectual gentility found himself sorely beset by a remarkably tenacious handful of very contemporary-looking harem beauties who clutched his legs, rode on his back, or just lolled about in overt paroxysms of desire.

These women obviously had completely forgotten that the true role of the clinging vine was that depicted by Herbert Draper in a painting called ''Day and the Dawn-Star,'' exhibited at the Royal Academy in 1906, in which he reestablished the proper pedagogic relationship between man (the sun) and woman (the moon, the satellite). It was yet another image of a strong, seraph-winged male supporting a fainting, clinging, wingless woman.

The verses which accompanied this painting told woman unambiguously what her role in life ought to be: "To faint in the light of the sun she loves, / To faint in his light and to die."

Clearly, notwithstanding all such high-toned directives, woman continued to be perversely unwilling to accept man's well-meant guidance. Solomon J. Solomon's gigantic (it measured eleven by seven feet) "Sacred and Profane Love" made this state of affairs quite explicit. As Henry Blackburn noted in 1889,

> Above on a mountain top an angel shelters with her wings an allegorical group of husband, wife, and child; below in the immediate foreground is a figure of beauty luring a victim to destruction, pelting him with roses while she drags him over a precipice. Between the two contrasting groups a cupid lies asleep. The foreground figures are in lurid light (*Academy Notes*, xix).

The angel of virtue in this painting is not so much feminine as androgynous—a being above material temptation—something that certainly cannot be said of the luridly lit beauty in the foreground, whose violent use of flowers is certainly a far cry from the ideal of the flower maiden who might have walked among or even upon flowers but who would definitely not have used them as missiles with which to stun man's virtue.

The link between woman and the flowers had, under the pervasive influence of Baudelaire, clearly taken on an ominous quality in the minds of fin-de-siècle males. The lily-toting virgin came to be replaced by rows of unruly and definitely unvirtuous—yet all-too-tempting—dandelions and daisies which seemed to blossom shamelessly on the street corners and alleyways of turn-of-the-century culture. To the French painter Marie-Félix Hippolyte-Lucas they seemed to be the daughters of Tantalus [VII, 11]—flowery temptresses of a hothouse variety, tantalizing adolescents, far more easily accessible but far less wholesome than the fruits Tantalus craved in Hades.

The theme of spleen and ideal also popularized by Baudelaire, had by the 1890s become the subject of a host of paintings. Baudelaire's writings, whose seductively beautiful language added immeasurably to the impact his rabidly misogynist point of view exerted upon his contemporaries, had glorified the image of the godlike male poet as early

VII, 11. Marie-Félix Hippolyte-Lucas (1854–1925), "The Daughters of Tantalus (Souvenir of Baiae)" (c. 1902)

as the 1850s. To break free and soar above a world populated with mean mothers and commanding wives was his abiding ambition. In the poem "Benediction," which introduces the "Spleen and Ideal" section of *The Flowers of Evil,* the poet manages to steel himself against the ruthless disgust of his mother. She is a crudely materialistic creature who claims that she would rather have borne "a knot of vipers" than to have weaned such a patsy of spirituality as her son. But

> Toward Heaven, where his eyes perceive a spendid throne,
> Quite unperturbed the poet lifts his pious arms,
> And the vast lightning-flashes of his lucid intellect
> Conceal from him the angry faces of the crowd.

The poet's wife, another virago, who has teamed up with his mother, has other plans for our poet-dreamer. As she stalks threateningly along the public thoroughfares, she cries,

> My nails, quite equal to a harpy's nails,
> Will scratch a ragged road into his heart.
>
> That heart, still trembling like a young bird's palpitating body,
> I'll rip, all red and dripping, from his breast;
> And—a mere morsel for my favorite pet—
> I'll fling it in disdain upon the ground.

Baudelaire, as should be obvious from the passage just quoted, had already gone considerably beyond representations of the clinging woman discussed thus far. His woman was an actual harpy, not merely a sensuous woman without a mind of her own, trying—for practical reasons—to hold her man down in order to keep his phallus available for the sort of worship depicted by Félicien Rops, who showed a nude young woman staring in ardent admiration at a gigantic erection enthroned in splendid disembodied isolation on the altar of her carnal desire.

In the eyes of many fin-de-siècle males, woman had become a raving, predatory beast, a creature who preyed on men out of sheer sadistic self-indulgence. She seemed the sort of regressive monster that in Carlos Schwabe's very Baudelairean "Spleen and Ideal" of 1896 had leapt from foam-crested waves, a poisonous animal with the dugs of an old hag, to twist herself snakelike around the body of a white-winged golden boy-poet who was just about to soar upward to the ideal.

For many of the intellectuals and artists of the years around 1900 it was not enough to portray woman as an empty-headed—or even empty-hearted—burden whose very existence was a regressive influence on man. They wanted to emphasize that she was, in fact, far more dangerous—that in her general characteristics and in the nature of her desires she was closely allied with the animals. If within the world of late nineteenth-century evolutionary advances she were to be allowed to continue to have a voice, progress might still be aborted, no matter how earnestly, in Yeats' phrase, "soul clap its hands and sing." The struggle between man and woman, the battle of the sexes, was a war between the forces of evolution and the emissaries of degeneration. Woman, the intellectuals wanted the world to know, was the Beast of the Scriptures, evil incarnate, an animal—and, worse, a veritable connoisseur of bestiality.

CHAPTER VIII

Poison Flowers; Maenads of the Decadence and the Torrid Wail of the Sirens

Beauty was the striving male's supreme temptation. A woman's downy skin was like the gently caressing echo of a yearning voice. The symphonic incantations of ever newly curving female bodies were like the choral movements of a satanic invitation to worldly abandon. Woman offered melodies of cradled melancholy to the laboring brain of sainted masculinity. Steely-browed and lean-loined Ulysses sailed past these aching calls, seeking financial self-sufficiency among the shoals of vice. The late nineteenth-century middle-class male already knew that Superman's ego was powered by gold. He feared that the Kryptonite of beauty could only weaken the essence of transcendent power he knew to be embedded in his seed. Even the thought of a strong woman with a will, a mind, and wishes of her own was enough to weaken the musculature of a selfhood nourished by the bitter herb of monetary gain. The ardor of man's will to power seemed to shrivel into insignificance before the tumescent homage of his body to the wonder of a woman radiant with life and unmoved by the commands of cash.

Much of the period's fascination with woman as the embodiment of evil was the logical outcome of a cultural environment in which the evolving male was expected to combine an attitude of socioeconomic belligerence with an ideal of personal continence in the service of worldly success. Thus, distrust of all others in the area of personal achievement came to be conjoined with a glorification of the virtues of sacrifice and abstinence. What the middle-class male knew about women inevitably came first of all from observing his mother and, if he had any, his sisters. But that mother was all too often only a shadowy creature and frequently she seemed too fragile and ill, too saintly and weak, too "ideal" to be approachable by an awkward adolescent. She was indeed often the sainted creature adored by Zola's Father Mouret, the woman who became "a Vessel of Honor chosen by God, the elected Bosom into which he wanted to pour his being and sleep forever more" (77).

At the same time, the women of his own age he might encounter—his sisters, her friends—would not at all tend to live up to the ideal transmitted to him by his father, whose mid-century generation had created the image of the household nun. Frightened, fascinated, feeling strange, unjustifiable erotic stirrings which interfered with his school-work and even with his surreptitious reading of the sensationally informative writings of the new breed of naturalists and bio-sexists, he would, more often than not, enter adult-hood full of strange scientific knowledge about the perversions to which woman's nature was prone, without ever having had more than a halting conversation with any of these creatures of mystery.

In his *Diaries* Paul Klee has given a candid record of what, in 1900, it was like to be a twenty-one-year-old single male with idealistic aspirations. Temptation and renuncia-tion struggled very fiercely within him. "I tortuously reasoned myself free of woman, but could not free myself from the dreamy look of young girls" (51). In 1920, the novelist Arnold Bennett described the male intellectual's plight in dealing with women: "He has too frequently either avoided them in a regrettable cowardice or an equally regrettable disdain, or has conducted his sexual relations with gross clumsiness and deplorable lack of comprehension. Titans of intellect whose labours influence the destiny of the human race have never kissed, and when confronted with a pretty woman have not known what on earth to do" (*Our Women*, 14).

During their adolescence, these men were generally not told anything at all about sexual relationships by their parents. If such instruction did take place, it most likely took the form of cautionary elaborations on the parameters of Nicholas Cooke's arithmetic of justifiable erotic exercise: "From once to thrice a month may be stated as a fair average frequency for the indulgence during the comparative youth and health of both parties, and when no circumstances exist to render abstinence a necessity" (*Satan in Society*, 150). Ironically, while they were being told these pious bits of hygienic wisdom, they were also, in the same books, reading all about the gruesome perversions to which woman was prone. And then there were Baudelaire, Zola, and the delectable ideal nudes of Bouguereau to confuse matters further. Klee recorded the effects of all those contradictory impulses on an intelligent and basically sensitive young adult: "Sexual helplessness bears monsters of perversion. Symposia of Amazons, and other horrible themes. A threefold cycle: Carmen-Gretchen-Isolde. A Nana cycle. *Théâtre des Femmes*. Disgust: a lady, the upper part of her body lying on a table, spills a vessel filled with disgusting things" (emphasis in orig-inal) (55).

In response to the dizzying contradictions of the commonplace, intellectual mysticism became the order of the day throughout Europe. Theosophy, Rosicrucianism, millenarian philosophies, and even elaborate theories, inspired by Poe, concerning the immortality of the soul as a material entity attracted the devoted attention of the educated middle classes and artists everywhere. Man's ambition was to rise to levels of divinity never dreamt of before, and many thought these levels might be achieved by their own generation.

In addition, the doctrines of Spencer and Darwin had fused to reflect the imperialist mentality in politics. The struggle for life was everywhere seen as a license to dominate the weak. Nature was no longer viewed as a benevolent, guiding force but as a rather primitive stage of experience to be overcome by those in the forefront of the evolutionary struggle. Nature's procedures for encouraging evolutionary development were, after all, at best haphazard and wasteful. In his famous essay of 1881 entitled "Sociology," William Graham Sumner had declared that "life on earth must be maintained by a struggle against nature" (*War and other Essays*, 173). The human brain, the most advanced product of evolution, was in a position to select the environmental conditions best suited to its own further development. In making this selection, it would literally be necessary for man to struggle to overcome the limited world of natural reality by creating in its stead a better-than-natural environment which was itself an emanation of the more perfect, integrative capacity of the human intellect.

Des Esseintes, the hero of Joris-Karl Huysmans' influential 1884 novel *A Rebours* (a title itself sometimes translated as *Against Nature*), remarks that "nature has had her day." There is, he continues, "no denying it, she is in her dotage and has long ago exhausted the simple-minded admiration of the true artist; the time is undoubtedly come when her productions must be superceded by art" (*Against the Grain,* 22). It is certainly true that Des Esseintes is here referring specifically to art, and is voicing the sentiments of the art-for-art's-sake aesthetes of the late nineteenth century. But the art-for-art's-sake aesthetic was a very deliberate expression of the evolutionary sentiment applied to the field of art. The artist, it was felt, needed to construct an environment of humanly created—artificial—beauty and intellectual purity which would provide an appropriate context in which the higher moral perceptions of the spiritually evolved vanguard of humanity could thrive.

Women had a role in the creation of this new, elevated sphere of beauty, but only insofar as they permitted their undeniable visual attractiveness to be used and, where necessary, restructured by the artist as part of the creation of an ideal aesthetic environment which could then serve as a stimulus for further explorations into the realm of perfect beauty. However, it was important to realize that, being deficient in the capacity for individuation, woman could only serve "as a receptacle for projected worthiness," as Weininger maintained. The discovery of ideal beauty in a woman was a creative act on the part of the artist, and was by no means expressive of inherent value in woman herself. It was simply a matter of man's "imposition of the ideal on her personality" (*Sex and Character,* 245).

If woman had only shown enough sense to remain content with her role as the passive human clay which man could mold according to his fantasies, to develop his perceptions concerning the structures of ideal beauty, everything would have been well. But she had remained a mere animal beneath the veneer of civilization with which the poetic spirit of man had covered her, and was, as we have seen, even prone to atavistic retrogression. Lodged in the earth, needing the earth—being, indeed, the very personification of that moist, fertile earth—woman was like a swamp, a palpitating expanse of instinctive physical greed whose primary natural function was to try to catch, engulf and, if possible, absorb the male and make him subservient to her simplistic physical needs. Woman, as the embodiment of nature, was therefore continuously at war with man, whose very purpose was to go against or beyond nature. No longer the personification of the warm, inviolate womb of domestic bliss, of motherly self-negation, woman became the womb of the earth, the all-encompassing, all-absorbing, indiscriminate receptacle of masculine vitality, the dark grotto of physical temptation opening mysteriously and wide before the terrified spiritual adolescence of man.

Woman was earth incarnate, and earth was the body of woman. According to Zola's rapacious imagination, the fertile earth was the deep cavern of carnal knowledge to which Albine lured Serge Mouret. The earth brought to this writer's art-filled mind the very image of the nymph with the broken back: "An immense green head of hair, ornamented by a storm of flowers, whose locks fell everywhere, escaped in insane dishevelment, created dreams of a swooning young giantess thrusting her head backward in the convulsions of her orgasm, a stream of gorgeous hair rising like a pond of perfumes'' (128–29). Zola made his tempted man of the cloth the vehicle through which he could express what were obviously his own most feverish yearnings, bringing together woman as moon, earth, and desperate temptress in what must surely be one of the purplest passages in all of literature:

> The vast plain spread out before him, more tragic under the moon's oblique paleness. Olive trees, almond trees, slender trees, stood as gray spots amid the chaos of the great rocks stretching to the dark line of the hills on the horizon, large splotches of shadow, broken ridges, bloody spots of ground where the red stars appeared to be staring at one another; white chalk spots became the scattered clothes of rejected women whose flesh was drowned in shade as they lay dormant in the country's hollows. At night, this ardent country assumed the tortured arch of a woman consumed by lust. She slept, but the covers had been thrown aside; she swayed, twisted, passionately spread her legs, exhaled in great warm breaths the powerful smell of a beautiful, sleeping woman dripping with sweat. It was like some strong Cybele fallen on her back, her breasts outthrust, her belly under the moon, drunk with the sun's heat, forever dreaming of impregnation (*The Sin of Father Mouret*, 90–91).

That, of course, was the real problem. Woman was Cybele, the goddess of fertility, mistress of brute nature, casual companion to the lion, and a creature to whom, in classical antiquity, the genitals of bulls had to be offered to assuage her insatiable hunger for seed. Even today she still was that ''bottomless pit of nature'' Maxim Gorki's Artamonov saw loom as he watched Paula Menotti dance out her ritual of seduction, that mythical ''patch of greensward in swampy forest country, where the grass grew particularly silky and green, but if you set foot on it you were lost, sucked into bottomless slime'' (*The Artamonov Business*, 207–8). Her hunger was endless, her fertility was all-encompassing, she was Diana of Ephesus, the idol with the countless breasts, portrayed time and again by Fernand Khnopff and other symbolist painters. She was the moon-circle, absorber of the sun, consumer of the souls and bodies of men. She was both source and ''symbol of the mystery of the cruel impassiveness and wastefulness with all her children—men and monsters, victors and beasts—of Nature, the natural fecundity of the earth,'' as William Walton, in *The Chefs d'Oeuvre of the Exposition Universelle* (1900), described Aristide Sartorio's sensational portrayal of the destructive, mind-blistering goddess [VIII 1].

These artists' depiction of Diana as the many-breasted idol of promiscuous, wasteful, earthbound fertility was again at least partially inspired by the scientific research of the evolutionists. Darwin, for instance, had attributed ''the not very rare cases of supernumerary mammae in woman to reversion.'' For him this was an excellent and interesting instance of the existence of a clear link between woman and her progenitors in the animal world. Thus once more science gave the artists of the turn of the century a chance to update the images provided by classical mythology. The many-breasted Diana of Ephesus, the pagan fertility idol, became the scientifically sanctioned symbol of woman's atavistic tendencies. She was the perfect visual image of feminine reversion, her ''additional mammae being generally placed symmetrically'' on her torso by the painters just as Darwin had prescribed (*The Descent of Man*, 41).

Flaubert had also provided useful material for the visual exploration of this theme; indeed he had served as the specific inspiration for Sartorio. In *The Temptation of Saint Anthony* he described how the idol appeared to his long-suffering hermit:

> Lions crouch upon her shoulders; fruits, flowers, and stars cross one another upon her chest; further down three rows of bosoms exhibit themselves, and from the belly to the feet she is

VIII, 1. Aristide Sartorio (1860–1932), "Diana of Ephesus and the Slaves" (ca. 1899)

caught in a close sheath, from which sprout forth, in the centre of her body, bulls, stags, griffins and bees. She is seen in the white gleaming caused by a disc of silver, round as the full moon, placed behind her head (241).

Walter Pater's *Marius the Epicurean* (1885) described how Diana had come to be seen—not so much in the mock-classical environment which provides the setting for the novel as in Pater's own late nineteenth-century intellectual context: "She is the complete and highly complex representative of a state, in which man was still much occupied with animals; not as his flock, or as his servants (after the pastoral relationship of our later orderly world), but more as his equals, on friendly terms or the reverse—a state full of primeval sympathies and antipathies, of rivalries and common wants." The "humanities" of the worship of Diana, says Marius, are "all forgotten today," and in Pater's conjoining of the classical world and that of the 1880s, he remarks that in "ritual," in the art of the time, one might say, "it was as a Deity of Slaughter—the Taurian goddess, who requires the sacrifice of shipwrecked sailors, thrown on her coasts—the cruel, moon-struck huntress, who brings not only sudden death, but *rabies,* among the wild creatures, that Diana was to be presented" (emphasis in original) (I, 255).

It was thus as the new, cruel Diana, the creature of animal impulse, that woman sought to separate the male from his godlike capacity for spiritual transcendence and drag him back into her regressive realm of mere physical existence. The only part of male being

VIII, 2. Thomas Theodor Heine
(1867–1948), "The Flowers of Evil"
(1896)

that woman truly understood was his sexuality, and it was therefore through this least effectively evolved element of his being that she tried to reach him and drag him back down to her own low level on the evolutionary scale. Thus woman became a nightmare emanation from man's distant, pre-evolutionary past, ready at any moment to use the animal attraction of her physical beauty to waylay the late nineteenth-century male in his quest for spiritual perfection. She was nothing but a golden fly, a blind force of nature whose very manner of doing things was likely to betray her link with the animals. Zola's Nana, for instance, has a "mane of loosened yellow hair" which "covered her back with the fell of a lioness." She is a "lewd creature of the jungle," possessing an "equine crupper and flanks" (223). As a matter of fact, she has such a close sense of her connection with horses in general that when she goes to the races and a horse with her name is about to win, she literally becomes the horse. "On the seat, without realizing what she was doing, Nana had started swaying her thighs and hips as if she were running the race herself. She kept jerking her belly forward, imagining that this was a help to the filly" (*Nana,* 376)

Since woman, being a tool of nature, had a voraciously predatory hunger for man's seed, since she was a veritable syphon of regression ready to gorge herself on that "great clot of seminal fluid" of man's brain, it is not surprising that she appeared everywhere, and in every imaginable guise, to tempt man into perpetual and desperate tumescence. That is why the favorite mid-century association of woman with flowers, which had originally been intended as a tribute to woman's virginal fragility and purity, now began to take on the tantalizing ambiguity of Hippolyte-Lucas' street-corner daisies. Baudelaire had already subverted the analogy's positive symbolism by referring to women as flowers of evil, and now the link between the flowers and the earth opened the chasm of sexual hunger between woman and the ideal of her floral inanity. In 1896 Thomas Theodor Heine added a racist dimension to this sexist equation by showing a black man as the motivating force behind a white woman's desire to pluck the sterile flower of sensuality. Wishing to make its link with the thought of Baudelaire explicit, he called this painting "The Flowers of Evil" [VIII, 2].

Man began to realize that to surround himself with flowers was to surround himself with the temptation of woman's orgiastic potential. Indeed, it became clear that some women—dared one mention it?—were fertilized by and could even be said to copulate with flowers! Take Zola's Albine:

> Under her fingers, it was raining roses, wide tender petals possessing the exquisite roundness and barely blushing purity of a virgin's breast. Like a living snowfall, the roses already hid her feet crossed in the grass. Roses came to her knees, covering her skirts, drowning her to her waist; while three stray rose petals, blown to the beginning of the valley of her bust, seemed to be three bits of her adorable nudity (*The Sin of Father Mouret*, 120).

Once again the restless search of turn-of-the-century painters for decorous ways to express—and feed—the overheated erotic imagination of the well-to-do middle-class male found a newly fetishized outlet: Woman as flower became a nightmare vision of woman as a palpitating mass of petals reaching for the male in order to encompass him, calling to him to be drained by her pistils yearning for fertilization. This idea was characteristically expressed in 1899 by Maximilian Lenz in his in intention deeply serious but, in retrospect, unquestionably hilarious portrayal of a thoughtful turn-of-the-century intellectual strolling through flowery fields in springtime, his mind quite obviously occupied with the exalted future of man, only to be beset by fiercely frolicking maidens and, worst of all, a bevy of transparently smocked young ladies carrying—indeed, becoming—huge stalks of blossoms. These decidedly daffy damsels personified tempting floral finitude as they urged the serious thinker to lose himself among their panting petals [VIII, 3].

VIII, 3. Maximilian Lenz (1860–1948), "A Daydream" [or "A World"] (1899)

The fin de siècle's fascination with the eroticized flower, expressed in countless images of this sort, had been fed by a number of sources, not least among them—as Claude Phillips pointed out in discussing Georges Rochegrosse's painting "The Knight of the Flowers" in *The Magazine of Art* in 1894—"the scene in Wagner's *Parsifal,* in which the boy-hero victoriously resists the allurements of the flower-maidens." Phillips pointedly called Rochegrosse's bevy of petal-headed, pistil-packing mamas "land-sirens," thereby clarifying the link between the painters' bouquets of bare-breasted landlubbers and the period's favorite denizens of the sea.

Zola, in his description of the erotic encounter of Serge Mouret and Albine among the flowers, was another principal source for the petal-pushing painters of the years around 1900. The novelist's account of this event succeeded in turning a bower of flowers into a house of ill repute:

> The living flowers opened like naked flesh, like bodices revealing the treasures of breasts. There were yellow roses shedding the gilded skin of savage girls, straw roses, lemon roses, roses the color of the sun, all the subtle shades of necks bronzed by burning skies. Then the flesh softened, tea-colored roses hinted at exquisite moistness, spread hidden modesty, displayed the parts of the body which are not shown, possessing silken smoothness, made slightly blue by the body's veins. Next the laughing rose-life blossomed out: rose white, barely tinted by a spot of lake, white like a virgin's foot dipping into the water of a spring; pale rose, more discreet than the hot whiteness of a half-glimpsed knee, than the light with which a young arm illumines a wide sleeve; frank rose, blood under satin, naked shoulders, naked rosebuds of a woman's breasts, half-open flowers of her lips, exhaling the smell of warm breath. And climbing rosebushes, large bushes raining white petals, dressed all these roses, all this flesh, in the lace of their clusters, in the innocence of their thin chiffon; while here and there, wine roses, almost black, bleeding, broke into this bridal purity with a passionate wound. Marriage of the scented forest leading May's virginity to the fertility of July and August; first ignorant kiss, picked like a flower on the wedding morning (*The Sin of Father Mouret,* 122).

It should be obvious that when a flower could become a temptress, when the virginal lily could turn into a hot-petaled rose of desire, the flower of evil was most likely only the initial manifestation of woman's chameleonic eroticism.

The sex-obsessed turn-of-the-century male could draw little solace from the remarks of a progressive thinker such as Edward Carpenter, who tried hard and, on the whole, honorably to serve as intermediary in the war between men and women. Carpenter, who at least advocated a world of experience consisting of more variation of thought and action than the prevailing dualistic mindset would allow, nonetheless helped fan the fires of masculine disdain and fear of the feminine by perpetuating the bio-sexists' myths about woman's nature. Carpenter, it is fair to say, was himself negatively influenced by Havelock Ellis' willing acceptance of the notions of Carl Vogt and Charles Darwin about woman's primitive, retentive character. Ellis had echoed the notions of these worthies in his historically important and, at the time of its first appearance, unusually liberal-minded *Man and Woman* (1894). Carpenter, having read Ellis, reiterated in his influential book *Love's Coming of Age* (1896) notions he had already expressed more expansively two years earlier in his pamphlet *Woman, and Her Place in a Free Society.* In addition to a number of rather sensible suggestions for reform, it also contained the by now standard contention that woman lived exclusively for and through sex. Said Carpenter:

> In woman the whole structure and life rallies more closely and obviously round the sexual function than in man; and as a general rule, in the evolution of the human race, as well as of

the lower races, the female is less subject to variation and is more constant to and conservative of the type of the race than the male. With these physiological differences are naturally allied the facts that, of the two, Woman is the more primitive, the more intuitive, the more emotional. If not so large and cosmic in her scope, the great unconscious processes of Nature lie somehow nearer to her.''

For Carpenter, woman was also much less capable of differentiating between spiritual passion and lust, an ability which he considered characteristic of males ''and which causes them to be aware of a grossness and a conflict in their own natures.'' Woman, in general, was blissfully unaware of such moral conflicts, ''since in a way she is nearer to the child herself, and nearer to the savage.'' Trying to be liberal but, by his own admission, sounding awfully like Michelet (though with the flavor of the jungle added for spice), Carpenter concluded that woman should be seen as the point of equilibrium to which ''Man, after his excursions and wanderings, mental and physical, continually tends to return as to his primitive home and resting-place, to restore his balance, to find his centre of life, and to draw stores of energy and inspiration for fresh conquests of the outer world'' (*Love's Coming of Age*, 39–40).

With even such well-intended would-be allies as Edward Carpenter confirming the turn-of-the-century male's suspicions concerning woman's primitive nature and her similarity to the savages, it is no wonder that many men began to suspect her of indulging in strange sylvan rituals, of having, for instance, an insatiable urge to dance. Few of Carpenter's contemporaries could see her any longer as the weary man's safe harbor, for she seemed too intent on flinging herself about at every opportunity in ecstatic forms of primitive movement. The scientists naturally had a good explanation for woman's insatiable need for rhythmic motion. Harry Campbell, for instance, had discovered in dance still another ''form of bodily activity in the liking for which the civilized woman more resembles the child than the civilized man.'' Woman's predilection for this form of movement was not surprising since ''the child and the savage are both very fond of dancing.'' Hence it ''constitutes a form of excitement in harmony with woman's emotional temperament.'' Getting suitably technical, Campbell suggested that

> the dancing epidemics of the Middle Ages are obviously related to the war and other ceremonial dances of the savage. The movements of such dances tend to excite strong emotion and to arouse enthusiasm; for just as the various emotions tend to call forth particular movements, so, contrariwise, do the movements expressive of an emotion tend to call forth that emotion. Thus the movements of these wild dances imperceptibly shade off into the co-ordinate movements of the hysterical fit—into ''hysteroid'' movements, to use a term originated, I believe, by Sir W. Roberts, and adopted by Gowers. Hence it is possible that the love of dancing, so peculiarly strong among women, is the outcome of a nervous organisation affording a suitable soil for hysteria. The dancing affords an outlet for pent-up energy (*Differences in the Nervous Organization of Man and Woman*, 169).

The culprit behind woman's uncontrollable urge to dance was her nervous organization, her tendency to hysteria. Now there certainly were few educated readers who did not know that hysteria was woman's way of giving vent to the chronic ''hyperaesthesia of the senses'' to which she was prone. As Arthur Symons had declared, ''the modern malady of love is nerves'' (*Selected Poetry and Prose*, 48). Even the term *hysteria*—the Greek word for womb turned into the designation of a malady—was linked to feminine sexuality. And although Havelock Ellis, in *Man and Woman*, cautioned against ''the mistake of supposing that there is some special connection between hysteria and the sexual organs,'' he nonetheless posited that the error had ''probably arisen from the undoubted fact that in women the organic sexual sphere is of greater extent than in men.'' Ellis conceded that the tendency of earlier scientists to associate hysteria with ''sexual irritation in any crude form, or any gross disease of the sexual organs,'' was probably exaggerated, but he insisted that ''many of the symptoms of hysteria can be traced back to a sexual origin,'' and he quoted with approval Clouston's definition of the malady as ''the loss of the inhibitory influence exercised on the reproductive and sexual instincts of women by the higher mental and moral functions'' (326). If anyone still doubted the link between dance and hysteria established by Campbell (and many others before him), Ellis made certain it would become obvious. ''One reason why women love dancing is because it enables them to give harmonious and legitimate emotional expression to this neuro-muscular irritabilty which might otherwise escape in more explosive forms'' (355).

Dancing women, then, were clearly sex-crazed women. The hordes of girls who hopped about decorously in all those painted meadows featured at the yearly exhibitions had a transparent secret: They were overheated hysterics letting off steam. Often they danced quite decorously, almost as if they were simply boarding-school girls on an outing who were letting their energies run free to prevent anything more untoward from happening on their return to their dormitories. Clad in flounces and flowers, their tendency to skip in circles around a maypole to acknowledge their allegiance to Maia, goddess of spring and fertility—a practice actually institutionalized by a number of progressive fin-de-siècle schools for girls—was, in such cases, the only outward indication of the dangerous fire in their loins.

Frequently, however, the painters appear to have encountered these young ladies in dark corners of the forest or in otherwise deserted fields, after they had unceremoniously taken off their clothes in tribute to Mother Nature. In such cases the ever-amorous Pan (symbol of the regressive, bestial, Dionysiac residues in the male) was likely to be found

VIII, 4. F. Humphry Woolrych (1864–1941), ''The Music of Pan'' [detail] (ca. 1900)

VIII, 5. Johann Victor Krämer (1861–1949), "Nymphs Dancing" (ca. 1895)

among them, having taken up his flute to urge his eager female followers on to bouts of reckless dancing. Few women, apparently, were able to resist such musical enticement. They would respond to Pan's notes with an uncontrollable panic of movement, as the flamboyant Australian-born American painter F. Humphry Woolrych documented in a richly textured visual—yet almost audibly musical—exploration of this theme [VIII, 4].

Not all established fin-de-siècle painters were as successful in combining elements of sound and fury as Woolrych. Most, like Johann Victor Krämer, relied on the lessons of the academy to convey the swirling, suspect passions of women in the throes of dance. In a painting of nymphs frolicking [VIII, 5] Krämer intimated that woman's proper place on the evolutionary scale was with the goat-footed boy-satyr of ancient myth. These are very contemporary fin-de-siècle women, lewd and out of control, who have chosen to be the companions to a single, half-bestial child-male representative of the pre-evolutionary stage of man's development.

As Krämer's painting also indicates, the nymphs of the woods and streams usually chose not to dance solo in Woolrych fashion. Most frequently they caroused hand in hand, forming a circle, as if to emphasize their coextensive nature and their fundamental uroboric lack of differentiation, which testified to the basic, primitive self-sufficiency of their erotic fervor. In Maximilian Lenz's study of the sorely beset intellectual strolling in a meadow, these women can be seen in the background, in various stages of transport, one flinging herself so fiercely backward that she might be in danger of pulling a muscle.

A similar mishap—or, worse, a possible broken back—seems to be in the offing for the impetuous nymph who abandons herself quite unwisely to the lure of movement in Robert Blum's mural for the Mendelssohn Hall Glee Club in New York [VIII, 6]. In Emil Vloors' "The Ring Dance" [VIII, 7] such abandonment has caused one of the ecstatic

VIII, 6. Robert Blum (1857–1903),
"The Dance" (detail of a mural decoration) (1890s)

participants to become completely airborne. Henri Matisse's famous paintings of women
dancing in a circle [VIII, 8] unquestionnably represent an innovation in style, but in terms
of subject matter they remain decidedly in the mainstream of the period in which they
were painted.

Once again the lure of the adolescent led to the adaptation of the theme to the representation of prepubescent girls, as in Francesco Gioli's once much-admired contribution to
the Venice Biennale of 1907 [VIII, 9]. It is probably also a factor in Cléo de Mérode's
success as a dancer that she had childlike features. In such productions as her ballet "Phryné,"
which in its very title promised the titillation of adolescent erotica, she exploited this
element quite knowingly. Indeed, turn-of-the-century men adored the stage spectacle of a
woman who lapsed into self-induced fits of orgiastic transport—and all in the name of art.
What could be more intriguing than to watch a woman, safely isolated from the audience,
revert publicly to the "savage" source of her being? In 1902 Paul Klee recorded his
impressions of "La Belle Otéro," another of the famous dancers of the period:

> At first she sang in a rather poor voice, posing in exquisite attitudes. When she started playing
> the castanets she seemed unsurpassable. A short, breathless pause, and a Spanish dance began.
> Now at last the real Otéro! She stands there, her eyes searching and challenging, every inch a
> woman, frightening as in the enjoyment of tragedy. After the first part of the dance she rests.
> And then mysteriously, as it were autonomously, a leg appears clothed in a whole new world
> of colors. An unsurpassably perfect leg. It has not yet abandoned its relaxed pose, when, alas,
> the dance begins again, even more intensely. The pleasure becomes so strange that one is no
> longer conscious of it as such.

Klee thought that from this dance "of an orgiastic character" a budding artist could learn
much about "the complication of linear relations" (*Diaries,* 85), and it is certain that the

VIII, 7. Emil Vloors
(1871–1952), "The Ring
Dance" ("La Ronde")
(ca. 1914)

VIII, 8. Henri Matisse
(1869–1954), "The
Dance" (1909)

VIII, 9. Francesco Gioli
(1846–1922), "Ring
Dance on the Tyrrhenean"
(ca. 1907)

visual legacy of art nouveau would be sorely diminished were it not for its many drawings and sculptures of dancers in ecstasy. These creations were usually based on the famous serpentine dances of the American-born Loïe Fuller, who thrilled the Paris of 1900 with her butterflylike stage perambulations, strategically lit by variously colored spotlights. Klee's teacher Franz von Stuck saw evidence of woman's mirrorlike reduplication in it all and painted images of dancers whose movements echoed their lack of differentiation. These images, in turn, inspired the linear lucubrations of his pupil. Isadora Duncan moved around onstage diaphanously dressed in Alma-Tadema–inspired classical garb, carrying palm fronds in her hands and balancing a basket of flowers on her head. She was known to bring the equally diaphanously dressed adolescent pupils of her dance school with her onto the stage. What ecstasy! What art! The movements of the celebrated modern dancers brought their audiences an entirely new spectrum of creative expression, but to the lubricious males of the period their dances gave further evidence of the hysterical eroticism and childlike nature of woman. While the women who danced rightfully saw themselves as artists liberated at last from the trivialities of woman's fate, Arthur Symons and his friends saw ''The Maenad of the Decadence'' (*Lesbia,* 28) in the stage dancer. In men's eyes modern dance was indeed the ''Sarabande of the Barbarians,'' as Paul Renouard entitled an engraving of a group of madly circling ballerinas.

The inevitable outcome of all this frantic dancing was to the scientists a foregone conclusion: sheer physical exhaustion—the same sort of exhaustion as that which was thought to be the result of hysteria and neurasthenia—the ''congenital exhaustibility of women,'' as Ellis put it, following the French pathologist Féré. Said Ellis, ''as among children, savages, and nervous subjects [women's] motions and their emotions are characterized by a brevity and violence which approach to reflex action (*Man and Woman,* 358–59). Thus, not even woman's ecstatic immersion in the supposed autoerotic pleasures of the dance was much more than the motor spasm of a mobile automaton. No wonder the dancers in such exalted documents of antique erotic activity as Sir Lawrence Alma-Tadema's mock-classical machine entitled ''The Women of Amphissa'' (1887) became virtual catalogues of late-Victorian fantasies about the nature of feminine sexuality. Alma-Tadema's painting featured, in Cosmo Monkhouse's words, ''a wandering troupe of Bacchantes lying in every attitude of exhausted nature in the market-place'' (*British Contemporary Artists,* 208). Oscar Seyffert, whose *Dictionary of Classical Antiquities,* first published in 1882, became a major iconographic source for the late nineteenth-century's representation of classical themes, had described these prostrate lovelies as ''inspired women, the *Maenades* or *Bacchantes,*'' followers of Dionysus, god of the earth, who engaged in rites which ''were wild even to savagery,'' and who ''wandered through woods and mountains, their flying locks crowned with ivy or snakes, brandishing wands and torches, to the hollow sounds of the drum, and the shrill notes of the flute, with wild dances, and insane cries and jubilation'' (191–92). It must have been quite a relief to the Victorians to be informed by science that if these women—indeed, all women if they escaped from the armed camp of civilization—tended to become savage and wild, and prone to fierce bouts of dancing, they were at least also prone to rapid exhaustion, thus making it possible to depict them, as in Alma-Tadema's opus, in decorously passive poses rather than while they were still desperately determined to pursue the dignified doctors of evolutionary theory into their libraries.

It therefore now becomes clear that the nymph with the broken back was most likely a maenad of the decadence who had fallen victim to the dangers of dances in which she had engaged primarily to relieve her unstilled, hysterical hunger for man's precious essence. The term *nymphomaniac,* to designate a woman with uncontrollable sexual desires, is one of those typical discoveries after the fact about the inherent characteristics of women with which nineteenth-century psychological research is so rife. First male desire created the image of the insatiable nymph, then the scientists discovered her existence as a medical fact.

Not surprisingly, the term came into general use during the 1860s to describe the "abnormal" interest of certain women in sexual gratification. At a time just before the cultural rediscovery of feminine sexuality, the term became a means for the scientific community to explain the astonishing atavistic reversion of contemporary women, whose sexual instincts should properly have atrophied in the midst of an advancing civilization and according to the laws of evolution. Instead, these women were returning to practices and desires chronicled in detail only by the historians of pagan societies. Nymphomania became a convenient and graphic term to describe a "syndrome within the sphere of psychical degeneration," as Krafft-Ebing defined it in *Psychopathia Sexualis* (322), which, in the words of Bernard Talmey, his enthusiastic American follower, was characterized "by irresistible exaltation and an insatiable appetite for sexual gratification." Warned Talmey; "At the mere sight of any man the nymphomaniac woman gets into such a state of excitement that without any tactile manipulation whatever she may experience real orgasm" (*Woman,* 113).

We also learn from Talmey that masturbation was a common cause of this frightening disease. The nymph with the broken back, it turns out, was obviously also a boarding-school inmate who, after some modeling sessions for paintings of the collapsed woman, had wandered into the primeval woods. No longer satisfied with a course of solitary self-gratification, she had gone in search of whatever male she might happen to come across, presumably dancing while she wended her way. As chronicled by Talmey, even the thought of a man could drive her to such ecstasy that, failing anything better, she would simply fling herself upon the forest floor in the throes of desire, twisting and turning with such abandon that in her very excitement her spine would snap.

Such incidents were by no means thought to be beyond the realm of possibility. Following the late nineteenth-century researchers' incestuous habit of quoting other authorities liberally and unquestioningly in matters for which they could not cite incidences of personal observation, Krafft-Ebing reported on a case involving a young girl who "became suddenly a nymphomaniac when forsaken by her betrothed; she revelled in cynical songs and expressions and lascivious attitudes and gestures. She refused to put on her garments, had to be held down in bed by muscular men (!) and furiously demanded coitus. Insomnia, congestion of the facial nerves, a dry tongue, and rapid pulse. Within a few days lethal collapse" (322). And from another source Krafft-Ebing culled a no less astonishing example: "Miss X., aged thirty; modest and decent, was suddenly seized with an attack of nymphomania, unlimited desire for sexual gratification, obscene delirium. Death from exhaustion within a few days" (*Psychopathia Sexualis,* 322–23).

August Forel, in *The Sexual Question,* defined nymphomania as "sexual hyperaesthesia," and also pointed to the autoerotic origins of this terrible condition. "I have observed," he said,

> in women two very different varieties of sexual hyperaesthesia. In one, true nymphomania, the subjects are attracted toward man bodily and mentally with an elementary force; in these the whole brain follows the appetite in quite a feminine manner. Other women, on the con-

trary, are driven to masturbation by a purely peripheral excitation; they have erotic dreams with venereal orgasms which torment rather than please them; but they do not fall in love easily, and may have difficulty in the choice of a husband. Their mind alone remains feminine, full of tact and delicacy in its sentiments, while their lower nerve centers react in a more masculine and at the same time more pathological manner.

Though it was not at all easy to spot a nymphomaniac at first sight, there were nonetheless clear public indications: "Nymphomaniacs often have polyandrous instincts, and they then become more insatiable than men." Indeed, warned Forel, "when a woman is possessed by passion she often loses all sense of shame, all moral sense and all discretion, as regards the object of her desires" (266–67).

Under such conditions, as Horace Bushnell had already noted in 1869, the veneer of civilization would flake off women's elegant surface with blistering speed. "Women are a great deal more violent, constitutionally speaking, than men; the very delicacy of their nature makes them so," Bushnell insisted with true dualistic consistency. "When the charities of a womanly nature are burned out, and nothing is left but spleen or frenzied passion, we have a spectacle both sad and frightful" (132–33). In truth, he argued, "women often show a strange facility of debasement and moral abandonment, when they have once given way, consenting to wrong. Men go down by a descent—*facilis descensus*—women by a precipitation" (*Women's Suffrage,* 142)

"Nymphomania transforms the most timid girl into a shameless bacchante" (296), warned Lombroso and Ferrero. Clearly, the nymphs, sirens, and maenads of late nineteenth-century art were the visual expression of a heady mixture of wish-fulfillment fantasies, fear, horror, hope, and revulsion crowding the nineteenth-century male mind. This mélange of elements also spoke loudly through the bio-sexists' disquisitions upon the bottomless pit of woman's sexual nature. At the same time, the painters' pursuit of these mythical, sex-starved creatures legitimized the average male's fantasies about "the wild women" who assumed a masculine, aggressive sexual role and were suspected by everyone of prowling about even in the drawing rooms of society—those outrageous feminists about whom one heard so much and of whom, unfortunately, one saw so precious little.

Still, the lurid masculinizing, amoral desires of the feminists, those frightening viragoes, as well as the never completely submerged "degenerative" tendencies of women in general, indicated that one might find these strange, powerful, inviting, and fearfully eager temptresses almost anywhere. Often these creatures seemed at first sight to be no more than perfectly normal young ladies who had put on comfortably free-flowing classic clothes—as was the current fad among progressive young women of the Isadora Duncan sort—and who had simply gone into the woods to cull a garland of wild roses. The young lady in Charles Amable Lenoir's painting "The Flower Girl" of 1900 [VIII, 10] is just such a specimen. But as Lenoir indicated both in the subtly serpentine stance of his young beauty and in the peculiar tension with which she clutches the roses she carries at her breast, it is clear that she is highstrung and taut with suppressed passion—a passion which also speaks quite straightforwardly in the slightly malevolent edge the painter has given to her smile. Her sidelong glance and the "come hither" glint in her eyes, hauntingly deep-set within the fashionably feverish, dusky rings left by dissipation and too many sleepless nights, are telltale signs of our otherwise apparently virginal beauty's less than virtuous past.

Lenoir, a close follower of Bouguereau and perhaps this master's most technically accomplished pupil, had made single-figure compositions of this sort his specialty. By the time he painted this work, Lenoir had unquestionably surpassed his—at this point already

VIII, 10. Charles Amable Lenoir (1861–
1932), ''The Flower Girl'' (1900)

quite elderly—teacher in several respects. Bouguereau had always been singularly incapable of expressing psychological depth in his work. The bodies and features of his women
inevitably resembled the exquisitely delineated outlines of wax puppets. Whether they
were meant to be regal and austere or voluptuous and tempting, they nonetheless remained
forever devoid of evidences of inner life. Lenoir, on the other hand, while technically
every bit as accomplished as Bouguereau, in addition often succeeded brilliantly in expressing multiple levels of pychological tension in his delineation of the neurotic, skittish
tensility which echoed disturbingly through the bodies and faces of the young Parisian
women he liked to paint. Although Lenoir continued to disguise these psychological portraits as classical subjects, the best among them, like ''The Flower Girl,'' document in
extraordinary fashion the manner in which the inner conflicts of the women of the fin de
siècle found expression in gesture and stance, glances, ambiguous smiles and falsely cheerful faces, as well as in the hesitant grace of imperfectly controlled bodies.

In a visual world populated by images of women with faintly flickering or already
extinguished eyes, the sexual woman, with her malevolent but spirited glance actually
comes as a relief, a sign that woman still existed as a thinking entity - even if all she could
think about—as in Andrea Carlo Lucchesi's supremely sensuous and superbly modelled
and composed sculpture entitled ''The Myrtle's Altar''—was the destruction of man [VIII,
11].

VIII, 11. Andrea Carlo Lucchesi
(1860–1924), "The Myrtle's Altar,"
sculpture (ca. 1891)

The eyes of Lucchesi's young lady, snakelike and piercing, are no longer turned inward but display a hypnotic, aggressive quality. She sizes up the (male) viewer with a licentious intensity calculated to produce in him, in Max Nordau's words, "the morbid state of degeneracy which renders a man woman's plaything and the victim of his own temperament" (*Paradoxes,* 258).

In an oblique manner, images of this kind returned to woman a quality of force and of active existence, thus contradicting the ideal of fatuous inanition. In consequence, it is perhaps not surprising that in the years around 1900 it became fashionable among society ladies to have themselves painted as femmes fatales. They would let themselves be pictured looking malevolently at the viewer from under partially lowered eyelids—as in Albert von Keller's portrait of "Baroness B." [VIII 12]. Franz von Lenbach went so far as to paint himself and his whole family in this fashion; the three female figures, especially the youngest, glare at us in a veritable symphony of malice.

Diabolical women with the light of hell in their eyes were stalking men everywhere in the art of the turn of the century. For instance, Arthur Hacker's Sir Percival, so fresh and intellectual that his spirituality surrounded him like a saint's halo, was being stalked by a lady of catlike mien whose only wish was to dissipate our hero's manly virtue [VIII, 13]. John William Waterhouse's Hylas, meanwhile, found himself surrounded by alluringly naked young water nymphs who wound their soft arms around his so that they might

VIII, 12. Albert von Keller (1844–1920), ''Baroness B.'' (1905)

drag him brusquely into their watery bower, where they could be expected to have their fill of him even as he drowned. Although he had been tempted since the beginnings of Christendom, St. Anthony found scores of new sympathizers among turn-of-the-century painters. Inspired by Flaubert's exotic restatement of the saint's struggle against the temptations of the flesh, they searched under every rug and basket to ferret out the temptresses who seemed to lie in wait everywhere to assault the spirit.

While the story of St. Anthony's temptation was pursued internationally, it was one of the few subjects which allowed for clearly delineated national modes of representation. French versions tended to find the saint beset by one or two voluptuous figures who, walking on flowers, would try to confuse the saint in his attempts to differentiate between

VIII, 13. Arthur Hacker (1859–1919), ''The Temptation of Sir Perceval'' (ca. 1894)

VIII, 14. Ferdinand Götz (b. 1874), "The Temptation of St. Anthony" (ca. 1900)

the realms of heavenly beauty and Eros. Carolus-Duran produced several versions of this sort. Otherwise French artists were inclined to work on the saint's fetishistic tendencies, as Adolphe Léon Willette did in his playfully obscene version of the theme, exhibited in 1911, in which the poor man was forced to concentrate on voluptuous breasts and buttocks as analogous sources of perdition. Generally speaking, French temptresses tended to be airy figments of the saint's imagination, as in several versions by Henri Fantin-Latour.

German painters, however, preferred to depict realistically painted women—through with bodies and features of exaggerated protoexpressionist coarseness which loudly proclaimed their bestial natures—who had a habit of crawling about all over the place, and even all over the saint himself, as they insisted on doing, for instance, in Ferdinand Götz's entry to the Munich Secession exhibition of 1901 [VIII, 14]. In terms of style and mode of representation, Götz had clearly been influenced by Lovis Corinth, who, in a painting of 1898, showed an even denser mass of undifferentiated crawling female flesh besetting a quite understandably panicky saint [VIII, 15]. Corinth, having hit upon a good thing, was to return to this theme repeatedly.

Obviously, religious subjects involving temptresses were especially opportune since they allowed the painters to show images of extreme lubricity without having to defend themselves against charges of prurience. But mock piety of this sort did not delude stern critics such as Max Nordau and Leo Tolstoy, who insisted that woman should be regarded as little more than a specialized piece of baby-making machinery. Tolstoy, in fact, vented his fury against a "Temptation of St. Anthony" [VIII, 16], exhibited at the Royal Academy in 1897 by the British painter John Charles Dollman (who had a thing for monkeys) and also shown at the Universal Exposition of 1900 in Paris. Looking through the volume of *Royal Academy Pictures* in which the painting had been reproduced, Tolstoy became justifiably angered by its visual dishonesty: "The saint is on his knees praying. Behind him stands a naked woman and animals of some kind. It is apparent that the naked woman

VIII, 15. Lovis Corinth (1858–1925), "The Temptation of St. Anthony" (1898)

pleased the artist very much, but that Anthony did not concern him at all, and that so far from the temptation being terrible to him (the artist) it is highly agreeable. Therefore if there be any art in this picture, it is very nasty and false" (*What Is Art?* 224–25).

Tolstoy's remarks are, of course, the key to virtually all the countless representations of religious and mythological scenes of temptation produced at the turn of the century. They might, for instance, be applied with equal justice to Cézanne's version of the St. Anthony theme [VIII, 17]. Cézanne is without question the most innovative and most influential artist of the fin de siècle. But for him as much as for his more conventional colleagues the true subject of the temptation of St. Anthony is the very contemporary war between the sexes. Cézanne's painting is a fascinating mixture of an extremely individual stylistic sophistication and a perfectly conventional interpretation of standard late nine-teenth-century subject matter. Cézanne's approach to this subject matter is indicative of his entrenchment within the dominant ideological structures characteristic of his epoch. The latter makes it more understandable why this master painter, who was so largely responsible for introducing much of what is most brilliant—most humane—in twentieth-century art, should have continued to take the trouble, year after year, to cart his paintings over to the salon, only to have them rejected by disdainful juries.

VIII, 16. John Charles Dollman (1851–1934), ''The Temptation of St. Anthony'' (1897)

VIII, 17. Paul Cézanne (1839–1906), ''The Temptation of St. Anthony'' (ca. 1880)

Cézanne's bacchanalia represent a minuscule portion of his work, and they are by and large the least successful of his productions. They cannot compete with his breathtaking visual reformulation of our perception of landscape in his depictions of Mont St. Victoire and the Bibemus quarries—paintings which have forever altered and deepened our understanding of the colors, textures, and shapes of rock and soil. Yet it is important to note that the same elements of style that have given us new insight into the beauty of the material world also served to add power and negative force to his image of woman as a towering Venus of earth, who oppressively threatens a cowering St. Anthony, egged on by Satan and surrounded by putti who seem more like resentful child-acolytes of the devil than the gentle fruits of domestic bliss.

Whether he was Cézanne or an academic hack with the most marginal pretensions to talent, for the painter of the fin de siècle religious or mythological subjects, as Tolstoy pointed out, had become little more than a convenient excuse to preoccupy himself with representations of what he thought woman was—and what he might think she wanted. For most it was quite obvious that what she really wanted was to encompass and destroy man's soul with the devilish beauty of her body. It was, of course, not by any means necessary to invoke the temptation of St. Anthony to depict woman's predatory desire for man's gray matter. The most primitive face-to-face confrontations between man and woman could serve just as well, as the French painter Jean-Baptiste-Augustin Nemoz' salon entry of 1890, ''At the Edge of the Abyss,'' demonstrates [VIII, 18]. Woman and a snake conspire to make an end of a young man who is naively seeking for no more than a kiss from a beautiful lady. As the grip of this serpent-eyed temptress' left hand and the locking hold of her right arm around the unsuspecting suitor's neck clearly indicate, this evil creature is about to send a young man crashing down into the deepest and coldest crevasses of Mother Earth's regressive body.

VIII, 18. Jean-Baptiste-Augustin Nemoz (ca. 1840–1901), ''At the Edge of the Abyss'' (ca. 1890)

Sirens and mermaids were clearly also an especially urgent problem facing the late nineteenth-century explorers of the soul. These daughters of the sea seemed to be virtually everywhere. Aggressive and predatory, driven by the ceaseless sexual hunger of the nymphomaniac, they should not be confused with that other group of watery creatures, the *ondines,* or "wave women" we have already encountered as they were being washed up, helpless and broken-backed, in the work of the same painters who also hurried to portray their predatory sisters [see IV, 23]. The *ondines,* though unquestionably creatures of nature and sexually active, were the very personification of appropriate feminine passivity—as Armand Silvestre pointed out in *Le Nu au Salon des Champs Elysées* for 1896. In his description of an *ondine* with an extreme case of spinal deformation by the painter Pierre Dupuis, Sylvestre took pains to emphasize that "this is not the cruel siren who with her perfidious chants, lures the sailors, only to drag them down into her deadly grottoes, but rather the merely playful *ondine,* who does not look for victims and who is the innocent flower of the sea's deep garden."

Instead, the siren was quite the opposite of the eminently ravishable *ondine.* She had allowed the masculine force of the bisexual primal state to resurface; hence she personified the regressive, bestial element in woman's nature. She was not the cultured pearl of modern, passive femininity but the dangerous, brutal, atavistic child of the sea's cold watery womb. In 1891, waxing lyrical while discussing a painting of sirens by Louis Adolphe Tessier, Sylvestre warned, "They have it in for man, these daughters of the sea, who hate all men because they still remember how Ariadne was abandoned by Theseus, man of courage and intellect, only to be rescued by Dionysus, god of the earth and erotic abandon:

VIII, 19. Herbert Draper (1864–1920), "The Sea-Maiden" (1894)

VIII, 20. Fernand LeQuesne (b. 1856), "The Legend of the Kerdeck" (1890)

That is why their cold eyes have the sparkle of iron
By which from afar man's breast can be pierced—
And, like two mirrors in which their thought catches fire,
They reflect the flames of eternal despair."

No wonder that, their insatiable urges forever unstilled, they perched everywhere. They could be encountered in the nets of fishermen [VIII, 19] or on the rocky cliffs along the ocean. They were recorded by the most illustrious as well as the most obscure turn-of-the-century painters. The American Thomas Moran showed them yearning in a setting which was something of a cross between Green River, Wyoming, and a South sea island. The Hungarian Lajos Mark, with characteristic Eastern European demonstrative insistence, exhibited a painting at the Universal Exposition of 1900 showing a nest of sirens stylishly coiffed in turn-of-the-century hairdos, all remarkably naked, receiving offers of gold and jewelry from a massed throng of madly desirous—who could blame them?—males of all ages. It was abundantly clear from Mark's image that the impetuous fervor of their male admirers was likely to end in a fall from the precipitous cliffs on which they were huddled in the throes of temptation. At the same Universal Exposition of 1900 the French State proudly exhibited a purchase of 1890 representing a dramatization of the Breton legend of the Kerdeck by Fernand LeQuesne [VIII, 20], in which a veritable crush of delectable

VIII, 21. Adolphe La-
Lyre (b. 1850), "A Nest
of Sirens" (c. 1906)

sirens is about to pounce upon a decent gentleman in Breton dress—a "new and very modern St. Anthony" in Silvestre's words—in order to drag him down among them into a watery grave, all this under a "silver light, virtually lunar in its vibrant white limpidity." These sirens, said Silvestre, were "a most certain evocation of the pagan world," these women "with their pearly, roseate throats, and the pale azure running in their veins, with their arms that seem to yearn for voluptuous embraces, and their chubby cruppers, whipped by the water's amusing anger" (*Le Nu Au Salon des Champs Elysées,* 1890).

Some painters, such as Adolphe LaLyre and Gustav Wertheimer, made the depiction of sirens virtually the single focus of their careers. LaLyre's "A Nest of Sirens" of 1906 [VIII, 21] is a characteristic example of the range of seductive poses this painter liked to have his models strike in order to depict the never-stilled need for gratification and the

VIII, 22. Reinhold Max
Eichler (1872–1947),
"Festival of Nature"
(1904)

VIII, 23. Constantin Makovski (1839–1915), "The Roussalkas" (1890s)

autoerotic fervor of his atavistic, nymphomaniacal sirens. LaLyre's symphonies of feminine desire were populated with exactly the same sort of creatures so vividly described by Talmey as likely to be found roaming the drawing rooms of turn-of-the-century polite society: "In their wild passion, casting all moral and social consideration aside, they throw themselves into the arms of sin. The more they abandon themselves to the gratification of their lust, the greater is the desire of their morbidly irritated sexual nerve-centres for lecherous satisfaction" (*Woman*, 115).

It must have seemed to the visitors to the yearly painting exhibitions that these creatures were virtually everywhere. Even such a usually sedate and decorous painter as Childe Hassam encountered them quite frequently. He showed one at Cincinnati in 1915, elegantly posed at the very symbolic vulval opening of her deep, dark, primeval grotto. They might be seen crawling threateningly toward us, as in Edvard Munch's "The Lady from the Sea" of 1896. They could be disguised as water sprites, mountain sprites, or Loreleis. In fact, the German painters, true to their largely landlocked geography, tended to paint the siren as a creature—part woman, part mermaid, and part snake—who inhabited the rivers and woodland ponds, as in Reinhold Max Eichler's "Festival of Nature" [VIII, 22]. These German *Nixen* tended to spend their afternoons clawing at hapless fishermen in rowboats.

In Britain they might suddenly appear to the startled mariner in the form of a kelpie, as in Thomas Millie Dow's painting of that title [see VII, 8], glowering ominously at her prospective prey while posing voluptuously on her rock to lure him into her arms no matter what his reason might advise him. In the guise of a naiad she could be found stalking her prey near the rivers and springs, and as the "russalka" of Eastern Europe she crowded the nightmares of the Russian painter Constantin Makovski [VIII, 23]. The ill-fated Ioulka of Prosper Mérimée's story "Lokis" (1869) explains the function of these creatures: "A

VIII, 24. J. Humphreys Johnston (1857–1941), ''The Mystery of the Night'' (ca. 1898)

VIII, 25. Jean-François
Auburtin (1866–1930),
''The Echo'' (1911)

VIII, 26. Otto Greiner (1869–1916), ''Ulysses and the Sirens'' (1902)

russalka is a water nymph. There is one in each of the pools of black water which adorn our forests. Never go near them! The russalka comes out, even lovelier than I, if that is possible; she drags you down to the bottom, where in all probability she gobbles you up!'' Whereupon the story's narrator exclaims, ''A real siren!'' (157.) Like the russalkas, the dryads, nymphs of the forests and trees, and the oreads, nymphs of the hills and mountains, could turn even the most steadfast idealists into fauns of the decadence as they fluttered past them in a dense fog of flesh and feverish desire [See IV, 8].

Mostly, however, these callously beautiful creatures liked to dwell along rocky coasts whose rough cliffs were expressive of the stony agonies of the materialistic hell awaiting any man who yielded to the outward softness of their bodies. Waiting in a state of mummylike self-involvement for their prey—as in the American J. Humphreys Johnston's striking painting ''The Mystery of the Night'' [VIII, 24]—or moaning in a tight, communal circle of homoerotic attachment among a world of seagulls and fog—as in Jean-François Auburtin's ''The Echo,'' his contribution to the Salon des Beaux-Arts of 1911 [VIII, 25] —she represented nature in its most regressive, most elemental state, nature expressive of death and destruction. Gustave Moreau liked to depict the sirens as virtually melting into the rocks, an integral part of the rotting, decomposing organic matter which obstructed the ideal male in his search for transcendence.

Sometimes the painters, especially if they were British or German and had been given a classical education, felt called upon to adhere to historical accuracy by referring to the story of Ulysses' encounter with the sirens. A painter such as John William Waterhouse might show Ulysses bound to his mast while being beset by sirens, who had shed their seductive outward femininity and had become the predatory flying harpies they really were. Otto Greiner used the myth of Ulysses to match masculine Aryan muscle, ripple for ripple, with feminine heat [VIII, 26]. Others stressed the link between the mermaid and the siren—and the intense hunger of both for the male's gray matter. For most turn-of-the-century painters, however, the story of Ulysses was little more than an apposite historical allusion meant to give a contemporary scene added narrative depth, a general practice to

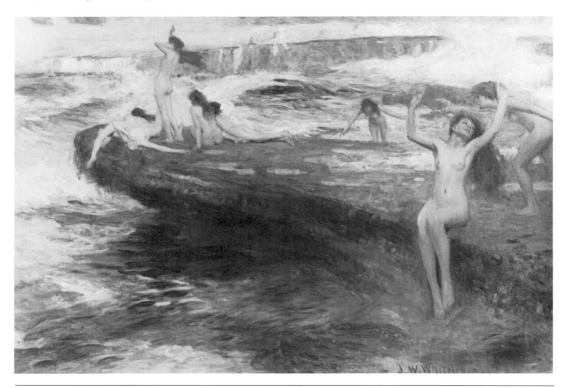

VIII, 27. John William Whiteley (active 1882–1916), "A Sail!" (1898)

which James Joyce was to give new life when he came to write the siren section of his own version of this pervasive cultural theme.

It was characteristic of these paintings to present the story of Ulysses as merely a convenient literary framework for a scene otherwise expressive of very modern cultural concerns. John William Whiteley's striking contribution to the Royal Academy Exhibition of 1898, titled, with supreme understatement, "A Sail!" [VIII, 27], is representative of this tendency. The title refers to the distant trireme under sail which makes its easily overlooked appearance at the extreme upper-left corner of the painting among the flaming, protoexpressionist pinks, reds, and aquamarines of a rocky coastline at sunset. This trireme, it may be assumed, represents Ulysses' vessel, which is about to sail into the dangerous realm of the sirens. It is these sirens who serve as the focal point of Whiteley's painting. Perched on a tonguelike rock whose monumental phallic outline dominates the composition, they seem like actresses fresh from London's West End who, having come to spend their summer holidays along the cliffs of Dover, were so seduced by the call of elemental nature that they decided to shed their clothes forever and take up permanent residence among the roaring waves of the sea. Insatiably desirous of physical gratification, these women, in their various poses, constitute a visual lexicon delineating the iconography of the dangerous woman in turn-of-the-century art. Yearning, calling, clambering, crawling, and ready to pounce—or, in the final throes of erotic transport, simply sitting and waiting in intense anticipation for whatever male might come along to be tempted— they symbolize the essence of material beauty ruling a world of primal passions whose hellish lure is fiercely expressed in the brilliant, broadly painted, intoxicating reds, purples, yellows, and blues of a fiery summer evening. Living under clouds and surf that seem like steam rising from the boiling cauldron of the elemental sea, these women represent that unabashed independence and elemental sense of freedom the men of 1900 feared, and found most fascinating, in the viragoes of their day. In the very directness of their passion and strength, these women embodied the paradox of the self-possessed and therefore hated, yet so very delectable and admirable New Woman, she who had thrown off the trappings of the household nun and and had toppled her weak and fainting mother's pedestal. In Whiteley's painting she appeared as a very real being of flesh and desire before the startled and fearful eyes of men, whose maudlin mythologies she had come to disdain in the triumphant self-sufficiency of her liberated personality.

It was the New Woman's insistence on power, equality with men, and an active life in which she and her sisters might, in Tennie C. Claflin's words, "boldly advance into the heat and strife of active business life, and show themselves competent to compete with the shrewdest and most experienced men" (*Constitutional Equality,* 66) which struck fear into the hearts of males, who were terrified of losing their own privileges. They yearned for the days of their fathers when women "at least had not begun to trespass upon men's ground," to use Max Beerbohm's phase ("The Pervasion of Rouge," 105). No wonder that many turn-of-the-century women were attracted to such paintings as that of Whiteley, and to the ambiguous image of the liberated female which they presented, for these women must have thought that it was better to be feared as a predator than to be disdained as a fool. In fact, it is clear that in their imaginative sweep, paintings such as Whiteley's obliquely paid tribute to the world of the New Woman, the world in which, in the words of Catherine Milnes Gaskell, "the age [had] gone by when weakness, physical or mental, passed for an attraction in women" ("Women of Today," 780). As such, these paintings may have added fuel to the belligerent determination of those "viraginous women" whose "unnatural" ambitions they were at least nominally designed to expose.

For many men, however, woman remained "The Witch," the title of a poem published by Theodore Strong in *The Smart Set* for August 1900, who in her "lure of moonlight," inevitably brewed "a soul's death" in "the falsehood and the smile / That disguised the wanton guile." (II, no. 2, 49) These witches, sirens, and predators were all the more dangerous because their deceptive appearance of activity only served to mask the engulfment of the striving male by the forces of passivity. In the popular lore of the years around 1900, the sea was ultimately passive, and woman was the creature of the sea, water being her symbol: totally yielding, totally flexible, yet ultimately all-encompassing and deadly in its very permeability. Her predatory sexuality was the tool with which nature tried to "draw back by cohesion and [refund] into the general watery surface" of the ocean of undifferentiated instinctual life the individualized "drop" representing the male intellect in Joseph Le Conte's scheme for evolutionary development.

The literature of the fin de siècle is, not surprisingly, no less densely populated with images of sirens and mermaids than the walls of the yearly exhibitions of painting. Theodore Dreiser's Carrie, longing to escape the monotony of her life as a laborer, "rocks to and fro" and sings under her breath, listening to what her own "siren's voice of the unrestful was whispering in her ear" (*Sister Carrie, 116*), and Edith Wharton's Lily Bart, in *The House of Mirth,* seems to Gerty Farish to be the siren of death itself. Gerty, the personification of the passive, steadfast, and undemanding clinging vine, sees her chance of cleaving to Selden, her heart's desire, vanish before Lily's confession to her of shared interest. Wharton comments, "The mortal maid on the shore is helpless against the siren who loves her prey: such victims are floated back dead from their adventure'' (193). Tennyson, Rossetti, Baudelaire—nearly any late nineteenth-century poet worth his salt water—grappled with sirens, mermaids, and their deadly desires. The highstrung Gerard Manley Hopkins hopped ''down that dank rock o'er which their lush long tresses weep'' (11) in his ''Vision of the Mermaids,'' and even T. S. Eliot's Prufrock, deacon of middleclass, middle-aged misery, thought he ''heard the mermaids singing, each to each'' while walking upon the beach, and fantasized that he had ''seen them riding seaward on the waves / Combing the white hair of the waves blown back / When the wind blows the water white and black'' (''Prufrock,'' 7).

As we have already learned, these sea maidens had a tendency to let themselves be caught quite deliberately in the nets of fishermen, as in Herbert Draper's painting [see VIII, 19] of 1894, based on a passage in Swinburne's *Chastelard*, where these toilers of the sea catch just such ''A strange-haired woman with sad singing lips, / Cold in the cheek, like any stray of sea, / And sweet to touch.'' Swinburne also indicates that it was a characteristic siren's trick to be caught in such a manner, because this made it all the easier for her to destroy the men beset by her in this apparently passive fashion: ''Men seeing her face, / And how she sighed out little Ahs of pain, / And soft cries sobbing sideways from her mouth, / Fell in hot love, and having lain with her, / Died soon'' (71–72).

The siren's physical allure spelled death to man's transcendent soul. She was, in Armand Silvestre's words ''the great vulture's sea-born kin,'' and would listen impassively to her victim's pleas for mercy. In an 1895 painting by Gustave Moreau titled ''The Poet and the Siren'' [VIII, 28] she is again seen as an integral part of organic matter, her diabolical eyes flashing as she pushes the boy-poet, the idealistic ephebe, who clamors miserably for mercy, implacably back into the primal mud whose emanation she is. With fine masochistic fervor Silvestre described the feelings of a siren's victim as he sang her his love song just before she pulled him down into a watery grave: ''When your lips touched mine, I shivered / My whole being tensing with the fear / Which courses through an animal who senses / That he must be eaten soon. / It was horrible and ravishing / To serve you as your prey. / My pain was equal to my joy in knowing / That I could let you wallow in my blood'' (*Le Nu au Salon des Champs Elysées*, 1897).

Clearly, the heavy weight of spiritual responsibility which the turn-of-the-century male felt elected to carry made his dread of the siren's enticements echo with disturbing elements of wish fulfillment and a yearning for freedom from the burden of responsibility. In consequence, the many painted documents recording the regressive woman's assault upon the ascending male, seeking to drag him down into the waters of physical indulgence, tended to place great emphasis on the typical female predator's very material charms. The male's fantasies of helplessness before the siren's physical enticements were not infrequently laced with a yearning to be seduced. Such a fantasy of seduction allowed him to combine the pleasures of indulgence with the innocent stance of the unwilling victim— thereby placing the responsibility for his weakness once again squarely on the shoulders of woman.

Mermaids, too, were not by any means the cute little housewives they have since become by Hollywood directive. They tended to be as vicious as the sirens. They might have fishy breath and flashy tails, but their bodies featured the same sort of classical perfection that made the sirens so delectable. Friedrich Heyser's perusal of the theme of ''The Fisherboy and the Water Nymph'' [VIII, 29], loosely inspired by Goethe's poem,

VIII, 28. Gustave Moreau (1826–1898), ''The Poet and the Siren''
(1895)

VIII, 29. Friedrich Heyser (1857–1921),
''The Fisherboy and the Water Nymph''

VIII, 30. Aristide Sartorio (1860–1932), ''The Green Abyss'' [or ''The Syren''] (c. 1895)

VIII, 31. Charles Shannon (1863/65–1937), "The Mermaid" (ca. 1900)

showed how sirens, mermaids, and nymphs could all serve the theme of the clinging vine whose perfect body had become a symbol of abject degeneration.

The fin-de-siècle painters liked to trace each step of the process by means of which the all-too-trusting and impressionable fisherboy came to be filed away in the watery locker of the mermaid's conquests. Aristide Sartorio, for instance, in his painting entitled "The Green Abyss" [VIII, 30], combined themes of moonlight, circularity, and the depths of forgetfulness in his depiction of a naked young man who let himself be tempted halfway out of his boat as he reached out to help a milky-breasted maiden, whose long fish tail—symbolizing her icy, predatory nature, as yet unseen by the young man—hovered in the dark waters behind her.

Charles Shannon depicted another mermaid who had already succeeded in taking possession of her prey with a solid wrestling hold, and who was just about to heave him unceremoniously into her watery domain [VIII, 31]. Charles Dana Gibson left his viewers no doubt that this predatory mermaid was none other than the girl next door, who, cool and collected and with disdainful eyes, remained in the swim of things, while her unfortunate suitor slipped helplessly below the water level of economic survival, his hands still vainly reaching out to his destroyer as he went down [VIII, 32]. In Burne-Jones' "The Depths of the Sea" [VIII, 33], finally, a woman with hypnotic eyes and a vampire's mouth has already completed her seduction and is carrying her prey—as if it were a huge, flowery bouquet of lost male morality—into the oblivion of her sensuality, where, we can be quite certain, he is to suffer the brain death which unfailingly accompanied the state of perpetual tumescence promised by the hollows of the siren's lair.

Masochistic fantasies such as these had a tremendous influence on the imagination of the intellectuals of the turn of the century, even if, in some cases, they were given a highly sophisticated form and couched in the language and knowledge of psychoanalysis. For instance, Arthur Schnitzler, in his novella "Mother and Son," used the theme of the siren

VIII, 32. Charles Dana
Gibson (1867–1944), "In
the Swim" (1900)

VIII, 33. Edward Coley Burne-Jones (1833–1898),
"The Depths of the Sea" (1885)

in delineating the slowly awakening realization of a previously virtuous mother that she had developed a clear sexual longing for her teenage son. He turned the climactic ending of his narrative into a scene which was a virtual reprise of the visual iconography of this theme as established by the painters. Mother and son, having rowed out at night to the middle of a mountain lake, are facing up to primal matters: "She drew him nearer to her, pressed against him, and an agony of longing came from the depths of her soul and flowed mystically over him." Inevitably Beatrice, the mother, and her son, Hugo, succumb to the demands of the flesh. Then, in the final paragraph of his novella, Schnitzler gives what he undoubtedly considered a deeply psychological, archetypal temptress-nymph denouement to this mother-son encounter:

> As she felt consciousness returning, she had enough strength of mind left to beware of a complete awakening. Holding both of Hugo's hands tightly in hers, she stepped to the side of the boat. As it listed, Hugo's eyes opened in a look touched with fear that bound him for the last time to the common lot of man. Beatrice drew her beloved, her son, her partner in death, to her breast. Understanding, forgiving, emancipated, he closed his eyes. But hers took in once more the grey bank rising up in the menacing dusk, and before the indifferent waves pressed between her eyes, her dying look drank in the shadows of the fading world.

Thus the "emancipated" son undergoes "the common lot of man" as he is pulled down into the water by his temptress-mother, who, even as she does so, still cannot help but long for the continuation of her material being.

Women who were so obsessed by the call of nature that they could not even leave their own sons alone were self-evidently no more than mere animals. As we shall see, mere animality was likely to revert to the most unblushing forms of dermatological degradation. When Eve consorted with the serpent, the scaly patterns that stuck to her skin tended to transform themselves into a fowl's feathers or, better still, into the mane of a predatory cat.

CHAPTER IX

Gynanders and Genetics; Connoisseurs of Bestiality and Serpentine Delights; Leda, Circe, and the Cold Caresses of the Sphinx

The male fantasies which led to the image of the evil woman, the seed-hungry temptress, that femme fatale in search of the perpetually tumescent male, created a fundamental conflict among turn-of-the-century intellectuals, a struggle between what one might call the wimps and the supermen. The supermen, in particular, despised the wimps for getting their kicks from their masochistic apotheosis of woman as the destroyer of souls, from turning a degenerate "freak of nature"—the viraginous woman, the feminist—into a creature with near-magical powers of seduction. The supermen regarded the wimps, those seekers after the sweet softness of the ephebe, the masculine substitute for the all-suffering household nun of their fathers, the lovers of the androgyne, and the fearful victims of the siren, as no more than the degenerate consorts of the "gynander"—the lesbian forager into the realm of the male soul. They were the decadents. Empire builders such as Max Nordau and William Graham Sumner, or a belated seeker after pastoral millennia such as Tolstoy, despised them with equal vigor.

Nordau, for instance, insisted that the decadents' fascination with the siren's song was little more than the wail of the wimp. "Strong-minded women," he said brusquely, "are attractive only to men of a feeble individuality, while men of clearly defined originality prefer to be and are attracted by the average type of womankind" (*Paradoxes,* 51). Statements such as this make it tempting to see in the doings of the decadents an oblique tribute to the powers of the feminine, but if this were so their tribute was cast in the form of an act of negation and had been shaped by revulsion, whereas the superman's derogation was cast in the smithy of arrogant disdain.

Joséphin Péladan, spokesman for the decadents, modish Rosicrucian, and author of the interminable, multivolume epic talkathon entitled *The Latin Decadence,* spelled out his party's position on the matter in the ninth volume, *The Gynander,* published in 1891 and therefore inevitably full of Darwinian clichés. "The androgyne," declared Tammuz, Pé-ladan's spokesman, "is the virginal adolescent male, still somewhat feminine, while the gynander can only be the woman who strives for male characteristics, the sexual usurper: the feminine aping the masculine!" To this wisdom delivered by Tammuz in true Socratic dialogue form his companion reacts by saying, "But of those two terms forming a pair, the one represents a positive concept, the other a negative." Of course Tammuz agrees: "The first originates in the Bible and designates the initial stage of human development; the Graeco-Catholic tradition has consecrated its use, whereas I have taken the other from botany, and with it I baptize not the sodomite but any tendency on the part of woman to take on the role of a man—and that includes a Mademoiselle de Maupin [Gautier's trans-vestite heroine] as well as a bluestocking" (43).

Tammuz seeks the mother's milk of feminine passivity to guide his soul to the Pla-tonic ideal, and since the women seem intent on withholding the comforting blanket of their Bible-designated helplessness, he sees it as incumbent upon him to redress the bal-ance of male-female harmony: "Since the gynander disguised herself as a man, Tammuz turned feminine; and by means of that maneuver he reestablished the conditions of normal contact. The result was that the superiority of his brain and his sex became apparent to the gynanders," making them cry out in disappointment, "We are nothing but imperfect, undeveloped versions of Tammuz, stunted in our growth, as unseasoned in thought as novices, and shapely of body only, not of mind" (133).

Running through the stages of ascension toward ideal love outlined by Plato in the *Symposium,* Tammuz thereupon argues that physical contact is representative only of the very lowest stage of love. "Even the pursuit of beautiful forms is not part of ideality unless one does so contemplatively." According to Tammuz, the male brings to a physical encounter only the excitation of his senses, his feminine element, but since sense experi-ence is the only measure of "soul" woman has, her self expires in the excitation of her body. Hence a physical encounter between males, unlike the encounter between a man and a woman, results in a strengthening of the male's higher faculties, of his "soul-force." This is inevitable, because "sodomy results in the atrophy of the sexual in either one or the other of opposing forces; and what is atrophied feminity but an incurable form of puerility?" Thus Tammuz concludes, tortuously and illogically, as Weininger was to a decade later, that homosexuality strengthens the intellectual power of male participants but that, given the laws of degenerative conjunction, same-sex relationships served to weaken women further into the condition of children: "Since she has no more reason or brain than a child of thirteen, the gynander will, in her conjunction with another woman, be no more than an idiot joined with a fool, and no elevation, no amelioration can ever be expected to come from such detestable mixtures" (145).

Tammuz, then, as spokesman for Péladan and the decadents, clearly does not offer any antidote to the muscle flexing of the supermasculine male. Instead, this man whom the supermen considered a wimp took pleasure in seeing himself as a sacrificial lamb to the masculine cause. With that part of his superior masculine being which was still in touch with the aboriginal feminine in humanity's bisexual origins he took on the "proper" feminine role which had been abandoned by the gynanders. He saw it as his mission to rescue the predatory relationship between the active male force and the passive feminine which must ultimately lead to the eradication of the feminine and the triumph of the mas-culine intellect, the soul. The gynander, on the other hand, became the symbol of complete degeneration. She was the predatory woman, the autoerotic or lesbian woman who consorted with males in a futile attempt to absorb or syphon off their masculine energies in order to "become masculinized," but who otherwise chose to conjoin herself only with other women in an orgy of degenerative, self-extinguishing regression into the absolute of femininity, a perverse journey back into the primordial earth.

As Nordau's remark about the preferences of "real" men shows, the supermen of the late nineteenth century saw their ideal of femininity in the prototype of what—courtesy of Hollywood—was soon to become the stereotype of the dumb blonde. On the other hand, as the gynander, the strong, self-assertive woman became the virago, the siren, the predator, the favorite villainess of both the wimps and the supermen. Both camps saw her as unnatural, doomed to self-destruction. But where the supermen saw the solution to her presence to be a sadistic gynecidal campaign of ruthless eradication which would keep the dumb blonde/household slave from being infected with the degenerative delusions of her self-assertive sisters, the decadents chose to wallow in agonized masochistic submission to the "unnatural acts" of the gynander in order to demonstrate their self-sacrificial virtue in the cause of masculine evolution.

In any case, it was clear to both the supermen and the decadents that since the intellectually evolved male was in a strong moral position, the number of victims in the full flower of their manhood that were available to the regressing woman was mostly limited to those men who were emotionally immature, sexually undeveloped, or radically degenerate to begin with. For, ultimately, only men who had not been able to keep pace with the evolution of the true male spirit were likely to be available as more than passing companions to the materialistic female. These men generally showed very clear evidence of their imperfect evolution. The proof of their failure, the shameful remnant of their bestial heritage was there for all to see. In the everyday world one encountered them as the lower classes, the workers and vagabonds, the unintelligent "fodder" of industrial civilization.

Certainly, there were among these masses many who through hard work and self-denial might attain to a small measure of intellectual dignity and hence obtain the right to be counted among the lower echelons of evolved male society, but the fact remained, in the words of William Graham Sumner, that "pauperism, prostitution and crime [were] the attendants of a state of society in which science, art and literature reach their highest developments." It was meaningless to try to elevate these creatures to the evolved level of their masters, for the processes of degeneration, that effeminacy of mind, had taken hold of their being. "If we should try," cautioned Sumner, "by any measures of arbitrary interference and assistance to relieve the victims of social pressure from the calamity of their position, we should only offer premiums to folly and vice and extend them further" (*War, and Other Essays,* 185).

It was not always quite clear what should be done with these creatures—for their own good, of course—but Sumner, for one, was charitable. "The sociologist is often asked if he wants to kill off certain classes of troublesome and burdensome persons. No such inference follows from any sound sociological doctrine, but it is allowed to infer, as to a great many persons and classes, that it would have been better for society, and would have involved no pain to them, if they had never been born." (187). As we now know, there were many "evolved" members of the master race who were to have no patience for such fine distinctions and who—good followers of Spencer, Sumner, and Weininger all—set out to help degenerates and effeminates, inferior groups and races, to become unborn again. Hitler's *Mein Kampf* in many respects slavishly echoes these theorists, especially

Weininger, whose *Sex and Character* found a place of honor on the shelves of the Reichs-führer's library.

Weininger, in fact, with his insistence on the need for an ultimate struggle between the masculine and the feminine principles, had put the matter in clear dualistic terms: "Mankind has the choice to make. There are only two poles, and there is no middle way" (329). Darwin, in *The Descent of Man,* had already expressed the opinion that even if considerations of humane concern demanded that in society "we must . . . bear the undoubtedly bad effects of the weak surviving and propagating their kind," it would nonetheless be very nice if evolving humanity, without intentionally neglecting "the weak and helpless," could somehow get rid of them. He suggested a eugenics approach, "namely that the weaker and inferior members of society do not marry so freely as the sound; and this check might be indefinitely increased by the weak in body or mind refraining from marriage" (152).

With such encouragement from the father of evolutionary science, it was not suprising that the turn of the century abounded with schemes for the appropriate pruning of the rosebush of evolving humanity. John D. Rockefeller, his head full of Spencer and Sumner, addressed a Sunday school class in the hope of instilling a future generation with an appropriate sense of selective purpose. Snuff out the weak in business and in society, he argued with fierce enthusiasm, and the future of the human race will be the better for it. "The American Beauty rose can be produced in the splendor and fragrance which bring cheer to its beholder only by sacrificing the early buds which grow up around it. This is not an evil tendency in business. It is merely the working-out of a law of nature and a law of God" (Ghent, *Our Benevolent Feudalism,* 28).

Rockefeller's instructive example of the "American Beauty rose," a flower widely used at the time to signify the vegetative glory and beauty of the American woman, is indicative of the manner in which metaphor came to be used to confuse and confound the issues of evolution, economics, sex, race, and class, producing a muddle of eugenics rhetoric which served to prepare the middle classes of Europe and America for the acceptance of the idea of genocide as a viable approach to the "problem" of potential evolutionary stagnation. Certainly the spectre of "devolution" so widely regarded as endemic to the natural tendencies of the feminine character, was a major factor in alerting the middle classes to the need for "social pruning." Warnings concerning the dangers inherent in letting the viraginous degeneration of woman proceed unchecked could, after all, easily be applied to the position of "effeminate races" in evolving Western civilization.

With their customary enthusiasm for visual extremes, late nineteenth-century painters set out with vigor to present their audiences with symbolic representations of this dread tendency to devolution which Nordau and the other imperial supermen had found to be endemic in the society of their time. The animal remnant in man was to be represented graphically. As it almost always did in turn-of-the-century art, mythology offered the visual means; it was clear that such half-bestial creatures as satyrs and centaurs were the perfect symbolic designation for those males who had not participated in the great evolutionary spiral, who had been content to be coddled in the bosom of effeminate sensuality. Since women, being female, were, as a matter of course, already directly representative of degeneration, there was no need to find a symbolic form to represent their bestial nature; their normal and preferably naked, physical presence was enough to make the point.

The archetypal representation of this conjunction of the normal female and the symbolic figure of the satyr, the man still partly animal, is unquestionably Bouguereau's "Nymphs

IX, 1. William Adolphe
Bouguereau (1825–1905),
"Nymphs and Satyr"
(1873)

and Satyr'' of 1873. [IX, 1] With this imposing canvas, the Grand Satrap of French salon
art did much to establish the provocative erotic-narrative framework which made such
depictions transcend the literary and cultural conventions of academic art. It is by no
means accidental that Bouguereau's innovations in the realm of mythological narrative
came at virtually the same time that the impressionists were rediscovering—admittedly in
a radically different manner—the narrative potential of the everyday. As early as 1879 Earl
Shinn had no difficulty in identifying the shift which had taken place in the narrative use
of mythology in Bouguereau's painting. "The trouble with the picture is that the people
are ladies, not Maenads or Bacchants. Their undressing is accidental or prurient, not ignorant. Look at any of their faces, and you feel that they need not insult your reason by
pretending not to write modern French and read the fashion-newspaper''(*The Art Treasures
of America,* I, 54).

For the next forty years there were always new versions of the contemporary woman disguised as nymph fooling around with a satyr, or groups of both, to regale and caution the visitors to the yearly Parisian salons. Cézanne's "Bacchanal" [IX, 2] is stylistically light-years removed from Bouguereau's primeval frolic. Yet it is only manner, not matter, which separates these two works. Even more than Bouguereau, Cézanne suggests, in his depiction of an encounter between a still half-bestial, emphatically hairy (and hence obviously as yet unevolved) group of men and an equal number of representatives of the Eternal Feminine, that rape is man's logical response to a woman who brings out the beast in him.

The omnipresent bacchante in the visual arts around 1900 was nothing but the graphic representation of the "bad" woman's facile renunciation of her maternal duties, which, as Lombroso and Ferrero had pointed out, were the only restraints holding back the release of woman's natural tendency to criminality. These two explorers of the dark side of the female soul had been in the forefront of the defense of civilization against the bacchante when they pointed to the degenerative effect of the sexual impulse in women. "Sensuality has multiple and imperious needs which absorb the mental activity of a woman, and, by rendering her selfish, destroy the spirit of self-abnegation inseparable from the maternal function" (*The Female Offender*, 153). Since, as Harry Campbell stressed in 1891, "self-control has come to be regarded as a *sine qua non* of good breeding," the unrestrained bacchante and her effeminate—because also unrestrained—men became the central figures in the bestiary of degeneration. Whether the painters called them nymphs, bacchantes, satyrs, centaurs, or fauns, they were always careful to show their links with the animals. The satyrs, especially, were given the "scientific" physiognomy of the degenerate, as delineated by Lombroso and others. They tended to have the caricatured features of the "bestial Jew," woman's closest degenerate companion in the otherwise civilized world.

When designing decorative statuettes for the Royal Copenhagen procelain factory, Gerhardt Henning even went so far as to make the physiognomy of degeneration an integral part of the nymphs and fauns destined for the mantel pieces of Europe's bourgeoisie

IX, 2. Paul Cézanne (1839–1906) "Bacchanal" (or "Amorous Struggle") (1885)

IX, 3. Gerhardt Henning (1880–1967), "Nymph and Faun" porcelain (ca. 1910)

IX, 4. Alfred-Philippe Roll (1846–1919), "The Feast of Silenus" (ca. 1885)

[IX, 3]. Alfred-Philippe Roll showed the effeminacy of the male company the bacchante kept by combining the boney, shrunken body of the satyr and the pudgy, effeminate baby flesh of the modern Silenus figure with three lusty ladies of the woods to create a veritable mountain of heaving flesh [IX, 4]. The Belgian sculptor Jef Lambeaux took his pointers from Rodin, who also liked to sculpt satyrs, fauns, and bacchantes in various stances of bestial transport. He showed the almost cannibalistic hunger of the bacchante as she played with a faun. This woman, up to nothing but animalistic pleasures, could be counted upon to bite the clumsy degenerate's ear the way a playful dog might, making the faun cry out in understandable consternation [IX, 5].

Lovis Corinth, never one to bother much about subtlety of statement, created an image perfectly representative of the colossal crudities of German symbolist art around 1900 by showing how the viraginous bacchantes could drive elderly businessmen from Munich to behave like satyrs. In his "Bacchanale" [IX, 6] he made one such victim hop around like a hairy, beer-bellied Silenus among clumsy Teutonic maidens. Corinth's bacchantes, with their characteristic circular dances, could only make this particular specimen dizzy and ridiculous as he tried to join the fun. Also featured in Corinth's fashionable image were "the childish black man," by nature fit to dance and hence much better at it than the pudgy accountant, and in the foreground the helpless, effeminate white male poet, being stomped lovingly into the ground among the daisies by a dream maiden straight out of the pages of Sacher-Masoch.

Corinth's painting, like virtually all the innumerable depictions of bacchanalian scenes

IX, 5. Jef Lambeaux (1852–1908), "The Bitten Faun," sculpture (ca. 1906)

which flooded the art market around 1900, was the sort of image that taught Anthony Ludovici, himself the son of artists, that "although Woman is constantly seeking to lure Man to specialize in his reproductive instinct, she never respects the man whom she thus succeeds in forcing to betray his other trusts," and that "all frivolous, superficial and pretentious women nowadays are to be found only shoulder to shoulder with degenerate man wherever and whenever he is 'enjoying himself,' and whiling away his empty existence in a whirl of still more empty pleasure" (*Woman: A Vindication*, 67).

The German painters, led by Arnold Böcklin, who was truly indefatigable in his pursuit of the licentious doings of barbarians, also had a special preference for the depiction of voluptuous young women who, while resting, bathing, or doing other assorted sylvan things, were being ogled or stalked by men who had remained decidedly animalistic. Böcklin's various depictions of fauns staring at sleeping nymphs are typical examples

IX, 6. Lovis Corinth (1858–1925), "Bacchanale" (1896)

[IX, 7]. These paintings are especially instructive representations of the theme because they vividly testify to Böcklin's Darwinian consciousness. He deliberately made his fauns into animalesque creatures with not only grotesquely caricatured Semitic or negroid features but the posture of monkeys as well—after all, as Darwin had shown, these simians were man's true ancestors. The fact that fully developed human women were the usual companions of these creatures was a telling indictment of the static nature of femininity; whereas once they might have been considered more human than man's ancestors, they had in the intervening centuries of male evolution proved to be incapable of evolving with him.

Böcklin's follower, Franz von Stuck, liked to show pale-skinned, fair-headed women gamboling with dark-skinned, hairy centaurs [IX, 8], or riding bareback—and bare-buttocked—on these atavistic holdovers at a full gallop. At other times he showed them having unmentionable fun with quite literally horny satyrs [IX, 9]. Maximilian Lenz painted a pair of high-strung and rather anorexic modern ladies taking a passionate tour of the flowery fields of summer on the backs of some rather unorthodox steeds [IX, 10].

The subject's potential for painters was markedly enlarged by the publication in 1893 of a sonnet cycle entitled *Les Trophées (The Trophies)* by José-Maria de Hérédia, a Cuban-born French poet. This work sold in phenomenal quantities, was translated into numerous

IX, 7. Arnold Böcklin (1827–1901), "Fauns and Sleeping Nymph" (1884)

IX, 8. Franz von Stuck (1863–1928), ''Centaur and Nymph'' (1895)

IX, 9. Franz von Stuck (1863–1928), ''Scherzo'' (1909)

IX, 10. Maximilian Lenz (1860–1948), ''Summer Air'' (ca. 1906)

languages, and became the turn-of-the-century intellectuals' favorite source of bestial lore. In his sonnet "The Centauress" Hérédia showed the agonizing demise of that mythical race, half noble horse and half human, as it fell victim to the unnatural passions of the outwardly fully human bacchantes. In centaur society, Hérédia implied, the female or mother-centaur had been a domesticating influence, but even in this bucolic world the viraginous, wild, fully human women were wreaking havoc. "From day to day," wails the centauress, "The cloud-begotten race, / Prodigious, slowly dies, as our embrace / Its sons abjure, for women desperate." Thus, roaming history in search of domestic societies to destroy, the modern bacchante demonstrated that she was the Eternal Feminine, causing the extinction of whole species.

Paul Klee, in discussing his etching "Woman and Beast" [IX, 11] in his diaries, explained the symbolic pattern which underlay the—for his time—characteristic juxtaposition of the fully human woman with males who had retained remnants of bestial anatomy: "The beast in man pursues the woman, who is not entirely insensible to it. Affinities of the lady with the bestial. Unveiling a bit of the feminine psyche. Recognition of a truth one likes to mask"(143). As Klee's etching clearly indicates, that truth was indeed the inherent affinity of woman with the bestial. Klee's woman not only has the physiognomy of the degenerate but also seems to be molded from the same organic matter as the phallic beast she lures toward her with the flower of her material passion. Growing out of the earth like the trunk of a tree, she is a frightening missing link between vegetal and animal existence.

The fact was that the women of the turn of the century were suspected by their men of having an extraordinarily clinical interest in the bestial remnant in man's nature. In 1913 Gaston la Touche exhibited a painting of a group of tea-drinking society ladies sitting around a table on which the body of a goat-footed, mustachioed faun has been placed;

IX, 11. Paul Klee (1879–1940), "Woman and Beast," etching (1904)

IX, 12. Albert-Joseph Penot (active 1896–1909), ''The Snake'' (ca. 1905)

before the ladies' avid eyes the faun is being dissected by an elegantly winged cupid. Fauns held a relentless fascination for these women and, as the painters of the period frequently hinted, that fascination could lead to genetic accidents of a peculiar nature. For instance, a painting by the British artist Charles Sims depicts two ladies of the leisure class enjoying a perfectly typical afternoon tea in the garden. Standing on the table, two children are reaching for the blossoms in a tree. One is a young boy of normal mien, while the other is an adorable little faun whose gamboling is watched over with maternal concern by one of the two elegant ladies.

Clearly, as late nineteenth-century science had conveniently discovered for the eagerly receptive artists and intellectuals, though one might wish to separate woman from the animals by socializing her and pretending that she was adaptable to the intellectually evolving world of man, ultimately one was likely to discover that it was impossible to take the animal out of the woman—that woman and animal were coextensive. Simply being in the vicinity of animals was enough to bring out the beast in woman. No matter how she struggled against the hegemony of the sexual instinct over her being, sooner or later she would have to succumb to its tyranny.

Sue Bridehead, Thomas Hardy's emancipated woman in *Jude The Obscure,* spends her life trying to escape from the sexual drive, going to extremes to avoid physical contact with the men who love her. In the end, however, ''the self-sacrifice of the woman on the altar of what she was pleased to call her principles'' (381) only succeeds in driving her first into hysteria and catastrophe and finally back into the ''civilized'' woman's most narrowly circumscribed self-image as a virtuous wife, a child-bearing device for her husband. On the verge of expiring, Jude, her true mate and natural lover—for he is, notwithstanding his intellectual aspirations, the offspring of generations of ''devolving'' men—contemplates her behavior and concludes, ''Strange difference of sex, that time and circumstances, which enlarge the views of most men, narrow the views of women almost invariably'' (414). In Hardy's Sue we see the degenerate principle in woman graphically at work. The one woman in Hardy's novel who literally has the last word, the person who continues to live blithely and unperturbed in the material world, is Arabella, who near the

IX, 13. Charles-Amable Lenoir (1861–1932), ''Poetry'' (ca. 1890)

outset of Jude's odyssey flings a butchered hog's penis into his face to catch his attention, and who is truly a modern Circe. She is described by Hardy as "a complete and substantial female animal—no more, no less"(44).

Painters were often as direct as Hardy in depicting the link between women and animals. Often they sought ways to show the generic similarity between a woman's physical presence and the appearance of various four-legged creatures. For instance, when Fernand Khnopff drew his study of "Déchéance," of degeneration in woman, he simply posed his model in such a fashion that the very curves of her body, the very tilt and tension of her torso, unmistakably suggested the outline of a cat. Similarly, Albert-Joseph Penot discovered in the delectable curves of a young woman the dangerous spring and curvature of a snake [IX, 12]. Penot, moreover, thought he saw in the peculiar indentations of the skin at the corners of his subject's lower lip hints of a viper's (or perhaps a vampire's) fangs. The woman's eyes clearly seemed to him to have the dark, hypnotic powers of a cobra.

In a much more subtle fashion, Charles Amable Lenoir hinted at the atavistic remnants in modern woman of her ancestors' all too intimate knowledge of the primeval faun's prodigious feats of sustained physical exertion. His "Poetry" [IX, 13], notwithstanding her modest downward glance, was clearly more a Dionysiac than an Apollonian muse, more representative of the self-destructive universality Nietzsche saw in Dionysiac "magic" than indicative of that justification of the individuated world he considered the essential environment for Apollonian art. To help us understand the nature of his "Poetry," Lenoir posed her, deerlike and graceful, in such a manner that her leg suggested the curve, spring, and tension of an agile animal, while the exaggerated space between her toes hinted at the cloven hoof of the faun. In this striking work, painted early in his career, Lenoir combined stylistic traits from the work of Courbet with the academic lessons of his teacher, Bougu-

IX, 14. Gustave Courbet (1819–1877), "Woman in the Wave" (1868)

ereau, again demonstrating that the distinctions among art movements, so confidently established by art historians, can only work on the basis of determined acts of historical exclusion rather than in conjunction with a thorough analysis of the full spectrum of a period's actual artistic production. The weighty shoulders and strongly molded breasts of Lenoir's nymph, for instance, have much more in common with such works as Courbet's "Woman in the Wave" of 1868 [IX, 14] than with the wax confections of Bouguereau. Courbet's version of the wave-born woman, as a matter of fact, is based on a similar sort of emphatic assertion of animal contours in the muscular tensility of his bather's breasts as is to be found in the body of Lenoir's nymph.

The suggestive images of woman's physical affinities with animals painted by Courbet, Lenoir, and Khnopff represent visual analogues to the sort of descriptions Zola liked to give of the "natural woman." This is his portrait of Father Mouret's sister Désirée (here seen through Albine's eyes): "She envied those strong arms, that hard chest, that completely carnal life in the impregnating heat of a flock of animals, that purely bestial expansion which made the plump child a peaceful sister to the red and white cow. She dreamed of being loved by the tawny rooster and of loving as the trees grow, naturally, without shame, opening every one of her veins to spurts of sap. It was the earth which made Désirée grow when she lay on her back" (*The Sin of Father Mouret*, 236).

Such elements of brute nature as were apparent in Désirée's dreams of strange, bestial couplings were also evident in the undignified doings of viraginous women. For instance, under the guise of documentary realism, the French artist Jean Veber exhibited a painting of "Women Wrestling in Devonshire" [IX, 15] at the Salon des Beaux-Arts of 1899. Against a background of haglike women with grotesque faces who are screaming encouragement, it shows two nude working-class women, with harsh faces and pulpy bodies, wrestling with each other for the coins which have been thrown in a grimy plate visible in the foreground. A few years later, in 1903, Veber's compatriot Emmanuel Croisé exhibited a painting which documented the nude wrestling contests of the infamously viraginous women of ancient Sparta [IX, 16]. It prefigures the late twentieth-century male's apparently boundless appetite for seeing women "debase themselves to the level of animals," as their ancestors of a century earlier would have said, in such ongoing instances of cultural devolution as female mud-wrestling contests and the like.

Although the extreme privileging of impressionism as an art movement in our own century has made it virtually impossible to look with objectivity at the productions of such a painter as Degas, it seems reasonable to take heed of the misogynist element in his work, which ideologically allies many of his productions with the forgotten prurient depictions of bestial women by such painters as Veber and Croisé. Maurice Hamel's comments on Degas—those of a critic who in 1890 declared that impressionism had "saved French painting from the trivial concerns of academic art" and who therefore cannot be dismissed as an unreasoning enemy of the movement—deserve consideration in this respect. In Degas' depiction of the feminine, specifically in the painter's famous studies of women in their circular bathtubs, Hamel discerned the "bestial poses," the "froglike crouch" of creatures seen as inherently degenerate. Hamel asserted that in his many representations of harshly footlit café singers [IX, 17], actresses, and dancers Degas was often not driven by sympathy or a desire to gain analytic understanding of the painful realities life held for these women, as they were forced to struggle for recognition in the realm of the popular arts. Instead, the painter's attitude was dominated by the ruthless curiosity of a bourgeois gone slumming. These images were the work of a man

IX, 15. Jean Veber
(1868–1928), "Women
Wrestling in Devonshire"
(ca. 1898)

IX, 16. Emmanuel
Croisé (b. 1859), "The
Girls of Sparta" (ca.
1903)

soured by disgust for the platitudes and ugliness of his subjects, for the base, cud-chewing, vegtative members of humanity, who merely undergo the pressures of daily life and do not react. He notes, with cruel amusement, the deformations they have undergone as a result of their professional life, whether these be physical or moral—and as a logical consequence he seeks out the artificial as an escape from the heavy blows of reality and has created an art in which the theater of life is turned into a fantasy world through insistent understatement, technical refinement, and an ever more supple vocabulary (*Salon de 1890,* 41).

These remarks hardly seem the obtuse observations of a typical late nineteenth-century critic doggedly insistent upon idealized representations of women, and consequently unwilling to recognize the fearless realism many of today's most prominent critics have come to see as characteristic of Degas' art. Instead Hamel, in these brilliant and subtle remarks, would seem to be anticipating the opinion of these twentieth-century critics (who, of course, had nineteenth-century counterparts with a similar point of view). He seems to be arguing against their interpretation not because of a desire to attack the concept of realism in art but because he wishes to expose a fantasy world masking itself as a documentary explo-

IX, 17. Edgar Degas (1834–1917), "La Chanteuse Verte," pastel (1884)

IX, 18. Fernand Khnopff (1858–1921), "The Meeting of Animalism and an Angel" (1889)

ration of the real world. Degas, Hamel argues, in effect used new elements of style and texture to make his very typical late nineteenth-century fantasies of feminine degeneracy appear to be realistic depictions of ordinary women. The fact that Hamel came from Degas' own intellectual milieu should make us think twice before dismissing his remarks. He may have been able to read the message of Degas' paintings in a far less mediated fashion than we are able to today.

In literature, as in the realm of the visual arts, fantasies concerning women's resemblance to animals increased steadily in frequency, ranging from simple comparisons ("catlike grace") to elaborate psychological characterizations. Suggestions of woman's animal nature could be found even in the most bourgeois American households. William Dean Howells, usually rather sedate in his presentation of natural phenomena, clearly saw the animal stirring in some very ordinary young ladies. In *A Hazard of New Fortunes* (1890), Angus Beaton, a rather foolishly self-confident artist who is charmed by the beauty of a millionaire's daughter, Christine Dryfoos, dares to toy with her affections. Having seen "from the first that she was a cat" (405), he nonetheless is fascinated with the idea of "holding a leopardess in leash" (404). He soon finds out, however, that such women are dangerous playthings:

> She felt him more than life to her and knew him lost, and the frenzy that makes a woman kill the man she loves, or fling vitriol to destroy the beauty she cannot have for all hers, possessed her lawless soul. . . . As he put out his hand to Christine, she pushed it aside with a scream of rage; she flashed at him and with both hands made a feline pass at the face he bent toward her (426–27).

Beaton manages to escape the "wildcat's" attack, but he has learned his lesson about the nature of incompletely domesticated female animals.

Lombroso and Ferrero, in *The Female Offender,* their nasty exercise in phrenological fanaticism, tell the lurid tale of a woman who murdered her husband, apparently because she was predestined to do so by the asymmetrical structure of her face and because her jaw happened to be "enormous with a lemurian appendix" (89). Similarly, several other murderesses demonstrated conclusively that they had criminal tendencies because they had "gigantic canine teeth" or eyes which were "wild in expression." The "big feline eyes" of one criminal and the "very ferocious countenance" of another were conclusive proof of their innate predisposition for evil. Even when these intrepid explorers of bestial elements in the female physiognomy could not find anything particularly unusual in the structure of their subjects' faces on which to hang their theories concerning the bestial impulse underlying the actions of the female offender, they would persist until they had found at least a hint of the animal. For instance, when they encountered a child murderess whose physiognomy was "relatively good, in spite of the subject's licentious tendencies, which age could not eradicate," they discovered with considerable relief "the cleft palate and fleshy lips which betray a luxurious disposition" (93).

In line with the evidence presented by the members of the women's movement, those "wild women" Mrs. Lynn Linton was so concerned about, Lombroso and Ferrero found that female criminals were, on the whole, clearly women with "virile traits," whose faces were habitually hard and cruel. Because of this, they obviously found it necessary to ask fundamental questions about the future of womanhood in general. Everything again pointed to the likelihood that degeneration was built into the very nature of woman. "A morbid activity of the psychical centers intensifies the bad qualities of women and induces them to seek relief in evil deeds; when piety and maternal sentiment are wanting, and in their place are strong passions and intensely erotic tendencies, much muscular strength and a superior intelligence for the conception and execution of evil, it is clear that the innocuous semi-criminal present in the normal woman must be transformed into a born criminal more terrible than any man." All this, moreover, was magnified by the fact that "women are big children," children whose childlike brains were now operative in adult bodies. In consequence, "their evil tendencies are more numerous and more varied than men's, but generally remain latent. When they are awakened and excited they produce results proportionately greater" (151).

Fernand Khnopff, in a drawing of 1889 entitled "The Meeting of Animalism and an Angel" [IX, 18], depicted the inevitable confrontation between woman, animal, and the dignified intellectual male, here shown as encapsulated in the armor of his angelic soul, a true androgyne of the Péladan variety, containing—yet, like this author's Tammuz, infinitely transcending—the realm of passive acculturated femininity. In contrast, primal woman, that tiger of aggressive sexuality, that voluptuous sphinx riddled with bestial needs, would

IX, 19. John Charles Dollman (1851–1934), "The Unknown" (ca. 1912)

seem to be only marginally responsive to the controlling spiritual strength of her master.

Few were the men who were as confident as Khnopff's determinedly upright masculine monad that they might prove capable of coping with these primal passions when unleashed in woman. Driven as she was by animal desire and animal instincts, it was not surprising that woman found she could get along better with animals than with men. Among the creatures most often frequented by woman in her passionate state, the monkey figured prominently—not surprisingly, given its shocking new prominence as the precursor of evolving humanity. John Charles Dollman, for instance, showed how monkeys and woman, equally childlike in their ignorant astonishment, tried to cope with the concept of fire in the primeval world [IX, 19]. Monkeys also took to visiting women in their dressing rooms, as in Otto Friedrich's painting "Vanity" [IX, 20], serving as subtle reminders of the promiscuous nature of the sex. After all, as scientists such as Harry Campbell emphasized "both the gorilla and the baboon were polygamous" (*Differences*, 45), and woman had the same uncivilized tendency. Forel, it will be recalled, had asserted that "nymphomaniacs often have polyandrous instincts, and they then become more insatiable than men" (*The Sexual Question*, 227). To this one needed only to add Carl Vogt's observation that "whenever we perceive an approach to the animal type, the female is nearer to it than the male," and that in any such male/female comparison "we should discover a greater simious resemblance if we were to take the female as our standard" (*Lectures on Man*, 180), to complete the outlines of a new feminine perversion. It consequently surprised few well-informed late nineteenth-century men to hear that, as Havelock Ellis was careful to point out in his discussion of bestiality, "Moll remarks that it seems to be an indication of an abnormal interest in monkeys that some women are observed by the attendants in the monkey-house of zoological gardens to be very frequent visitors. Near the Amazon the traveller Castelnau saw an enormous Coati monkey belonging to an Indian woman and tried to purchase it; though he offered a large sum, the woman only laughed. 'Your efforts are useless,' remarked an Indian in the same cabin, 'he is her husband' " (*Studies in the Psychology of Sex*, vol. II, pt. 2, 84–85).

Given the vigorous "scientific" bandying about of such provocative tidbits of biological folklore, we can begin to understand why such a sculpture as Emmanuel Frémiet's "Gorilla", [IX, 21] exhibited at the Salon of 1887, was given a medal of honor instead

IX, 20. Otto Friedrich (1862–1937), "Vanity" (ca. 1904)

IX, 21. Emmanuel Frémiet (1824–1910), "Gorilla," sculpture (ca. 1887)

of being howled unceremoniously out of the slightly soiled halls of turn-of-the-century High Art. Who, among the intellectuals, would have felt called upon to doubt, given woman's scientifically proven predilection for simian sensations, that a forward-looking ape might not also have decided that a shapely young woman was more delectable than a mate of his own species? King Kong was born in the jungle of nineteenth-century evolutionary science, and the details of his story were fleshed out in the hothouse of turn-of-the-century art.

Another prodigious polygamist who frequently appeared alongside women in late nineteenth-century art was the lion, who had been singled out by Darwin himself for his orientalist predilections. "As I hear from Sir Andrew Smith, the lion in S. Africa sometimes lives with a single female, but generally with more, and, in one case, was found with as many as five females; so that he is polygamous. As far as I can discover he is the only polygamist among all the terrestrial Carnivora, and he alone presents well-marked sexual characters" (*The Descent of Man*, 247).

While the comparison of women with cats, already a time-honored tradition, had become virtually endemic by the 1890s, José-Maria de Hérédia's *Trophées* added a graphic descriptive dimension to a fantasy whose details had until then remained relatively hard to imagine. Hérédia took up the story of Ariadne after her abandonment by Theseus on the island of Naxos. Pursuing the suggestion that she has been rescued by Dionysus and made to serve as a priestess of his cult, Hérédia linked her to Cybele, the great mother, goddess of fertility and wild nature, to whom the lion was sacred. Inevitably her worship was supposed to have been accompanied by sinful dances and orgies, and she herself had

IX, 22. Léon Victor Solon (b. 1872),
"Bacchanale" (ca. 1903)

presumably been wont to ride around on the back of a lion. In his sonnet "Ariadne," Hérédia described "The Queen, reclining nude / On a great tiger's back," as she entered the scene of an orgy devoted to Iacchos (Dionysus). He then sketched, in language which it still extraordinary in its frankness, the details of the queen's imagined ritual copulation with the great cat she had been riding:

> The royal beast, by her caresses wooed,
> Arches his loins and spurns the yellow plain,
> Roars amorously as she drops the rein,
> And champs his flowery bit in servitude.
> Upon his flanks her unbound tresses steaming
> Like clustered amber amid dark grapes gleaming,
> The Spouse, his growls of protest heeding not,
> Lifts her wild lips, flushed with ambrosial bliss,
> Her outcries stilled, the faithless one forgot,
> And smiles to Asia's Conqueror for his kiss (27).

Contrary to misconceptions still current, the leaders of fin-de-siècle society did not shy away from descriptions of any sort of sexual activity, provided such descriptions remained appropriately couched in the trappings of mythology. Hérédia's poems therefore became popular as graphic documents of the nuances of woman's atavistic nature. Inevitably they also inspired the painters to find ways in which to depict the events described by the poet. Léon Victor Solon provided a graphic representation of the details of Ariadne's experience [IX, 22] in an image which was apparently considered so true to nature that it was promptly reproduced in *The International Studio*. Artists such as Fidus habitually showed women encountering or huddling with lions or tigers. The British animal specialist Arthur Wardle painted dryads frolicking with leopards, or an enchantress lolling about with them in a playful mood, a satisfied, drowsy look on her face. One of Wardle's most successful canvases showed a bacchante roaming through the fields with a whole flock of the amorous beasts [IX, 23].

Angelo Graf von Courten, in his painting "Love and Strength" [IX, 24], showed that he knew very well how to play the game of visual double entendre which the intellectuals of the turn of the century used incessantly as a rather transparent veil with which to shield the icons of their libido against the intrusive censorship of the moral majority. Clearly, in Courten's painting the lion, traditional symbol of masculine strength, has been brought to the point of somnolent exhaustion by the amorous attentions of the spirited young woman who is hugging him in this well-posed snapshot of unorthodox domestic bliss. The young woman in question may not yet be wearing the pants in this family, but she is quite clearly in possession of the *flèche phallique,* the arrow of love, which in the lion's more manly days of mastery would have rightfully belonged to him. Now woman, usurping cupid's task, has taken to injecting even the king of the animals, the majestic polygamist of whom even Darwin had stood in awe, with the unmanning sedative of her insatiability.

In the United States Frederick Stuart Church, who had made the documentation of ambiguous encounters between women and animals a lifelong specialty, showed himself

IX, 23. Arthur Wardle (1864–1949), "A Bacchante" (ca. 1907)

IX, 24. Angelo Graf von Courten (1848–1925), "Love and Strength" (ca. 1894)

quite the equal of Courten in providing delicate images of women's prodigious sexual powers for the perusal of men whose neuroses were to lead the psychoanalysts to their preordained discoveries. His outwardly so delicate and properly feminine "Enchantress" [IX, 25], who in 1911 stalked undauntedly through the pages of *The Century*—one of America's favorite monthly magazines—was accompanied by a brace of tigers whose growling jaws suggested the *vagina dentata* which turn-of-the-century men feared they might find hidden beneath this raven-haired beauty's decorous gown.

In 1913 Constantin Starck exhibited a sculpture of a nude woman riding on the back of a lion. Starck, like Lenoir, mindful of Nietzsche's disquisitions on the Dionysiac nature of the feminine, entitled this work "Lyric Poetry." In the United States Daniel Chester French produced a group of sculptures representing the continents to adorn the facade of New York City's Customs House. He chose to portray Asia, fashionably, as a swarthy woman with a Buddha in her lap, a platform of skulls under her feet, benighted nations on one side and a gigantic tiger sidling up to her in amorous admiration on the other.

Sometimes—for the sake of variety, no doubt—artists tried to link woman amorously with grizzlies or polar bears. But whether the companion they chose to depict was a bear, tiger, lion, or domestic cat, artists made certain to emphasize the affinity between woman and her pet, as Charles Chaplin did, for instance, in his "Portrait of Miss W." [IX, 26],

IX, 25. Frederick Stuart Church (1842–1923), "The Enchantress" (1907)

IX, 26. Charles Chaplin (1825–1891), "Portrait of Miss W." (1889)

IX, 27. Franz von Lenbach (1836–1904), "Miss Peck" (ca. 1890)

in which the lady in question and her cat are quite obviously perfectly matched. Franz von Lenbach went Chaplin one better by having the "Miss Peck" of his version of this theme [IX, 27] hug her cat in the sort of "glued together" tête à tête which, as we have seen, was usually reserved by such painters as Cassatt and Sargent for the depiction of a more parental sort of affection. Lenbach's Miss Peck glowers at us with a more catlike intensity than her feline companion, whom the painter has for once, portrayed—with an admirable attention to realistic detail—as suffering (quite understandably, given the circumstances) from acute physical discomfort.

Hans Makart, in search of appropriate symbols with which to freight his depiction of two marble-hearted, luxuriant, sinfully cigarette-smoking, gossiping, and more than slightly suspect modern women, could find no better analogue for their lazily predatory ways than the cat snuggled comfortably on the pillow against which one of his cool seductresses has also chosen to lean her head [IX, 28]. Finally, as was only to be expected, Fidus found a way to combine the welter of suggestive implications worked out by his contemporaries, concerning the link between females and felines, in a drawing of a contented lion who has nestled his muzzle in a young woman's lap. This instructive image he captioned "My Little Dog" [IX, 29], thereby adding to his already none too subtle creation a reference to still another of the turn of the century's favorite subjects for the establishment of amorous links between women and the animal world. For dogs, like cats, were highly sought after by artists as the not-so-symbolic companions of choice in woman's exploration of the pleasures of bestiality. Even before Darwin, Proudhon had already insisted, that in times of decadence women "returned to their bestial nature" (*La Pornocratie*, 269), and Zola's Serge Mouret got an eyeful of evidence for this when he saw Rosalie, the promiscuous farmer's daughter, "sprawled under an olive tree with Voriau; the dog was licking her face and making her laugh. Her skirts flying and her arms beating the ground, she shouted to the dog, "You're tickling me, you dumb thing! Stop it" (30). Earlier, Mouret had seen the same dog "run into the petticoats" of Rosalie. Zola was clearly implying—and the men of the turn of the century would certainly have suspected—that the dog's affectionate

IX, 28. Hans Makart (1840–1884), "Marble Hearts" (ca. 1880)

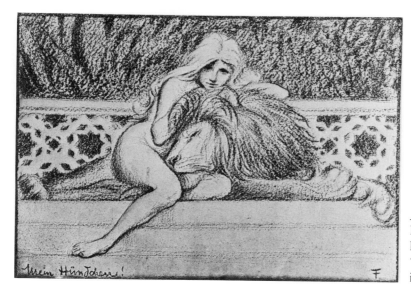

IX, 29. Fidus [Hugo Höppener] (1868–1948), "My Little Dog," drawing (ca. 1897)

tribute to her face might, in private, be repeated by dogs everywhere in admiring homage to less accessible areas of a woman's anatomy. As Havelock Ellis noted, Darwin had again planted important seeds of new knowledge. "The possibility of sexual excitement between women and animals involves a certain degree of excitability in animals from contact with women. Darwin stated that there could be no doubt that various quadrumanous animals could distinguish women from men—in the first place probably by smell and secondarily by sight—and be thus liable to sexual excitement" (*Studies,* vol. II, pt. 1, 85).

The accounts of the German bio-sexist Moll—who was relentlessly interested in the perversions to which women were prone, and who was ubiquitously cited by others at the turn of the century—gave Bernard Talmey a chance to bewail the harm done to scientific exploration by the fact that it was very difficult to surprise young girls "in the very act" of bestial excitement, even though, according to Moll, they were "fond of taking male dogs to their beds and practicing lustful improprieties with them" (162). Talmey brought together a veritable compendium of lubricious dog stories, again citing Moll as his authority; for example, he wrote that many unmarried females were known to "keep dogs for their sexual gratification, training the animals to practice cunnilingus" (*Woman,* 162). Most of these stories reappear in virtually identical form in the writings of Havelock Ellis, who, on their basis concluded—very scientifically, of course—that "when among women in civilization animal perversions appear, the animal is nearly always a pet dog. Usually the animal is taught to give gratification by cunnilinctus. In some cases there is really sexual intercourse between the animal and the woman" (*Studies,* vol. II, pt. 1, 83).

A number of the turn-of-the-century bio-sexists had insisted that, in Ellis' words, "bestiality . . . resembles masturbation and other abnormal manifestations of the sexual impulse which may be practiced merely *faute de mieux,*" and often a link was made, as Ellis did, between girlhood explorations of same-sex relationships and a later predilection for encounters with animals. From the case histories—virtually all based on hearsay—cited by these researchers, it would appear that in the 1880s and 1890s the more exotic brothels in such cities as Paris, Berlin, and San Francisco sometimes featured staged performances of encounters between prostitutes and dogs to accommodate male curiosity concerning

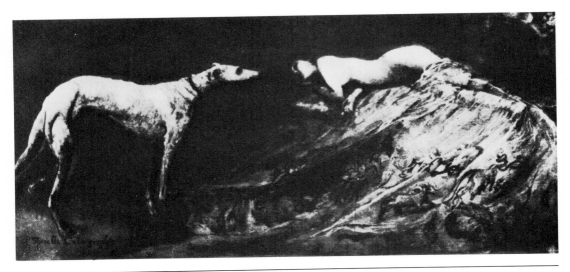

IX, 30. Aristide Sartorio (1860–1932), "The Child of Pleasure" (ca. 1889)

IX, 31. Jakov Obrovsky (1882–1949), "A Sultry Evening" (ca. 1911)

IX, 32. Fritz Erler (1868–1940), "A Grey Day" (1902)

women's degenerative sexual impulses. An anecdote recounted by Irving Rosse in an article entitled "Sexual Hypochondriasis and Perversion of the Genesic Instinct," which appeared in the *Virginia Medical Monthly* of October 1892, was widely used to illustrate woman's perverse, self-reflective tendency to prefer pets to the men in her life. Rosse had quoted a San Francisco prostitute involved in such encounters staged for male audiences as having insisted, in Ellis' words, "that a woman who had once copulated with a dog would ever afterwards prefer this animal to a man" (*Studies*, vol. II, pt. 1, 83).

In art these suggestions of a new outlet for feminine lubricity and disrespect for the male's reproductive responsibility certainly had something to do with the late nineteenth-century's focus on the society woman and her dog. Not merely ordinary lapdogs, these creatures often took on formidable proportions. It is unclear whether the vogue among the upper classes for the keeping of long-nosed, lean-bodied varieties of dogs, such as greyhounds and wolfhounds, helped spawn the male fantasy of what women did with these dogs, but it is certain that the artists of the period did everything they could—with propriety, of course—to feed that fantasy. A drawing by the infamous fin-de-siècle pornographer Franz von Bayros, Austria's answer to Félicien Rops, in which a nude woman with a riding crop in her hands leans back on an elegant sofa while a greyhound nestles its nose between her legs, gives more than ample indication of its very conscious nature.

Only slightly less blatant than von Bayros' drawing was the engraving which Gabriele D'Annunzio had Aristide Sartorio design in 1889 as a promotional stunt for his novel *Il Piacere (The Child of Pleasure)*. Sartorio produced an engraving in which a nude woman, elongated on a platform or bed, appears nose to nose with a truly formidable greyhound [IX, 30]. D'Annunzio and his entourage, in fact, had a fetish for large, pure-bred dogs which helped feed the fashion for the depiction of women accompanied by extremely nosy pets. For instance, the painter Jacov Obrovsky, much in the spirit of the writings of the internationally famous Italian author and poseur, depicted "A Sultry Evening" in which two women had been joined by one of the period's favorite canines to form a pyramidal structure of morosely reposing flesh. Obrovsky appears to have wanted to stress that if womanhood was going to the dogs, it was most probably because woman and dog were roughly equal on the evolutionary scale. For, in the same spirit of generic comparison which had motivated Chaplin and Lenbach, he managed to give the woman who sits at the

IX, 33. Louise Breslau (1856–
1927), "The Pensive Life"
(ca. 1908)

top of his pyramid a face whose structure, drowsy expression, and prominent proboscis closely echo the features of the dog sitting in the lower left corner of the painting, which in body pose as well as expression finds its analogue in the woman above [IX, 31].

A similar fascination for the exploration of similarities between a woman and her dog seems to have inspired the production of such affecting works as Fritz Erler's "A Grey Day," in which the elegant lady portrayed is also gifted with a nose of extraordinary prominence almost matching that of her dog [IX, 32]. Louise Breslau's very peculiar contribution to the Salon of 1908, entitled "The Pensive Life" [IX, 33], featured as its central personage a truly gigantic specimen of canine breeding that seemed to exhibit a remarkable proprietary affection for the region of his lady's lap. Arturo Dazzi's sculpture of the Countess Marie-Jeanne de Berteaux, exhibited at the Venice Biennale of 1914 [IX, 34] is characteristic of the stylish perversity associated with the women in D'Annunzio's circle.

Perhaps the most striking and odd among these suggestive fin-de-siècle fantasies about women and their dogs is Wilhelm Trübner's "At the Swimming Hole" [IX, 35]. The period's art of double entendre narrative is here brought to a new level of perfection. Ostensibly this is meant to be no more than an affecting dog story of the sort British painters such as Sir Edwin Henry Landseer and Briton Rivière had been so fond of chronicling. A faithful dog patiently stands guard over his mistress' clothing as she whiles away a sunny afternoon in just the fashion late nineteenth-century painters seem to have been convinced all attractive women spent their days: bathing in the nude in the open air. But there is in Trübner's painting a strangely suggestive element in the conjunction the artist has established between the dog, with its tensely alert musculature, and the feminine garments—delineated with a fetishistic attention to detail—over which it stands guard. The glare in the animal's eyes, it tautly straining limbs, linked with the suggestion of a woman's vulnerable nudity—unseen, but therefore all the more powerfully present—conjoin in this image to play upon the most prurient elements of the fin-de-siècle male's fantasies about rape, sexual aggression, and bestiality.

A quick glance at the pages of such a magazine as *Jugend* will convince anyone that the artists of the turn of the century had read their Darwin and the writings of the biosexists very diligently. Darwin had said that "the Asiatic *Antilope saiga* appears to be the most inordinate polygamist in the world" (*The Descent of Man,* 246), and, sure enough, *Jugend* began featuring images of nude women riding at breakneck speed on the backs of precisely such long-horned antelopes. Other animals singled out by Darwin for their philandering tendencies also attracted the attention of the ubiquitously unclad lasses of what has been considered the German artistic renaissance. These young women consequently played with wild boars, rode elephants as well as lions, and even naughtily nudged seals. Artists also favored the conjunction of women with extremely long-beaked fowl. One of these well-endowed creatures paid his hopeful respects as a suitor to an impressively unclad young lady in Richard Müller's "Lover's Quest" [IX, 36], which was reproduced in *Jugend* in 1914. Max Klinger, Gustave Moreau, and Félicien Rops, among many others, also tapped this source of artistic inspiration. With the period's characteristic sense that more of a good thing is always likely to be better, the theme was finally pursued by Auguste Matisse, who, though in actual terms only a few years older, can be regarded, at least insofar as

IX, 34. Arturo Dazzi (1881–1966), "Marie-Jeanne de Berteaux," sculpture (ca. 1914)

IX, 35. Wilhelm Trübner (1851–1917), "At the Swimming Hole" (ca. 1900)

IX, 36. Richard Müller (1874–1954), "Lover's Quest" (1913)

IX, 37. Auguste Matisse (1866–1931), "In the Gold of the Evening" (ca. 1905)

subject matter is concerned, as the far more famous Henri's spiritual painter-father. Auguste specialized in a form of decorative art considered "ideal" in its focus. This permitted him to produce such suggestive scenes of feminine life as his symphonic opus of 1905 entitled "In the Gold of the Evening" [IX, 37], which was composed of a swamp-sized congregation of delectable ladies and long-necked, long-beaked birds. The painting featured, in the foreground, among a cluster of nymphs, one young lady who understandably appeared to be concerned about her rounded exposure to the pointed attentions of an approaching cormorant. To suggest, as cultural historians have been wont to, that paintings such as these were naive, unconscious images of archetypal sexual symbols whose real meaning the psychoanalysts were in the process of uncovering at just this time represents a fundamental misunderstanding of the anything but unconscious prurience of the late nineteenth-century art world.

The intellectuals of the years around 1900 thought that woman's special predilection for animals was a logical concomitant of her inability to adapt to the conditions of the civilized world. Hence, men who did not keep their women on a tight leash and let them roam free ran dangerous risks. As we have seen, Beaton, the artist-aesthete of Howells' *A Hazard of New Fortunes,* discovered that fact to his edification. Another fictional artist who discovered the regressive capacity of woman at close range appeared in a little verse drama entitled "The Posing of Vivette," which played itself out in the pages of an 1897 issue of *Scribner's.* Profusely illustrated with engravings after illustrations by Albert B. Wenzell, it might be considered risqué even now if it were published in a family magazine. J. Russell Taylor, its author, set the scene for an artist's startling discovery of the ease with which pagan sentiments could reassert their hold over apparently civilized women. The artist has taken his youthful model, Vivette, out to the woods for a posing session as—what else—a bacchante. The young lady is wearing, as she herself describes it, "a smile, a garland, and a leopard skin / That doesn't fit: 'tis too decollete." Not to be numbered among the best-educated young women, the model barely knows what a bacchante is supposed to be, and, as a city-bred lass, she doesn't particularly enjoy being "in these stupid woods." She is uncomfortable and tells the artist, "I wish you'd hurry: this old tree / Is rough, my foot's asleep, and ants / Explore my back." Not exactly the most auspicious of circumstances, it would appear, for the establishment of a link with woman's "universal nature." And yet this is precisely what happens. For this thoroughly modern model, without the slightest knowledge of classical mythology or literature, suddenly and inexplicably falls into a trance and starts to spout what the author undoubtedly regarded as a reasonable cross between the poetic language of the Euripidean bacchae and Hérédia's poems about the orgiastic past:

> VIVETTE: The tiger, teased too far
> Last night (she sprawled an insolence of grace
> Along his cruel beauty, black and gold
> And sumptuous white), nipped
> Ariadne's hand.
> I had not guessed the cone concealed a blade,
> Though there I saw the Theban horror drip:
> She shrieked; the thyrsus glittered, pinned the brute
> By one gaunt striped flank unto the moss
> Where he writhed roaring till the god released
> And healed him. . . . Come, flushed ancient, we must stir.

O for the chanting dithyramb, to wake
The rain of pattering satyr-hoofs, and thrill
The purple ankles of the Hyades
Till every grape-kissed girl-foot taps in tune
And every nymph in Naxos floats her hair!
O Dionysus, drinker of raw blood!
O Dionysus, golden, honey-sweet!

[He (the artist) steps forward as she begins to dance, and touches her arm.]

THE ARTIST: Vivette—

VIVETTE: A man here at the Mysteries! Maenads, a man! [She leaps, one hand reaching his face before he can seize her wrists and hold her, struggling]

THE ARTIST: You wild-cat! Why, Vivette!
[She falls forward against him: he lays her on the grass.]

VIVETTE: Hullo, what's happened?
O, you hurt my wrists! What have I done—I fainted?

THE ARTIST: Died, I think.

VIVETTE: Your face is bleeding—aie!

THE ARTIST: I know, I know:
That blackberry caught me when I ran to help.

The artist has decided to let sleeping bacchantes lie, as it were, and not tell his model what has actually happened. Vivette now appears to be herself again, and the artist has learned his lesson, he indicates when he says to Vivette, "Come, child, we'll go. / I'll paint you serving tea-cups after this." The message of this little drama is never to give your women a chance to reconnect with their natural habitat, for if you let them they will find a way to scratch you to pieces.

The literary qualities of this piece are not of the sort to make it worth dwelling upon. What is quite remarkable, however, is the fact that it should have been published in a popular American mass-market monthly in the supposedly still puritan 1890s, and that it should have made so blatant and graphic a link between woman and her animal nature. It shows how commonplace the notion had become at this point. Caught up in a prosaic, boring, materialistic world of their own making, late nineteenth-century men, having built themselves up as the intellectual hope of civilization, were actually stuck in a rather unexciting world of habit and overelaborate etiquette which was quite unlike the realm of their transcendental aspirations. Being intellectual, they could not admit to erotic fancies of their own, but, being bored, they had plenty. Woman was there, conveniently, first to supply the fantasies and then to take the blame for them. The result was the pervasive myth of the animal-woman.

The sex researchers of the years around 1900 could not make up their minds about the frequency of actual bestiality among the female population. Talmey thought it to be "a rare anomaly or vice, in women" (*Woman*, 161), notwithstanding the proportionally considerable amount of space and scientific energy he devoted to it, while Forel, at virtually the same time, declared this "filthy, obnoxious and degenerative practice" to be "not rare in women" (*The Sexual Question*, 256). Given what Abba Goold Woolson had called, as early as 1873, "the common masculine talk about woman's instincts," all that was needed to convince the intellectuals, artists, and writers of the prevalence of erotic contact between women and animals was a mere hint of the existence of actual cases. For, as Woolson had pointed out with understandable resentment, all this talk about woman's animal instincts was nothing but "an attempt to deprive her of all credit for her actions, by allying her with bees, beavers, and other curious creatures, who work after a pattern set them in the Garden of Eden" (*Woman in American Society*, 22)

Indeed, the doings of Eve in paradise, and her being tempted by Satan, now came to be explained almost exclusively as a factor of her assumed general similarity, in terms of instincts and desires, to that most licentious primal length of living matter known to man, that seemingly perpetually tumescent creeper which had kept man's loins from being the peaceful bower of spiritual bliss he dreamt it might be: the serpent. Symbol of the uroboric primal earth, of woman as earth mother, it also represented, in Arthur Symons' words, "man's desire, / Godlike before, now for a woman's sake / Descended through the woman to the snake" ("Hallucination," *Selected Poetry and Prose*, 46)

In *Marius the Epicurean*, Walter Pater stressed the link between the serpent and the temptations presented to the male by those daughters of the household nun who had renounced their role as the white doves of domesticity and had ceased to be cradles to the male soul. Marius, who is no more representative of Roman culture than Pater himself, remembers his mother, who "became to him the very type of maternity in things—its unfailing pity and protectiveness—and maternity itself the central type of all love." Under his mother's guidance, Marius learns to see "the soft lines of the white hands and face" of the woman of the 1850s as the ideal of purity, and he remembers how his mother once told him that his soul was like a "white bird" which "he must carry in his bosom across a crowded public place." Growing up, Marius wonders whether he will be able to bring this white bird into "the hands of his good genius on the opposite side, unruffled and unsoiled." One day in early summer he encounters a nest of "snakes breeding" as he walks along a narrow road. It is clear that these snakes represent both his own awakening manhood and the women who cause it to stir. Pater effectively expresses the fear such stirrings wedged into the minds of the sons of the household nun:

> It was something like a fear of the supernatural, or perhaps rather a moral feeling, for the face of a great serpent, with no grace of fur or feathers, unlike the faces of birds or quadrupeds, had a kind of humanity in its spotted and clouded nakedness. There was a humanity, dusty and sordid, and as if far gone in corruption, in the sluggish coil, as it awoke suddenly into one metallic spring of pure enmity against him (I, 24–26).

Thus, the phallus, with its uncontrollable capacity to spring forward, which warred against men who carried the white bird of intellectual transcendence in their bosom, was indeed the satanic companion to the women whose presence caused the serpent to uncoil. In the evil, bestial implications of her beauty, woman was not only tempted by the snake but was the snake herself. Among the terms to describe a woman's appearance none were more overused during the late nineteenth century than "serpentine," "sinuous," and "snake-like." Only the "catlike" graces of woman provided any competition to considerations of reptilian sinuosity. Arthur Stringer, in an issue of *The Smart Set* published in July 1900, pondered the alternatives available to woman in the event that her soul might actually survive in "some other form" after her death. He could discover only two possibilities: "O sinuous being, lithe and strange," he cried, "Soft tiger-soul, snake-spirit shy, / Of you which will they make—/ A velvet padded hunter, or / A shadow-loving snake?" ("Metempsychosis," 90)

Thus, Eve and the serpent became coextensive. As Eva Nagel Wolf explained in *The*

International Studio in 1919, woman was "Lamia, the serpent goddess, with all the sinuous grace of the serpent and the tantalizing haunting memory of a beautiful woman" ("Elenore Abbott, Illustrator," xxvii) Maxim Gorki's Artamonov saw the primal serpent in Paula Menotti's movements: "She writhed as if trying to leap down from the piano, but unable to; her suppressed cries became more and more loathsome and evil; what was particularly disgusting to see was the way her legs writhed, together with the jerkings of her head which sent her hair flying first over her breasts, then lashing over her back again like the tail of a snake" (*The Artamonov Business*, 208). Gorki's description of Paula is almost a verbal equivalent of Penot's "The Snake"[see IX, 12]. The Russian novelist, writing in 1925, was describing, no doubt at least partially from personal memory, the sort of entertainment that would have been available to a businessman in just about the year in which Penot had first exhibited his painting. His description thus confirms a close connection between what the painters painted and what the entertainment industry was presenting to the general public. Period illustrations of stage shows also indicate that "serpentine dances" performed by young women, both with and without actual snakes in attendance, were extremely popular in theaters and circuses alike. It is therefore not surprising that Frank Wedekind's Lulu, as an archetypal performer, also represents "the primal form of woman," the snake "created for every abuse / to allure and to poison and seduce, / To murder without leaving any trace," who "coils herself with strong squeeze around the tiger" (*The Lulu Plays*, 10–11). And in George Meredith's cycle of poems entitled "Modern Love" the sobs the husband hears coming from his disaffected wife are "stangled mute, like little gasping snakes, / Dreadfully venomous to him" (*Poems*, 3).

In Rossetti's "Eden Bower," Lilith—Adam's reputed first mate and the original "wild woman" who did not have the appropriate qualities of abject passivity Rossetti valued in the household nun—claims to have been "the fairest snake in Eden." The nasty lady is all scales and no petals: 'Not a drop of her blood was human, / But she was made like a soft sweet woman" (*Poems*, 19). When John Collier painted the poet's Lilith, he showed her with a gigantic snake curled around her erect nude body, the snake's head cuddled like a puppy's in the luxurious hair at the crook of Lilith's neck—two lovers become one.

Being a true daughter of Eve, the animal-woman of the turn of the century thus had a special appreciation for the erotic abilities of the snake. She liked to be with serpents, use them in strange rituals, and generally become involved with them in the most dubious ways. As in so many instances, it was Flaubert who brought into focus this element of late nineteenth-century lore. In *Salammbô* (1862) he described what is clearly an erotic encounter between Salammbô and the serpent, her partner in the rituals she performs as a priestess of Baal, a god representing the male exterminating principle:

> Salammbô undid her ear pendants, her necklace, her bracelets, her long white gown; she unfastened the band round her hair, and for a few minutes shook it over her shoulders, gently, to refresh herself by loosening it. The music outside continued; there were three notes, always the same, headlong, frenzied; the strings grated, the flute boomed; Taanach kept time by clapping her hands; Salammbô, her whole body swaying, chanted her prayers, and her clothes, one after another, fell around her.

> The heavy tapestry shook, and above the cord holding it up the python's head appeared. It came down slowly, like a drop of water running along a wall, crawled among the scattered garments, then, its tail stuck against the ground, reared up straight; and its eyes, more brilliant than carbuncles, fixed on Salammbô.

> Horror of the cold or perhaps a certain modesty at first made her hesitate. But she remembered

Schahabarim's orders, came forward; the python fell back, and putting the middle of its body round her neck, it let its head and tail dangle, like a broken necklace with its two ends trailing on the ground. Salammbô wound it round her waist, under her arms, between her knees; then taking it by the jaw she brought its little triangular mouth to the edge of her teeth, and half closing her eyes, bent back under the moon's rays. The white light seemed to envelop her like a silver mist, her wet footprints glistened on the floor, stars shimmered in the depths of the water; it tightened round her its black coils striped with golden patches. Salammbô gasped beneath this weight, too heavy for her, her back bent, she felt she was dying; and with the tip of its tail it gently flicked her thigh; then as the music ceased, it dropped down again (174–75).

Predictably the painters and sculptors rushed in to illustrate Salammbô's lubricious encounter. Gaston Bussière's ''Scene of the Serpent'' emphasized the python's approach, while Gabriel Ferrier concentrated on the lady's pleasure [IX, 39]. The American Charles Allen Winter's ''Fantaisie Egyptiènne,'' exhibited at the salon in 1898, was a free adaptation of Salammbô's passionate communion with her appreciative lover. In general structure it was much like the sculptor Jean-Antoine-Marie Idrac's version of Flaubert's heroine, which found a place of honor at the World's Columbian Exposition of 1893 in Chicago [IX, 38]. Idrac's sculpture was, in turn, closely related to Collier's painting of Lilith, thus demonstrating how easily the literally hundreds of painted and sculpted versions of Lilith, Salammbô, Lamia, and assorted other snake charmers came to blend as generic depictions of Woman, the eternal Eve. For this first and, in the eye of the fin-de-siècle male, most perverse of women was also perpetually shown in intimate communication with serpents of virtually every length and girth throughout the forty-year period spanning the turn of the century. Henri Rousseau, creator of numerous jungle Eves and snake charmers un-

IX, 38. Jean-Antoine-Marie Idrac (1849–1884), ''Salammbô,'' sculpture (ca. 1890)

IX, 39. Gabriel Ferrier (1847–1914), "Salammbô" (ca. 1881)

doubtedly found it an easy and logical step to move from such images to the depiction of a very contemporary woman, whom he imagined in the process of copulating with a beast that looked like something of a cross between a bear and a donkey. And all the while, as she was being so formidably serviced, Rousseau's young lady quite unconcernedly insisted on holding above her head her oval hand mirror, symbol of the primal woman's narcissistic autoerotic self-sufficiency.

By using the symbol of the snake, the artists could afford to be somewhat less graphic and yet satisfyingly symbolic. Snakes were so horribly slithery that only the most tantalizingly perverse women could love the creatures enough to coddle and kiss them, as Kenyon Cox demonstrated in his rendition of Lilith [IX, 40], which was reproduced in *Scribner's* in 1892 for the edification of the American intelligentsia. Herbert Draper painted Lamia in such a fashion that no one could overlook the fact that this luscious lady with an asp on her arm had the soul and features of a serpent and the slithery body to match.

In 1920, as the heyday of the depiction of serpentine feminine bestiality was coming to a close, Arthur Symons summed up the symbolic meaning of this evil creature in his poem "The Avenging Spirit," identifying Lilith and Lamia as mother and daughter, united in evil:

> For in your body is the inevitable
> Sting of the Serpent made of the Snake's desire,
> The desire he had of Lilith, whose strange spell
> Woven around him made his breath respire
> The odours of no death, not damnable,
> But deadly when the blood that's mixed with mire
> Propagates evil. You the Insensible
> Beast of the Wilderness where root and briar
> Mix, and the ways thereof no man can tell,
> Jungles and forests, lion's lust and ire (*Poems,* II 301).

The link between Lamia and the late nineteenth-century feminists, the viragoes—the wild women—would have been clear to any intellectual reasonably well versed in classical mythology, since the Lamia of myth was thought to have been a bisexual, masculinized, cradle-robbing creature, and therefore to the men of the turn of the century perfectly representative of the New Woman who, in their eyes, was seeking to arrogate to herself male privileges, refused the duties of motherhood, and was intent upon destroying the heavenly harmony of feminine subordination in the family. The same was certainly true of Lilith, who, in her unwillingness to play second fiddle to Adam, was, as Rossetti's work already indicated, widely regarded as the world's first virago.

The Medusa, with her bouffant of snakes, paralyzing eyes, and bestial proclivities, was the very personification of all that was evil in the gynander, as Fernand Khnopff did not fail to emphasize in his "Istar" [IX, 41], a lithograph of 1888 inspired by Joséphin

IX, 40. Kenyon Cox (1856–1919), "Lilith" (ca. 1892)

IX, 41. Fernand Khnopff (1858–1921), "Istar," lithograph 1888)

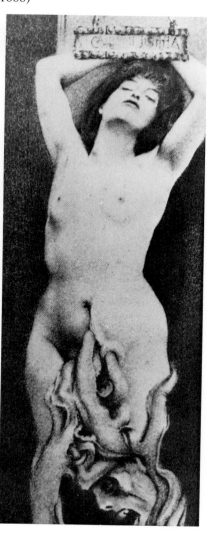

IX, 42. Carlos Schwabe (1866–1926), ''Medusa,'' in a frame designed by the artist, watercolor on paper (1895)

IX, 43. Joseph Müllner (1879–1968), ''Medusa,'' sculpture (ca. 1909)

Péladan. He showed a bestial Venus, arrogantly self-possessed even while chained in punishment to the walls of subterranean lust, while the polyplike tentacles of a giant medusa head, screaming in frustration, covered the feminine loins whose barren symbol, whose aggressive ''vagina dentata,'' the Medusa's head was widely thought to be, with its nearly masculine, phallic, yet hypnotically ingestive powers.

Many of the painters of the turn of the century depicted the Medusa frontally, with her mouth half-parted or, as in Khnopff's version, wide open in a silent scream, the snakes of her viraginity threatening the male viewer even as he was being lured by the enticements of bestial forgetfulness emanating from the cavernous vulval round of this ultimate siren. Carlos Schwabe's 1895 watercolor of the Medusa [IX, 42] is an especially striking example of this theme. With her claws poised, her cat's eyes staring, her mouth wide open, and her snake hair coiling into a bright bouquet of poison-toothed pink labia, this Medusa is a veritable nightmare visualization of woman as predatory sexual being. As if this weren't enough, Schwabe enclosed his monster in the circle of primal femininity and decorated his self-designed frame with a slithering chain of uroboric snakes biting each other's tails to form the eternal round of woman's materialistic preoccupations.

Other artists preferred less graphic representations of the Medusa's viraginous characteristics, emphasizing the almost masculine features of the creature's face, as the sculp-

IX, 44. Edward Robert Hughes (1851–1914), "Biancabella and Samaritana, Her Snake-Sister" (1894)

IX, 45. Franz von Lenbach (1836–1904), "The Snake-Queen" (1894)

tor Josef Müllner did [IX, 43]. He also made it a point to have the Medusa's serpentine coiffure appear as lifelike as possible.

Gustav Klimt, with his penchant for seeing women in water, naturally portrayed her in a number of works as a sea serpent. Edward Robert Hughes, in a painting exhibited at the Venice Biennale in 1895, made use of Italian fable to paint "Biancabella and Samaritana, Her Snake-Sister" [IX, 44]. The painting served as a good excuse to depict a very modern young lady in a Salammbô-like scene with the snake who is her sister, her mirror image, while she stands in one of those circular tubs so conveniently reflective of woman's same-sex, autoerotic self-involvement.

Franz von Lenbach's "Snake-queen" of 1894 [IX, 45] made woman into the tree of sensual knowledge around which the serpent wound itself. In "The Garden of the Hesperides" Lord Leighton succeeded in bringing together the theme of the collapsing woman with that of the encircling, uroboric snake (the detached phallus, the male turned into a sexual automaton, the dependent extension of woman). Leighton, like Schwabe, further emphasized the autoerotic self-sufficiency of the animal-woman by using a circular canvas for this depiction of a trio of blissfully inert daughters of night encircled by the body of their guardian snake, Ladon, which in its Teutonic enormity showed itself to be more than a match for Lenbach's impressive creeper.

IX, 46. Franz von Stuck (1863–1928), ''Sensuality'' (1897)

Other painters, like Franz von Stuck, explored the snake theme with maniacal intensity. The German painter variously portrayed Eve, the personification of sin, as wearing a gigantic snake over her shoulder or between her legs [IX, 46], or having a horizontal tête-à-tête (or, rather, torse-à-torse) with a snake whose proportions obviously stuck imposingly in the minds of impressionable youths such as Federico Fellini, who seems to have wanted to better von Stuck in the matter of snake size when he created the nightmare images in his film ''Juliette of the Spirits.'' Von Stuck was in the habit of repeating his compositions endlessly, as eager new clients demanded more images of evil women to hang on their walls as cautionary emblems. He also had in his repertoire an Eve offering Adam the fatal apple in a particularly close sort of cooperation with the serpent. He placed the offending fruit in the reptile's mouth and had Eve offer the apple, reptile and all to Adam. Von Stuck also liked to depict the horrific effects of woman's serpentine lubricity upon the future of man, placing men, women, and snakes, tightly packed, into the devil's dugout, thus giving his audience a convenient intimation of that netherworld in which a man might expect to end his days if he allowed himself to be dragged to ''the edge of the abyss.''

To return to the realm of the written word, though not, it should be said, to literature, Ayesha, the mysterious feminine emanation from the past who is the subject of H. Rider Haggard's still immensely popular, though abjectly written, potboiler *She* (1886), took care of the males around her in an equally snaky manner. We are told that she had the heart, the eyes, and the movements of a serpent; moreover, she had a ''terrible whisper, which sounded like the hiss of a snake'' (205). When she undressed, she showed herself ''shining and splendid like some glittering snake when it has cast its slough'' (198). Her body was ''instinct with a life that was more than life, and with a certain serpent-like grace which was more than human'' (161). Whenever she moved, we are told, ''her entire frame seemed to undulate, and the neck did not bend, it curved'' (149). Ayesha's eyes have a snake's hypnotic power, and in a scene dramatically illustrated in the novel's original edition, she uses them first to kill a woman who competes with her for the affection of the young British, golden-tressed, godlike hero of the narrative and then to hypnotize this clean-shaven young man ''and take hold of his senses, drugging them, and drawing the heart out of him.'' Her eyes, we learn,

> drew him more strongly than iron bonds, and the magic of her beauty and concentrated will and passion entered into him and overpowered him—ay, even there, in the presence of the body of the woman who had loved him well enough to die for him. It sounds horrible and wicked indeed, but he should not be too greatly blamed, and be sure his sin has found him out. The temptress who drew him into evil was more than human, and her beauty was greater than the loveliness of the daughters of men (238–39).

Painters eagerly responded to encouragement of this kind—especially to the rather more classy and mellifluent images produced by Flaubert and Baudelaire, who were wont to describe women in terms which, though certainly more elegantly formulated were not much different from those used by the journeyman English writer.

As had already become obvious in Flaubert's description of Salammbô's encounter with the python, it could no longer be denied that woman, more than merely liking to be seen with animals, was tempted to copulate with them as well. However, when dealing with the representation of straightforward cases of bestiality, turn-of-the-century writers and painters, as always, liked to make suggestive use of decorous myths rather than give too graphic or direct a representation of what men had rather worriedly begun to see as a major feminine fascination.

Anatole France's energetically sinful Thaïs, the heroine of his 1890 novel of that title—now probably more famous as the heroine of Massenet's opera—is a breathtakingly beautiful woman of rather disconcerting habits when we first meet her. For instance, she likes to enact in public "some of the shameful scenes which the pagan fables attribute to Venus, Leda, or Pasiphae" (31). Among this trinity of feminine connoisseurs of bestiality, Venus' principal predilection in the realm of this special sort of animal husbandry is well indicated by the object of the affections of one of Bouguereau's uninhibited bacchantes. The artist showed her as she frolicked with a frisky goat. This young lady's claim to being a true daughter of the great goddess of carnal passion would not have escaped a well-versed fin-de-siècle aficionado of academic art. However, it is not surprising that when Paphnutius, the monk who sets out to reform Thaïs, is haunted by nightmares of her, he dreams of her not as Venus but "as Leda, stretched out sensuously on a bed of hyacinths, her head thrown back, her eyes moist and sparkling, her mouth relaxed, the curve of her breasts revealed, and her arms cool as two streams" (32). For it was woman in the guise of Leda which fascinated the turn-of-the-century male viewer most. Not only did the painters have a long tradition of depictions of Leda—stretching from Leonardo to Delacroix— to fall back upon for thematic and iconographic justification, but, in addition, the long neck and snowy whiteness of the swan provided endless possibilities for elegant, sugges-

IX, 47. Albert-Valentin Thomas (active 1890–1905) "Leda" (1905)

IX, 48. Alfred-Henri Bramtot (1852–1894), "Leda" (1887)

IX, 49. Max Klinger (1857–1920), "Leda," sculpture (ca. 1900)

tive, and serpentine juxtapositions of woman and her bestial lover. Consequently, Ledas became legion in turn-of-the-century art.

Cézanne painted the swan-beset woman, as did Renoir, who took his cue from Léon Riesener, who in turn had found his inspiration in a composition traceable to Michelangelo. As the nineteenth century drew to a close, the representations of Zeus' swan-disguised exploits in pursuit of Leda became ever more explicit. Albert-Valentin Thomas, for instance, showed the godly swan as he triumphed in an especially aggressive and dominating fashion [IX, 47]. Alfred-Henri Bramtot, in his contribution to the Salon of 1888, appropriately placed Leda in the pose of the nymph with the broken back at the approach of her fowl-feathered rapist [IX, 48]. Max Klinger, one of the period's most relentless reporters of woman's bestial proclivities, carved a most graphic depiction of his suspicion that Leda had found a remarkably inventive new use for her nonaggressive and understandably submissive swan's conveniently stubby beak [IX, 49]. The symbolic role played by the swan's long neck was stressed more decorously, but no less obviously, by Klinger's compatriot J. M. Heinrich Hofmann, who painted a robust nymph as she appreciatively weighed the approaching cephalic splendor of her volatile suitor [IX, 50]. Félicien Rops, as usual, won the fin-de-siècle intellectuals' admiration for producing the period's most explicit—and still publishable—rendition of this theme [IX, 51].

Art lovers of the years around 1900 were so familiar with the carnal relations which had been established between the swan and the backsliding human woman that direct references to the Leda myth became entirely superfluous, as the title of Hofmann's painting already indicates. Similarly, Albert Besnard's innocuously titled "At Eventide" showed a woman, water, reflected light, and a swan sidling up to her—thus presenting information enough for any contemporary viewer to get his drift. Many painters habitually posed women at the edge of bodies of water in the vicinity of swans. Decorative and conveniently decorous, these graceful human creatures of nature could thus mingle with equally graceful fowl, which, as everyone knew, were their preferred companions in that world of simple

IX, 50. J. M. Heinrich
Hofmann (1824–1911),
"Nymph and Swan" (ca.
1884)

animal pleasures that had been transcended by man but was still central to woman's being. A painter such as Gaston la Touche seemed able to introduce naked women, water, and swans—often a veritable proliferation of the latter—in virtually every work he produced. Others, such as J. F. Auburtin, liked to combine primal woods, water, Chabas-like adolescent nymphets, and an impressive array of long-necked swans in order to provide their audiences with special insight into the world of woman.

In the years after 1900, hordes of stylistically derivative and qualitatively undistinguished painters sent works to the salon that coyly played upon their audience's knowledge of the Leda theme. A widely reproduced candy-colored confection sent to the Salon des Artistes Français in 1911 by H. C. Daudin featured two half-naked young women, straight from Montmartre, who were shown reclining in a classical architectural setting along a marble-edged garden pool. A swan was seen floating at their side, its elegant outlines ogled with enthusiasm by the denuded twosome. The title of the painting posed a wistful question: "And what if it were Jupiter?" Another painter, Floutier, had sent a similar production to the salon two years earlier, showing a trio of prepubescent girls frolicking suggestively with swans at the edge of a woodland pool. One of them in the center of the painting had been given the pose of the nymph with the broken back. Several swans were shown stretching their long necks expectantly toward her while the other two young girls watched what was going on with intense curiosity.

Once again it was the sterile, unproductive, and hence destructive and degenerative quality of woman's encounter with the swan which fascinated the intellectuals of the turn of the century. Maurice Barrès, in a section of his book *Of Blood, Delight and Death* (1894), given the title "The Evolution of the Individual," referred to "the embrace, so passionate but sterile, which Leda and the Swan exchange" (257). He equated the swan with Wagner's Lohengrin and emphasized the fowl's resemblance to the snake. "Serpent, bird, and fish at once, the swan is a composite, like nature itself." Barrès, future fascist and leader of the extremist French nationalists, saw in the swan's assertive act of rape the authoritative return of woman to her predestined position of abject submission to male

IX, 51. Félicien Rops (1833–1898), "Leda," etching (ca. 1890)

authority. The swan represented, for him, the essence of man's "pantheistic aspirations" (258).

A few years later Remy de Gourmont, in "A Woman's Dream" (1899), had a painter pose his model in a natural setting by a forest pool and had him play the masochistic counterpart to Barrès' sadistic conception. The model, displaying what Gourmont clearly saw as the lack of decorum characteristic of women, inevitably began to take the part of Leda seriously:

> When we returned, Leda, seated by the edge of the water, was feeding from her hand a large, shy swan which beat its wings at the slightest noise, flying over the water, returning like a galley to the young woman who stretched out her arms to it. Beside her the long snake stretched out, slid along her legs towards the withdrawn hand and sometimes, for a second, the bird covered with his large angel's wings the shivering body of a lover; one day the swan's golden eyes seemed enamelled and Leda closed hers, duped by her part, ready for the illusory marriage of mythological dreams.

The model subsequently confesses that "under the caress of the warm but dripping plumage she had indeed felt a vague desire of stupration; as the bird opened its wings above her leg she decided not to move, to let it happen" (I, 114–15). The painter in the story concludes, with a certain sense of relief, that such a woman is beyond male responsibility; he comes to realize that man can only participate distantly in the self-contained existence of these other-than-human creatures. "Nothing can be given to such a woman, scarcely even pleasure, which she receives disdainfully, almost as a compliment; a man is

not her lover, he is still her adorer when she forgets her divinity in our respectful arms''
(I, 116).

Certainly the most famous product of all these turn-of-the-century explorations of
Leda's bestial desires is William Butler Yeats' poem ''Leda and the Swan,'' first published
in 1923. Like virtually everything Yeats wrote, the poem's symbology is completely determined by the cultural atmosphere of the years around 1900. Yeats' Leda is a typically
brainless woman, the very personification of ''body,'' whose ''terrified vague fingers'' and
''loosening thighs'' can do nothing against ''the brute blood of the air.'' As a consequence
of her feminine inability to counter the destructive cycles of history with the force of her
soul, ''A shudder in the loins engenders there / The broken wall, the burning roof and
tower / And Agamemnon dead'' (211–12). Once again woman's bestial desires were to
breed a new cycle of evil, which would unleash the misery of war and murder upon the
world.

Just as popular as Leda's exploits with her swan were several of Zeus' other bestial
disguises in the service of woman's appetite for elemental pleasures. In 1889 Armand
Silvestre needed to take only one look at Alfred-Philippe Roll's ''Woman and Bull'' [IX,
52] to recognize its symbolic significance as a modern version of the rape of Europa. He
saw in the bull everything that was forceful and aggressive in ''nature swollen with the
sap of springtime, sweating from all its branches the fluids of obscure virilities.'' And,
said he, ''this beautiful, naked young girl—who with her fragrant hair caresses the rapt
muzzle of a young bull, warming her shoulder voluptuously against the tepid vitality of a
hide bristling softly with desire, who draws down against her own neck the caressing neck
of this animal and brushes her own rump against the moist down of his breast—this girl's
name, doubtless, is not Europa.'' Instead, declared Silvestre, being just an ordinary modern girl,

> she will, tomorrow, follow the shepherd who has told her the greatest number of lies. And he,
> the bull, bellowing in dense bewilderment, will buttress his mountainous body against the
> yielding loins of a heifer held captive between his knees. Each will end up following the bent
> of love naturally designated to their kind—the only form of love permitted in this era of false
> modesty. But for a single instant the spirit of the passing deity, that mortal yet altogether
> vivifying breath that brings kisses to everyone's lips, will hold these two, swollen with a vague
> desire and a delicious but useless melancholy. Together they shall have inhaled the same
> poison emanating from the same flowers whose petals caress their legs in the same manner''
> (*Le Nu au Salon—Champ de Mars*).

Silvestre's poetic flights unmistakably demonstrate that the so-called symbolic significance
of Roll's painting—and many others like it produced during the years around 1900—was
certainly not lost on his contemporaries, whose very graphic and anything but timid erotic
imagination has since largely come to be obscured by chronologically fuzzy assumptions
about the Victorians' supposed naiveté about sexual matters. If decorum did not permit
direct public enunciation of the baroque erotic fantasies which filled the minds of men and
women during this period, this was so to a large extent only because the very thinly veiled
antique symbolism of the writers and artists made their pornographic extremism far more
widely accessible, palatable, and lasting in influence than could have been accomplished
by crude immediacy of statement.

Nor is it possible in retrospect, to smile, as so many have, and wonder at the simple
naivete about sexual symbolism displayed by the artists and writers of this period, which,

so this mode of interpretation goes, made them expose their repressed sexual desires in images whose Freudian undercurrents have since been brought to the surface by more psychologically enlightened audiences. Instead, the writings and images presented here should make it self-evident that the artists and writers of the late nineteenth century who developed the sexual symbolism that was subsequently declared to be archetypal were in turn influenced by the prurient speculations and suggestive lucubrations of the scientists of the period. These scientists, in consort with the artists, planted a full-grown tree of sexual imagery in the minds of the psychoanalysts, who had grown up with these ideas and images all around them, and who now sought to prove the archetypal nature of these images by pointing to the unconscious symbolism of the very generation which had, to a large extent, quite consciously created that imagery.

Representations of humanity's links with the animal world have been legion in every period of cultural history. But no other age was as various in its very deliberate exploration of the role of animals in the satisfaction of woman's scientifically established hyperaesthesia as the fin de siècle; and no other age has ever made this theme as much into an elegant subject for playful parlor decorations destined for the homes of the ruling class as the age of Freud. A case in point is Thomas Theodor Heine's "Spring's Awakening" [IX, 53], which, through its title, directly linked itself to Frank Wedekind's hothouse play,

IX, 52. Alfred-Philippe Roll (1846–1919), "Woman and Bull" (1889)

IX, 53. Thomas Theodor Heine (1867–1948), "Spring's Awakening" (ca. 1910)

which in turn had taken as its theme the late nineteenth century's pervasive fascination with the inevitable awakening of adolescent womanhood to the call of nature and its reproductive (and destructive) needs. In his late Jugendstil work Heine showed two dainty yet spirited young ladies in the process of heeding the call of the wild by prancingly leading a hefty bull out of his pasture in a harness made of a garland of spring blossoms. Heine has given the bull an expression which indicates that it is slightly dubious about, but not entirely uninterested in, the peculiar doings of the young ladies. There is absolutely no question that the artist meant his viewers to observe this little scene and ask themselves the question also posed by Daudin in France at roughly the same time: "And what if it were Jupiter?"

Gustave Moreau was one of the late nineteenth century's principal creators of the sort of archetypal imagery which was to serve future generations of psychoanalysts. This painter was a master of the subtle, colorful, and elegant visual exploration of woman's interest in bestial liaisons. The exploits of Zeus in his various disguises therefore inevitably figure very largely in Moreau's oeuvre. Aside from numerous versions of the Leda theme—and as many explorations of the experiences of Europa and her bovine lover—this painter also explored lesser-known feminine liaisons with such mythical animals as unicorns, griffins, and the like. Moreau was especially fond of the story of Pasiphaë, perhaps because this woman's ambitions in the realm of amorous animalism represented an intensity of determination matched only by the enormity of her appetite for physical stimulation, easily making her the most uncompromising connoisseur of bestiality to entertain the fantasies of the awestruck intellectuals of the fin de siècle.

Pasiphaë, as any well-trained schoolboy of the time would have known, was the wife of the mythical King Minos of Crete, who was the product of the aforementioned liaison between Zeus, in his bovine guise, and Europa. Pasiphaë became enamored of a gorgeous white bull which Minos was supposed to have sacrificed in tribute to Poseidon but had kept for himself. He undoubtedly realized that this was not a wise action when Pasiphaë, taking to the bull, thought up an ingenious way for the bull to take her. The result of this truly monumental mythological get-together was that Pasiphaë became pregnant and gave birth to the infamous Minotaur. This creature, appropriately gifted with the body of a man and the head of a bull—and hence a living symbol of the effects of miscegenation—thereupon got into the habit of devouring various Athenian youths and maidens delivered up to him as he roamed around in the labyrinth built by Daedalus.

Moreau's representations of the bestial loves of women were never crude or graphic. The many sketches, watercolors, and paintings he devoted to Pasiphaë's courtship of the formidable animal she had come to desire are good examples of his understated approach to this hothouse material. Among them is a brightly colored watercolor in which Pasiphaë, having spotted her prospective lover in a field at sunset, can be seen undressing herself in the immediate vicinity of the specially constructed, life-size carpenter's facsimile of an attractive cow, into which she will climb within moments, to make the bull's access to her more convenient—and, one may assume, from the bull's point of view somewhat more orthodox. Daedalus, the cunning artificer who constructed this fascinating contraption for Pasiphaë, can be seen sitting somewhat pensively near his creation. It was a scene Moreau recreated, with minor variations, time and again, and one which, given his example, other late nineteenth-century painters pursued with equal diligence to warn men the world over about the dangerous proclivities of the female sex.

Given women's presumed regressive tendencies and their consequent interest in bestial

IX, 54. Arthur Hacker (1858–1919), "Circe" (1893)

relationships, it was also not surprising that the painters of the fin de siècle were especially eager to use Circe as a cautionary example of the eternal feminine. This Homeric witch's habit of turning men into swine was, after all, a clear indication of man's need to maintain his distance from the animal-woman. Arthur Hacker was there, as always, to provide a striking general representation of the theme, showing Circe as a voluptuous turn-of-the-century model sitting enticingly on a barren floor strewn with petals and cut flowers, which the artist actually succeeded in transforming into a strange sort of livid, fleshy debris suggestive of the—for her lovers—dire consequences of Circe's carnal appetite. Facing Circe and ogling her are members of Ulysses' crew in various stages of their transformation into perfect beasts [IX, 54].

In his commentary for the *Royal Academy Pictures* volume of 1893 (the year Hacker's opus was first exhibited), the well-known art critic M. H. Spielmann explained the cautionary nature of this painting with notable moral enthusiasm. The artist, he said, has portrayed Circe

> as a woman of arts and wiles, trusting to her own damning charms of body alone, with no bait of luxury or dazzling glory in her surroundings to entrap her half-willing victims. And to emphasise the moral of the story beyond, so far as I know, what any other artists have done, he has mingled with the enchanted pigs the men not yet metamorphosed, while retaining much the same expression on the faces of all. He has thus sought to accentuate the degradation of bestiality and sensual depravity, the depth of which is clearly sounded by the indifference of the human beings to the horrible change which is taking place around them.

John William Waterhouse combined the theme of the circular mirror with that of Circe's

IX, 55. Louis Chalon (b. 1866), ''Circe''
(1888)

carnality to show her offering her poisoned cup to a suspicious Ulysses, whose image is reflected in this mirror, while various swinish creatures surround the throne of the bestial witch. Waterhouse also painted a ''Circe Invidiosa'' in the process of poisoning the sea itself in order to infect the widest possible cross-section of males with animal desires.

In France Louis Chalon's ''Circe,'' exhibited at the Salon of 1888 [IX, 55], combined a foreground of groveling swine with decorative winged serpents and a throne graced by catlike creatures, topped by the jewel-encrusted circular halo of her materialistic self-containment. This image set the tone for numerous later depictions. Chalon's own inspiration had undoubtedly come, at least in part, from Villiers de l'Isle-Adam's description of Susannah Jackson, the ''Scottish Circe'' of his story ''The Eleventh-Hour Guest,'' one of his *Cruel Tales:* ''She is like quicksands: she swallows up the nervous system. She exudes desire. A prolonged attack of enervating folly would be your lot. She numbers several deaths among her memories. Her type of beauty, in which she has complete confidence, drives mere mortals to frenzy.'' Villiers' modern Circe, whose dream is to ''go and bury herself in a millionaire's residence, on the banks of the Clyde, with a handsome boy whom she would while away the time by killing at her leisure'' (87), is the sort of completely bestial female depicted by Alice Pike Barney in her pastel of a lion-maned woman, growling like an animal, her vampire's mouth ready to devour all comers, while she hugs the enormous head of a nasty-looking boar [IX, 56].

IX, 56. Alice Pike Barney (1860–1931), ''Circe,'' pastel (ca. 1895)

IX, 57. Elenore Plaisted Abbott (1875–1935), "Circe" (ca. 1918)

IX, 58. Félicien Rops (1833–1898), "Pornokrates," etching (1896)

Elenore Plaisted Abbott, who had an extremely successful career as an illustrator during the first two decades of this century, depicted Circe as a raving fury, flinging the fermented grapes of bacchic inebriation at a formidable host of boarish monsters [IX, 57]. It is difficult to detect any sort of reassessment or positive transformation of male stereo-types concerning woman's bestial nature in the images of either Barney or Abbott. Their work suggests that these artists had uncritically accepted the concept of the feminine estab-lished by the fin-de-siècle male establishment, and that they had made this acceptance the basis for personal fantasies of revenge in which they saw woman as gleefully hauling the male back into her nature-ordained prison house of degenerative materialism. Abbott's image, an illustration for an edition of Keats' poems, is also an instructive instance of the manner in which the great fantasy illustrators of this period were wont to infuse their work with prevailing notions of feminine evil. Even—one is tempted to say especially—such children's book illustrators as Arthur Rackham, Edmund Dulac, Walter Crane, and How-ard Pyle have contributed immeasurably to the perpetuation of antifeminine stereotypes among succeeding generations of young readers.

In any event, it is clear that by 1900 writers and painters, scientists and critics, the learned and the modish alike, had been indoctrinated to regard all women who no longer conformed to the image of the household nun as vicious, bestial creatures, representative of a pre-evolutionary, instinctual past, who preferred the company of animals over that of the civilized male, creatures who were, in fact, the personification of witchery and evil, who attended sabbaths and dangerous rituals astride goats, as the Spanish painter Louis Falero, among others, depicted them. Woman, in short, had come to be seen as the monstrous goddess of degeneration, a creature of evil who lorded it over all the horrifically horned beasts which populated man's sexual nightmares. She was the human animal viciously depicted by Félicien Rops as "Pornokrates" [IX, 58], ruler of Proudhon's "Pornocracy," a creature blindly guided by a hog, the symbol of Circe, the bestial representative of all sexual evil.

Lombroso and Ferrero had warned that the ferocious hatred for man which filled the criminally erotic woman "has no cause whatever, and springs from blind and innate perversity" *(The Female Offender,* 157). It was the result of a congenital primitivism in woman which was well beyond the male's capacity for understanding. It is clear that the painters agreed with this assessment. As a result of her instinctual affinity with the animals, woman had truly become "the idol of perversity" (see frontispiece), the livid-eyed, snake-encircled, medusa-headed flower of evil, whose aggressively pointed breasts were as threatening as the fangs of a devouring animal. This woman, if she were to breed at all, could only be expected to mother hordes of degenerative temptresses, treacherous sea creatures, predatory cats, snakelike lamias, harpies, vampires, sphinxes, and countless other terrible, man-eating creatures.

Indeed, given women's continuous concourse with animals of every possible shape and form, it was not at all surprising that their primal nature, their *Urgestalt,* as Wedekind called it, could only be grasped when they were depicted unflinchingly in those half-human, half-bestial states of being which characterized their nature. If the dangerous and vicious spider painted by Jules Verdier in 1905 was still outwardly wholly human, the creature's pose showed that its intentions were purely bestial [IX, 59]. Various other creatures who actively evinced the degenerative effects of miscegenation could be found crawling along the walls of the yearly art exhibitions, from homoerotic witches with cats' paws, horrible snake-bodied Eves, and tiger-tailed seductresses, to the Wagnerian "Swan-Women" of Walter Crane [IX, 60]. These monsters showed the logical effects of woman's repeated indulgence in close encounters with animals.

Among other such associations, Franz von Stuck chose to depict a formidably heterosexual sphinx—clearly not a creature any wholly human male should have tried to tangle with in the first place [IX, 61]. Males seeking relief from temptation could hang it on their walls thanks to an enterprising publisher, who had published a large photogravure of this timeless masterpiece suitable for framing. Von Stuck was also in the habit of painting nude women stretched out in sphinxlike poses in the night air, on slabs of marble which clearly belonged in a cemetery.

The point was, after all, that all women were sphinxes at heart, as Edvard Munch kept stressing in his paintings and lithographs. The Dutch painter Jan Toorop, on the occasion of a joint exhibition of his work and that of Van Gogh in Utrecht in 1898, gave the following instructive explanation (heavily influenced by Rosicrucianism) of his own version of the sphinx [IX, 62], a work on which he labored, off and on from 1892 to 1897. "They," he said,

> who are completely caught under the weight of the sphinx's claws, are unevolved beings. In the center of the painting man and woman (dualism), struggling toward ever higher evolution,

IX, 59. Jules Verdier (b. 1862), "Araena" (ca. 1905)

IX, 60. Walter Crane (1845–1915), "The Swan-Women" (ca. 1895)

IX, 61. Franz von Stuck
(1863–1928), "The
Sphinx's Kiss" (1895)

are chained to earth. Man, who has succeeded in enhancing his art with the ideal qualities (the veil) of woman, whose head is encircled with the aureole of eternal virginal beauty, whose breast heaves for the Rosa Mystica. To the right are those who have freed themselves from the sphinx's claws and who therefore constitute the propellant force of all spiritual labor. In the foreground of the painting one may find the higher, tender, understanding and intuition (song and the sounding of strings) of unsullied female souls floating onward to an ever more ethereal spiritual life. Two children (lower right) symbolize the tenderness of innocence. The inward-looking ascetic (left), the thinker lost in unworldly thought, represents the link between worldly struggle and the highest aspirations . . . *(J. Th. Toorop, De Jaren 1885 tot 1910, 44)*.

If in Toorop's mind the mysterious bestial sphinx and the ethereal household nun were still to be found in total dualistic opposition, many of his contemporaries gave the latter image short shrift, choosing instead to concentrate on woman as the predatory beast. Ivan Meštrovič's sculpture of a sphinx [IX, 63] concentrated on the creature's viraginous fury. With the hypnotic stare of a snake, the paws of a cat, lethally protuberant breasts, and the overblown musculature of a late twentieth-century female bodybuilder, this sphinx represented a masochistic male fantasy of the ultimate dominatrix, the goddess of stony bestiality.

But many turn-of-the-century artists pointed out that the contemporary sphinxes men encountered all around them did not need to have the bodies of lions or cats; their sphinx-

IX, 62. Jan Toorop (1858–1928), ''The Sphinx,'' crayon and wax pencil (1892–97)

like nature was revealed by the poses they assumed even in such ordinary circumstances as when sitting down to supper in a restaurant, as Léon Bakst was careful to show in a painting he exhibited in Munich in 1902 [IX, 64]. Regardless of how the sphinx showed herself, she certainly had a fierce appetite for manslaughter. Gustave Moreau's many sphinxes were always surrounded by the dismembered bodies of their prey [IX, 65]. Fraught with similar implications, Elihu Vedder's ''Sphinx of the Sea-Shore'' shows a young woman with long red hair, wide-open eyes, and a half-open mouth, her face distorted by elemental passion, nestling her cat's body flat against the soil. She is surrounded by the wreckage of ships, half-buried skulls, the skeletal remains of an arm and a hand still reaching for her from the sand. Like a cat playing heedlessly with the body of a mouse, she holds a man's skull between her forepaws, nestling it against her bosom, her nails digging into its bone. This painting, wrote a commentator in *The International Studio* in 1900, is perhaps Ved-

IX, 63. Ivan Meštrovič (1883–1962), "Sphinx," sculpture (ca. 1910)

IX, 64. Léon Bakst (1866–1924), "At Supper" (ca. 1901)

IX, 65. Gustave Moreau (1826–1898), "Oedipus and the Sphinx" (ca. 1888)

IX, 66. Fernand Khnopff (1858–1921), "The Supreme Vice," crayon (1885)

der's "most unique painting." He described it as "a sombre landscape, a chimera-like figure with a woman's head, her mouth open, the white teeth gleaming through the gloom—a conception of a scourge teeming with purport" (X, p. ii).

The purport was that the image of woman had for many men shifted from the dualistic absolute of the totally submissive mother-woman to that of the totally destructive man-eater. In Flaubert's *Temptation of St. Anthony,* the Chimera, sister to the Sphinx, exclaims: "Like a hyena in heat, I turn around thee, soliciting the impregnations the want of which devours me" (341).

The period's many representations of sphinxes and chimeras were characteristic examples of the nineteenth century's habit of juxtaposing—without synthesizing—extreme dualistic opposites in a single image. The sphinx, "half woman and half animal" as Oscar Wilde said, still had the outward appearance of the warm, all-yielding mother of the 1850s, for she had the nursing nun's tantalizing, milk-white breasts with which she lured her unsuspecting sons as with a promise of benevolent passivity. But once nestled at nature's fertile bosom, the helpless sons discovered that their mother, metamorphosed into her daughters, had dared to grow claws, to consort with the devil, to become a creature with demands. The Oedipal seed might have originally been cultivated in the family womb of the household nun, with "its nest-like peace and warmth, its jealous exclusion of all that was against itself and its own immaculate naturalness, in the hedge set around the sacred thing on every side" (II, 124), as Pater's Marius had experienced his family's motherly loins, but the flower of late nineteenth-century manhood all too often had to grow up to discover the lineaments of the sphinx in the nun's surrogates, her justly impatient daughters.

These daughters consequently came to stand for the mother-turned-evil, the ultimate monster, "The Supreme Vice" [IX, 66], depicted by Fernand Khnopff. In Khnopff's drawing, behind every man's vulnerable, statuesque, nude virgin-sister standing haloed on her pedestal there loomed formidably the evil sphinx of holy motherhood turned barren aggressor, her enticing breasts—the enlarged echo of the breasts of the virgin standing before her—guarded by the claws of polyandry and the death's-head of bestial passion.

In his sonnet "Une" (perhaps best translated as "One of Them") from his collection *In the Garden of the Infanta* (1893), the French symbolist poet Albert Samain expressed in words precisely what Khnopff had tried to portray in his engraving:

A Sphinx with emerald eyes, angelic vampire,
She hides in dreams under the cruel gold of wavelike hair;
Her mouth afire with red, hot like the hearth's slow embers,
And her eyes false, her heart false, her love still worse.

Her hard brow harbors a dim dream of empire.
She is the haughty, frigid flower of all ills.
And mortal sin, acrid in its slow unfolding
Rises in dark perfumes amid the venom of her flesh.

On her throne, which has been carved with dismal skill,
Immobile, distantly, she hears the howling:
Seas of poor hearts all bleeding from her evil will.

Lulled by these cries, she dreams, and sometimes
Singes with a look in which her lewd self smolders
The lily's virgin soul still dying slowly in her hand.

The war between the sexes, the war between male and female, between Apollo and

Dionysus, was a war between the godly future and the earthly past, between science and sorcery. It was ultimately a struggle between woman's atavistic hunger for blood—which she regarded as the vital fluid of man's seminal energies and hence the source of that material strength she craved—and man's need to conserve the nourishment that would allow his brain to evolve. Woman was a perverse instrument of the vampire of reversion, and by giving in to her draining embrace, men thought, they must needs bleed to death.

CHAPTER X

Metamorphoses of the Vampire: Dracula and His Daughters

The sphinx, soft-breasted mother and steel-taloned destroyer conjoined, was only one of many chimeras of womanhood expressive of the late nineteenth-century's extreme dualistic mentality. How these fantasms came to be created was spelled out very clearly in Vernon Lee's short story "A Frivolous Conversion," added to the second edition of her collection *Vanitas* in 1911. In that story Count Kollonitz, a man of violently absolutist proclivities and clearly a member of the generation born around 1870, proclaims his credo: "I am religious, in a way too," he remarks during a conversation. "Darwin, Haeckel, Nietzsche—that sort of thing? I used to read a lot of it when I had more leisure. You see all is founded on brute strength: the strong animal eats the food of the weak animal, and sometimes the weak animal itself" (10). Later on he exclaims, "Women! Why, of course, I believe in women! My mother was a saint. My married sister is a perfect angel—I must show you her photograph. I think that women are *far better* than men; I think women ought to be a kind of angels—and when they are not, why . . . You know how they used to treat vampires in my country—people who were corpses reanimated by devils and who sucked peoples' blood?" Mme Nitzenko, to whom these remarks are addressed, is all too familiar with Count Kollonitz's point of view. She attempts to reason with the young man, although she is well aware of the futility of her efforts:

> "As regards women, don't you think it's rather hard on them to divide them into angels and vampires only? It would be very unfair to divide all men into devils and curates. Why shouldn't a man expect from a woman just the same amount of strength or of weakness as from himself?"
>
> Kollonitz was sincerely shocked. "Good Heavens," he cried, "my dear lady, you talk like Ibsen! But you are jesting. A good woman like you cannot hold such opinions. Don't you see that you are destroying the *ideal*—and without the ideal in some form or shape—God, woman, beauty, chivalry—why, the world would be a pit of darkness."

Mme Nitzenko smiled sadly.

"Would it?" she answered. "Well—is it much else, with its population of female vampires of which you tell me, and its idealizing young men, believing that women ought to be angels, and dividing their time between—such angels—and the female vampires in question?"

At these remarks, the author notes, "Kollonitz merely laughed," and we take our leave of this typical representative of the idealizing young men of his time as he "paused on the terrace and philosophized over remnants of Schopenhauer and Nietzsche" (15–17).

The virgin and the whore, the saint and the vampire—two designations for a single dualistic opposition: that of woman as man's exclusive and forever pliable private property, on the one hand, and her transformation, upon her denial of man's ownership rights to her, into a polyandrous predator indiscriminately lusting after man's seminal essence, on the other. William J. Robinson, MD, chief of the Department of Genito-Urinary Diseases and Dermatology at the Bronx Hospital and author of a popular guide to *Married Life and Happiness,* published, appropriately, by the Eugenics Publishing Company of New York in 1922, spelled out the issue in clear and simple language. Wives, he said, who are "satisfied with occasional relations—not more than once in two weeks or ten days," are relatively reasonable in their demands and may be considered normal. However,

> there is the opposite type of woman, who is a great danger to the health and even the very life of her husband. I refer to the hypersensual woman, to the wife with an excessive sexuality. It is to her that the name vampire can be applied in its literal sense. Just as the vampire sucks the blood of its victims in their sleep while they are alive, so does the woman vampire suck the life and exhaust the vitality of her male partner—or victim. And some of them—the pronounced type—are utterly without pity or consideration (90).

Robinson was positing as a scientific commonplace a direct equation between woman's supposed hunger for seminal substance and her bestial blood lust. This blood lust was thought to be precipitated by her insatiable need to replenish the blood incessantly lost to her system as a result of her degenerative subjection to the reproductive function and its attendant sexual cravings. The equation in question was perhaps first directly expressed in literature by Baudelaire, in his poem "Metamorphoses of the Vampire," written around 1852 and deleted by the censors from the first edition of the poet's *Flowers of Evil* in 1857. In this poem a woman, obviously a prostitute and consequently a woman of "excessive sexuality," a polyandrous sphinx, a livid idol of perversity, "coiling like a snake / across hot embers," promises to provide with "fluid lips" unheard-of pleasures, guaranteed to make "old-fashioned conscience disappear into the darkest reaches of the bed." The narrator succumbs to the woman's enticements, but "once she'd sucked the very marrow from my bones," the creature seems to turn into "a slime-flanked mollusc full of pus." In horror, the narrator closes his eyes, and when he opens them again the protean monster has once more transformed itself: "Instead of that cadaver taut with force / Having gorged itself on blood, / Scattered pieces of skeletal remains, trembling in confusion / Squeaked without provocation like the sour night-cries of a weathervane" (*Oeuvres Complètes,* 143).

For the men of the second half of the nineteenth century—who strove to soar upward into the empyrean of intellectual transcendence upon the shoulders of their ever-pliant, gratefully suffering wives—it seemed that the pleasures of the body were to be paid for with death. The womb of woman was the insatiable soil into whose bottomless crevasses man must pour the essence of his intellect in payment for her lewd enticements. The hunger of the beast was in her loins, and the hunger of the beast was the hunger for blood. Woman's bestial couplings, her tendency to atavistic reversion, brought out the beast in man. The conjoining of bestial woman with the remnant of the beast in man could only spawn human animals, evil creatures from the distant past coming back to haunt civilization: hungry, half-human sphinxes, winged chimeras—blood-lusting vampires all.

Prosper Mérimée in his novella *Lokis* (1869) made the connection between blood lust and bestiality quite clear. The story takes place in Lithuania—in Eastern Europe, for the East was symbolic of the past, the morning of the world, the pre-evolutionary era, while the West could only be the *Abendland,* the land of fruition, ripened by the light of the sun, harbored in the bosom of evolutionary transcendence. There, in the primitive East, we encounter the Countess Szemioth, "a waxen figure" whose "wide-open eyes" stare "vacantly." We are to consider her mad. During the course of the narrative, we discover that she has been insane for at least twenty-seven years, and that she has suffered from violent, seemingly aimless outbursts of anger ever since the birth of her son, the present count. What has happened—and Mérimée, without ever precisely saying so, drops enough heavy-handed hints to make this quite clear to the reader—is that only two or three days after her marriage she was raped by a bear. That was her own fault, of course, since, like all Lithuanian ladies, she was a regular amazon, a veritable wild woman who had insisted on joining the hunt with her new husband. In the primal forests, "the great womb, the great nursery of the animals," Mérimée hints broadly, she strayed and was carried off by the offending bear. Later she was found "badly scratched, unconscious, of course," and with a broken leg. From then on she was mad—and, as it turns out, also pregnant. Everyone thinks this is her new husband's doing, but the reader knows better.

Enter Ioulka, Mademoiselle Iwinska, who is a tease because she has a sense of humor and a temptress because she has energy and a mind of her own. She flirts with the new Count Szemioth, who, unknown to all, including himself, is the son of our impetuous bear—his bestial nature still dormant because he has not yet been tempted. Ioulka—what else can one expect of a temptress, a "frivolous coquette"—loves to dance. She therefore dances barefoot, "at the risk of showing her leg," the dance of the russalka, that East-European water nymph previously encountered, who, if you are a man, likes to gobble you up. And, in a sense, she does. She marries the count, who has crumbled before her abilities as a dancer. Even before the wedding, however, the count, his bestial nature stirred, becomes extremely interested in stories about Uruguayan gauchos, who drink the blood of animals in times of need. He informs himself about that country's president, who, like other "white men who have lived for a long time with the Indians," acquired such a liking for blood that he "hardly ever missed a chance of gratifying that taste" (153). Indeed, "one day, when he was going to Congress in full uniform, he passed a *rancho* where a young foal was being bled. He got off his horse to ask for a *chupon,* a suck; after which he delivered one of his most eloquent speeches."

With information of this sort buzzing in his head, and with his animal nature aroused by an impudent woman, what is a bear cub count to do? Ioulka and he go through the marriage ceremony. They are festively sent off to their bridal bed. The silence of night. Strange, bestial noises. The sound of "a dark body of considerable bulk" passing by a window and falling "with a dull thud into the garden" (167). When, late the following morning, the happy young couple have not yet shown themselves to the wedding guests, the door of their room is forced open. "The young Countess was stretched out dead on her bed, her face horribly lacerated, her throat torn open and covered with blood. The Count had disappeared, and no one has heard anything of him since.'' One may assume that, the blood lust stirred in his veins by his luckless lady's primordial russalka nature, the count, having drunk his fill of woman, went off in search of sterner stuff among the denizens of his burly dad's primeval forest, a terminal victim to woman's ability to bring out the beast in man. For what happened was ultimately all Ioulka's fault, with her nymph-like dancing, her flashes of bare feet, and her frivolous materialism, just as the count's mother, with her amazonian proclivities, had brought upon herself what she had had to bear. Indeed, in a painting that would seem to be a virtual illustration of Mérimée's narrative, Henri Rousseau showed a bear enthusiastically ogling a lady whose nudity seems to have been deliberately designed to serve as a lure, so that an unsportsmanlike hunter can take advantage of the brawny animal's quite understandable distraction. Moritz Bauernfeind, no less attuned to the details of late nineteenth-century naturalistic lore than Rousseau, joined the substantial group of fin-de-siècle artists who, like Mérimée, had uncovered new, scientific motives for the encounters between women and bears found in folk tales and legends. The painter chose to show a quite literally enchanting young woman who had been able to take the bite out of a formidable bear and turn the beast instead into a simpering, love-struck sissy, an oversized lapdog [X, 1].

The supposed invigorating nutritive qualities of blood, so well documented in *Lokis*, made it particularly easy for men to suspect women, with their generally anemic constitution—women's blood was thinner, more watery than men's, as Havelock Ellis pointed out in *Man and Woman*—and their inevitable periodic blood loss, of having a constitutional yearning for this tonic. In his *Six Moral Tales* Jules Laforgue described an ''ideal maiden on her deathbed,'' wracked with consumption, who still has ''a little mouth that is greedy

X, 1. Moritz Bauernfeind (1870–1947), ''From the Cycle 'The Three Sisters Chronica,' central panel'' (ca. 1908)

even though it is bloodless." As a daughter of the moon, woman could hence easily become an actual vampire, that "white bloodless creature of the night, / Whose lust of blood has blanched her chill veins white, / Veins fed with moonlight over dead men's tombs," as Arthur Symons characterized her in "The Vampire" (*Lesbia*, 1). After all, did not woman's very nature make her a vampire, a blood drinker? If she did not allow herself to be domesticated by becoming a submissive wife—and, as soon thereafter as possible, a mother—the veneer of civilization would soon peel away.

In the same year that Mérimée wrote *Lokis,* Horace Bushnell described the sort of woman who would become the norm if one stripped her of the leavening dignity of motherhood. The appearance of a woman who was permitted to work, to get involved with politics, and to develop a mind of her own was likely to be as follows: "The look will be sharp, the voice will be wiry and shrill, the action will be angular and abrupt, wiliness, self-asserting boldness, eagerness for place and power will get into the expression more and more distinctly, and become inbred in the native habit." Such women who did violence to their single role in life were apt to have "thin, hungry-looking, cream-tartar faces, bearing a sharper look of talent, yet somehow touched with blight and fallen out of luster" (*Women's Suffrage,* 135–36).

In 1900 the French novelist Rachilde (the pseudonym of Marguerite Eymery-Vallette, wife of the editor and founder of *La Revue Blanche*) published a short story called "The Blood Drinker" in which she meshed the themes of the Eternal Feminine, blood lust, and the degenerative effects of female sexuality. Rachilde was held in high regard by the French symbolists; she was described by her champion, Maurice Barrès, as "Mademoiselle Baudelaire" because of her aptitude for the representation of "cerebral perversity" in such works as *Monsieur Vénus* (1884), a role-reversal novel in which a woman sets up a "male-mistress" in an apartment, treating him in very much the same way as nineteenth-century males kept their mistresses, watching him grow fat, slothful, and passive in response to her aggressive-possessive viraginous behavior.

Monsieur Vénus is characteristic of Rachilde's work. Its role-reversal theme was not meant as a serious critique of the state of late nineteenth-century male-female relationships. Instead, the narrative is an early example of the unthreatening reversal games which, during the past century, and especially among the French intelligentsia, have too often taken the place of serious social criticism. Rachilde's novels, with titles such as *The Marquise de Sade, Madame Adonis, The Female Animal, Poison Tail,* and so on, milked this vein extensively to provide cheap thrills to shock the bourgeoisie. Not surprisingly, they had a considerable vogue among bourgeois intellectuals.

"The Blood Drinker," like *Lokis,* ingeniously and obliquely plays upon a rather gruesome sidelight of late nineteenth-century medical theories on how to cope with the anemia which plagued and weakened so many middle-class women, and which was deemed a principal source of the weakness of effeminate males. What better way to strengthen one's blood, it was reasoned, than to drink the blood of others—not that of humans, to be sure, but the blood of strong animals like oxen. This blood should be as fresh as possible. In consequence, slaughterhouses everywhere began to attract "blood drinkers," anemics who came to ingest their daily cup of ox blood. Exhibiting the characteristic flair of certain late nineteenth century salon painters for anything sensational and anecdotal (in many ways they were the forerunners of the twentieth century's newspaper photographers), Joseph Ferdinand Gueldry painted an extraordinary record of this phenomenon and exhibited it at the Salon des Artistes Français in 1898 [X, 2]. The painting caused a sensation and was

X, 2. Joseph-Ferdinand Gueldry (1858–after 1933), "The Blood Drinkers" (1898)

widely reproduced in such periodicals as *The Magazine of Art,* whose editor described the work as follows in his report on the Paris salons of 1898:

> One of the most popular pictures of the year is undoubtedly Monsieur Gueldry's gorge-raising representation of "The Blood-Drinkers," in which a group of consumptive invalids, congregated in a shambles, are drinking the blood fresh from the newly-slain ox lying in the foreground—blood that oozes out over the floor—while the slaughterers themselves, steeped in gore, hand out the glasses like the women at the wells. What gives point to the loathsomeness of the subject is the figure of one young girl, pale and trembling, who turns from the scene in sickening disgust, and so accentuates our own (495).

Given such horrific would-be medical solutions to the problem of anemia, the actual prevalence of anemia among women, and the period's preoccupation with the conflict between civilization and brute nature, it was all too easy to see in the actions of those who drank blood for medical reasons an indication that, like the fictitious president of Uruguay in Mérimée's tale, one could actually acquire a taste for such practices. It began to seem by no means farfetched to suspect the existence of vampires, and especially vampire women.

Rachilde, in her emphatically symbolist story "The Blood Drinker," positions herself in the interstices between the reality of the late nineteenth-century cures and the psychological fascination of her contemporaries for the notion of the bestial vampire woman. Her blood drinker is none other than the moon—the feminine principle—beamed in upon herself in "the eternal desperation of her own nothingness." In her story we learn what effect the moon—"domineering, imperious, open and round like a golden hole," a goddess whose "hydralike face" is admired by "fascinated sand snakes"—has on the actions of a young peasant girl of fifteen. The moon, we are told, in "her silent vampire's flight" searches the earth for votaries, and "drunk, yet desirous to drink more, she draws to her, she sucks in, she swallows up into her open, golden pit-mouth everything that is of the spirit or of blood." Soon she spots below her a tiny black creature. "First it is an insect, then an ant standing erect, a snake undulating on its tail, then a bird stepping and dragging its wings, and finally it is a woman." The young girl to whom we are introduced by means of this clumsy, would-be phylogenetic traversal of her evolutionary prehistory is, we soon discover, wracked by the indistinct longing of a creature in heat. She has therefore been sent out-of-doors by her grandmother, who herself is "a woman submerged in folly," to encounter the moon, the presumed source of her discomfort, the medusa head of the devil, "the blond, severed head, still searching eternally for the dispersed blood of her former body."

The young girl's encounter with the moon has an immediate effect on her. "Her legs become numb; she's overcome by an infinite languor. She looks at the moon, and the moon must have seen her; her moon face, with its transparency of a woman who has died of consumption, has become even darker, even more livid with corrupt blood" (72–74). Overcome by the stare of that "golden monster," the young girl collapses into a sensuous sleep during which she dreams that "she's embraced by the moon, that the moon is a mouth of honey." When she awakens, the moon has disappeared, but the girl has become her avid disciple. A woman now, a genuine votary of the moon, her periodicity, we are to understand, must henceforth coincide with the cycles of that moon. "Hurting, she turns away, a small shadow, leaving spots of shadow on the pale road." And above her, we are told, "the moon snickers, the moon, flower of fire, who lives off the blood of women!" (75)

When next we encounter the young girl she's become a true bacchante, frolicking with a boy under the greedy eyes of the moon, "who watches them as the pupils of a hefty cat might watch a pair of mice." Once again the girl is to shed blood in service to the moon, this time as her companion demonstrates to her the error of her teasing remark to him "that he would not seem to be a man." Inevitably thereafter, when we encounter the young girl for the last time, "she has a belly so big that it forever precedes her down the road." It is at this point in her narrative that Rachilde pulls together all the ribbons of cliché about the Eternal Feminine which she has so laboriously unwound. She makes the story end in a symbolic scene of ritual murder which must have satisfied even the most avid woman haters among her male admirers on the editorial staff of *La Revue Blanche.* Lying down in a ditch, the girl calls to

> all the evil angels to deliver her! And *one* comes, who with his pointed nail rips open her belly. Then she is furious, drunk with anger; she takes him by the throat—he is very small, but the worst demons are all small—and she strangles him without realizing, the poor girl, that she is strangling her own child.
>
> She climbs out of the ditch, covered with filth, covered with mud, bent double, grown old all of a sudden, and she imitates her grandmother, the insane one; she shakes her fist at the heavens, at the Moon.
>
> The latter, victorious over the black cloud-squadrons of death, scintillates, grows into a white flame, pure, pure as the light of a white taper. The Moon shines; the Moon, flower of fire, who lives off the blood of women! (80)

Rachilde's primary concern in everything she wrote was for spectacle. But precisely because of this she could afford—more easily than writers concerned with balance or artistic coherence—to play directly upon the fears and fantasies of her readers—and, of course, her own. As such, ''The Blood Drinker'' is an extraordinary document of the level of self-hatred reached by some turn-of-the-century women. As a result of the relentless, strident indoctrination by the men around them, many had developed a fierce distaste for the mere fact of being a woman, and of having to cope with what men everywhere were loudly proclaiming the ''inherent'' physical disabilities and ''animal requirements'' of women. Rachilde does not show the slightest inclination to doubt the male cliché of her time that held woman to be an intuitive animal, unaware of, yet totally involved in, her own unindividuated participation in the bestial cycles of nature—even if, having entered into civilization, she was incapable of responding as automatically as an animal might to the cycles of fertility and gestation. The clash between her instincts and her imperfectly individuated will thus caused woman to wreak upon the world the sort of unreasoning, murderous havoc described in ''The Blood Drinker.''

Having been given a touch of imagination but no intelligence in the course of the evolutionary process, woman, Rachilde argued, literally became the moon, the very inversion of the creative impulse, a negative reflection of the will to power. Left in her primal environment without guidance, she reverted to the predatory nature of the animals. Removed from the gestatory influence of the sun, from the civilizing male intellect, she became the moon's tool, an instrument of degeneration, a child murderer, a vampire, a werewolf, one of those creatures who could live only by night, under the auspices of the moon, in the realm of the senses, the realm of darkness, of sex, of bestial desires, that realm which the turn of the century had declared to be woman's true realm because it was the moon's realm; a barren desert in which the warming, ''civilizing'' rays of the male sun could no longer guide and protect feeble (and feebleminded) womanhood.

This desert of materiality, in which women are stalked by untold horrors and become beasts in the absence of men, assumed its modern form in the nineteenth century in a concatenation of symbols clearly based upon the psychology of social domination. The image patterns developed then may since have lost their overt ideological underpinning. Yet when we go to horror movies and are thrilled to watch terrified women—for they are always terrified *women*—being stalked by vampires and werewolves under the full moon, we are still unconsciously waiting for the beast to get them, and for the beast to be *brought out* in them. In these films men are always absent at such crucial moments. One need not wonder why presumably intelligent and otherwise overly protective males always seem bent on leaving their women alone *at night*. After all, in the popular entertainments of the twentieth century night is still not the realm of men. In the evening the male must leave—like the sun—and be present only in the thoughts of the terrified woman. Indeed, it is always woman's principal function in narratives of this sort to keep the light of her protector's former presence flickering in the mirror of her helplessness. And when the hero, the true male, the *intellectual* male—he is usually a scientist or a philosopher—comes back to release her—*if,* that is, she has been well trained in the ways of civilization and has been able to maintain the vestiges of his light—he always brings the sun with him. He slays the vampire by exposing it to the light of day, for he must overpower the beast of night, the minions of the moon. He must exorcise the inherently regressive, degenerative susceptibilities of woman with the broad sweep of his superior, light-born male intellect. The writers of contemporary Gothic novels, the makers of vampire movies, as well as the many men and women who are virtually addicted to these narratives, pronouncing them harmless fun or simply campy entertainment, are still unconsciously responding very directly to an antifeminine sensibility established in its modern form and symbolic structure by the sexist ideologues among the nineteenth-century intelligentsia.

The continuing popularity of the two most successful late nineteenth-century vampire narratives, Joseph Sheridan Le Fanu's *Carmilla* and Bram Stoker's *Dracula,* is a case in point. Both have been made into movies repeatedly and both are central documents of the late nineteenth-century war on woman, direct expressions of the dualistic sentiment which made Nicholas Francis Cooke declare in 1870—at the same time that Le Fanu was writing *Carmilla:* "The temperament of woman exposes her to the most singular inconveniences and inconsistencies. Extreme in good, she is also extreme in evil. She is inconstant and changeable; she 'will' and she 'won't.' She is easily disgusted with that which she has pursued with the greatest ardor. She passes from love to hate with prodigious facility. She is full of contradictions and mysteries. Capable of the most heroic actions, she does not shrink from the most atrocious crimes." Indeed, concluded Cooke, women "are more merciless, more bloodthirsty than men" *(Satan in Society,* 280–81).

Carmilla marks one of the first appearances, center stage, of a female vampire in modern fiction. Previous vampire narratives, such as Polidori's *The Vampyre,* Maturin's near-vampire *Melmoth The Wanderer,* the gruesome and endless potboiling nastiness of *Varney the Vampire,* had all still featured male predators, and even Gautier's Clarimonde had to be satisfied with playing second fiddle to the schizophrenia of her living lover Romuald. Carmilla, however, was born right on schedule among the daughters of the household nun. As a creature of moonlight, she, like most of the late nineteenth-century's crop of female vampires, is not permitted any direct vampire power over men. Instead it becomes her role to prey on other women. She is delivered into the house of the family whose daughter she is to deprecate by a "hideous black woman, with a sort of colored turban on her head," the veritable personification of woman's animal nature emanating from the evil East, the predatory past. Becoming fast friends with the daughter of the house—who is also the narrator of the story—Carmilla soon shows herself possessed of strange passions.

> She used to place her pretty arms about my neck, draw me to her, and laying her cheek to mine, murmur with her lips near my ear, "Dearest, your little heart is wounded; think me not cruel because I obey the irresistible law of my strength and weakness; if your dear heart is wounded, my wild heart bleeds with yours. In the rapture of my enormous humiliation I live in your warm life, and you shall die—die, sweetly die—into mine. I cannot help it; as I draw near to you, you in your turn will draw near to others, and learn the rapture of that cruelty, which yet is love . . ." (291).

It becomes clear that Carmilla, even if she is real, is a mirror image, the photographic negative of Laura, the fashionably invalid young narrator. She is Laura's erotic primal nature made flesh. So completely opposite is she from Laura that the latter for a moment even suspects that her new companion might actually be male: "What if a boyish lover had found his way into the house, and sought to prosecute his suit in masquerade, with the assistance of a clever old adventuress?" Thus, Carmilla, the vampire, whose real appearance is that of "a sooty-black animal that resembled a monstrous cat" (304), has all the viraginous qualities of the bestial woman. Underlying the realistic setting of the story is a symbolism every bit as deliberately developed as that in Rachilde's "The Blood Drinker." When Carmilla murmurs to Laura, "You and I are one forever," she is enunciating a truth which goes beyond the ostensible resolution of the story in the inevitable ritual exorcism of the bestial creature, for the evil in this narrative is the never-ending evil of all women—their blood link with the animal past. Not only are both the demon and her intended victim women, but they are significantly related on the mother's side. Carmilla is a Karnstein; she represents the personified atavistic reappearance of the vile trait of sensuality characteristic of this "bad family," whose members even "after death continue to plague the human race with their atrocious lusts" (327). But Laura is a Karnstein, too, and that is why Carmilla has come back to haunt her. "I am descended from the Karnsteins," Laura admits, "that is mamma was" (299). Moreover, another notable figure among Carmilla's victims, the ill-fated daughter of the general who helps exorcise the monster, was also "maternally descended from the Karnsteins" (318). In addition, the monster of unspeakable passions and blood lust can return to possess them all the more easily because both girls have lost their mothers, who normally would serve as a protective buffer of acquired civilization against the return of the bestial past.

The vampire Carmilla, then, is the eternal animal in woman, desperately struggling with the forces of civilization to reenter the body from which it has, in the course of history, been expelled. Toward the end of the story the matter is quite specifically spelled out. Laura's bestial mirror, Carmilla, her name itself mirroring the family's Ur-mother Mircalla, whose emanation she is, infected their blood with the mark of the vampire as a result of a passionate love affair, thus forever branding succeeding generations of women born into this family with the mark of the beast, the lust for blood. Only a stake through the heart—symbolic of the death of the woman whose sexual sense has been stirred, who, like Mérimée's Count Szemioth, has acquired a taste for blood—can drive the vampire back into her bestial past and finally give Laura that heavenly peace to be found in the condition of chastity, which, in the words of Pater's Marius, "is the most beautiful thing in the world, and the truest conservation of the creative energy by which men and women are first brought into it" (II, 124).

The cultural preoccupation around 1900 with the struggle of evolutionary progress against the forces of bestiality and degeneration was dramatized most coherently and consistently in what is certainly the masterpiece of vampire literature, the ever-popular *Dracula* (1897) by Bram Stoker. In Stoker's novel virtually all elements of the dream of future evolutionary possibility and all aspects of the period's suspicions about the degenerative tendencies in women have been brought together in such an effortless fashion that it is clear that for the author these were not so much a part of the symbolic structures of fantasy as the conditions of universal truth. Stoker's work demonstrates how thoroughly the war waged by the nineteenth-century male culture against the dignity and self-respect of women had been fought, and how completely the ideological implications of the dualistic struggle between the angels of the future and the demons of the past had entered into that semiconscious world which nurtures the cultural commonplaces governing the average person's perceptual environment. Stoker clearly was a man of limited intelligence, typical of the fairly well-educated, fairly well-off, middle-minded middle class. But he had a remarkably coherent socio-logical imagination and a brilliant talent for fluid, natural-sounding, visually descriptive prose. Together these qualities made it possible for him to write, perhaps without ever completely realizing what he had done, a narrative destined to become the looming twentieth century's basic commonplace book of the antifeminine obsession.

It is certainly true that Dracula, the narrative's pivotal vampire, is a male, but the world in which he operates is a world of women, the world of Eve, a world in which reversion and acculturation are at war. Dracula himself is merely an updated version of the art world's vile, unevolved, grossly bestial satyr, whose inability to control his desires is only further illuminated by his effeminate aristocratic airs. Dracula may not officially have been one of those horrid inbred Jews everyone was worrying about at the time Stoker wrote his novel, but he came close, for he was very emphatically Eastern European, and hence, like du Maurier's "filthy black Hebrew," Svengali (*Trilby,* 52), a creature who had crawled "out of the mysterious East! The poisonous East—birthplace and home of an ill-wind that blows nobody good" (377). Like Svengali, Dracula approaches his victims "with a terrible playfulness, like that of a cat with a mouse—a weird, ungainly cat, and most unclean; a sticky, haunting, long, lean, uncanny, black spider-cat, if there is such an animal outside a bad dream" (*Trilby,* 83)

When Jonathan Harker crosses the Danube on his way to the Carpathian Mountains, he has the impression that "we were leaving the West and entering the East" (11). He soon discovers the regressive ignorance and inefficiency of Eastern life: "The further east you go the more unpunctual are the trains. What ought they to be in China?" (13) It is a hair-raising question for any efficient Britisher. To travel eastward is to travel into the past. The road to Castle Dracula is "hemmed in with trees, which in places arched right over the roadway till we passed as through a tunnel" (21). Harker has to travel into the bowels of man's past to find Dracula, and when he finds him the count has all the phrenological characteristics of the sensuous satyr: "His face was a strong—a very strong—aquiline, with high bridge of the thin nose and peculiarly arched nostrils; with lofty domed forehead, and hair growing scantily round the temples but profusely elsewhere, his eyebrows were very massive, almost meeting over the nose, and with bushy hair that seemed to curl in its own confusion." If that weren't enough, he has the "peculiarly sharp white teeth" of an animal, ears "at the tops extremely pointed," and hands which, like a monkey's, "were rather coarse—broad, with squat fingers. Strange to say, there were hairs in the center of the palm" (27). Centuries old, and therefore not surprisingly looking like an old man when Harker first meets him, Dracula, the personification of the past, feeds on the blood of young girls to grow young again, for the bestial past lives in the blood of woman.

Having had his fill of feminine coagulants, he becomes even more Semitic in appearance, "a tall thin man, with a beaky nose and black moustache and pointed beard" (179), a virtual visual reprise of du Maurier's Svengali, who was also possessed of "bold, brilliant black eyes, with long heavy lids, a thin, sallow face, and a beard of burnt-up black, which grew almost from his under eyelids; and over it his moustache, a shade lighter, fell in two long spiral twists" (*Trilby,* 10). Stoker felt called upon to show off his up-to-date knowledge of scientific fact in his characterization of the fiend of reversion. After he has his scholar-hero Van Helsing mumble in broken English that Dracula represents the regressive criminal type, is genetically "predestinate to crime," and has a "child-brain," he lets Mina Harker, for good measure, affirm that "Nordau and Lombroso would so classify him" (346).

Once this devil of blood lust and regressive, effeminate sensuality has been transported to England, the predatory forces of atavistic bestiality symbolized by Dracula can begin their challenge to the evolutionary acculturation of British womanhood. For it is woman, after all, that the chilly count is after—a fact that soon becomes obvious. Of the truly "male" men in the novel, none, not even Harker himself, the man who ventured into the very lair of the beast, becomes a victim to the creature. Dracula, in this respect, demonstrates a very myopically heterosexual bent. The two women on whom the count sets his bloodshot eyes, Lucy Westenra and Mina Harker, represent the success and failure of modern man's arduous attempts to acculturate woman to the civilized world. They are the two faces of Eve.

Lucy, as any well-read Victorian reader would have guessed from the very start, does not stand a chance before the onslaught of the demon of bestial hunger, for Stoker makes it quite clear that in her the attempt at acculturation has failed. We soon learn that Lucy has all the immodest, aggressively eager, viraginous sensuality of "a horrid flirt," as she appropriately characterizes herself. Without even knowing it, she bears the degenerative stamp of the new woman, for in rapid succession she is confronted with three different men—Arthur Holmwood, John Seward, and the American Quincy Morris—and horror of horrors, she falls in love with all three. In a shocking admission of her degenerate, bestial, polyandrous instincts, she exclaims, "Why can't they let a girl marry three men, or as many as want her, and save all this trouble?" Lucy admits that "this is heresy, and I must not say it" (68), but the predatory cat is out of the bag, and it is no wonder that Dracula heads straight for Lucy upon his arrival in England.

On a symbolic level, Lucy is indeed to have her wish of marrying "as many as want her," for it is obvious that she was specifically created to demonstrate how the polyandrous woman becomes man's conduit to the primal beast. As Lucy, increasingly "languid and tired," becomes a true collapsing woman under the draining ministrations of the thirsty count, she receives, in rapid succession, blood transfusions from each of the three men she loves. Stoker leaves not the slightest doubt about the fact that these transfusions should be equated with a sexual union between Lucy and her donors. Van Helsing, the Dutch scientist and student of the occult who knows all about such rituals, calls upon each of Lucy's lovers to undergo this sacrifice of masculine essence as Dracula continues to drain Lucy's blood. Arthur is called upon first. "You are a man," says Van Helsing, "and it is a man we want"—for Stoker is determined to keep things heterosexual and "clean"—"to transfer from full veins of one to the empty veins which pine for him" (130). After the transfusion has been completed, Van Helsing refers to the "much weakened" Arthur fondly as Lucy's "brave lover"—and Lucy contentedly notes in her diary: "Somehow Arthur feels very, very close to me. I seem to feel his presence warm about me" (137).

Dracula continues to sip, and Doctor Seward is next to offer Lucy his blood. That done, Lucy's new conquest exclaims with all the ecstasy of a satisfied lover: "No man knows, till he experiences it, what it is to feel his own life-blood drawn away into the veins of the woman he loves." But Van Helsing has Seward swear to keep quiet to Arthur about his happy experience, for "it would at once frighten him and enjealous him, too." Given Lucy's fondness for the Dutch scientist ("I quite love that dear Dr. Van Helsing" [141] she writes in her diary), it is no wonder that he, too, is drawn into the risqué ritual of transfusion. With Dracula continuing to do his thing, and Lucy sliding rapidly into the condition of the undead, Van Helsing cries out in desperation, "What are we to do now? Where are we to turn for help?" Nor does he fail to notice that there are robust, healthy women around who also happen to have blood in their veins. But even with Lucy on the point of death, the decorum of heterosexual transfusion must be maintained, no matter how flimsy the excuse. "I fear to trust those women, even if they would have the courage to submit," the good doctor mutters (156).

Of course, Quincy Morris, the American suitor, appears just in time, and Van Helsing is pleased as punch. "A brave man's blood is the best thing on this earth when a woman is in trouble. You're a man, and no mistake. Well, the devil may work against us for all he's worth, but God sends us men when we want them" (157). Thus Lucy gets her wish—the blood, the symbolic semen—of "as many men as want her." Later Arthur, who is still unaware that his friends have all shared his lustful lover, declares that ever since the transfusion he has felt "as if they two had really been married and that she was his wife in the sight of God" (181). Whereupon Van Helsing chuckles—though to Stoker this is obviously no joke—"if so that, then what about the others? Ho, ho! Then this so sweet maid is a polyandrist" (182).

And so Lucy was indeed. Dracula, in Lucy's case, had done his sipping virtually all offstage—for it was central to Stoker's purpose to expose the evil drain placed upon true manhood by the bestial polyandry of the unacculturated primal woman. Fed by the seed of four men, Lucy turns into a wild woman, one of those horrible creatures who prey upon that central symbol of the future potential of mankind: the child. Woman's misplaced viraginity, that masculinizing force which in real life encouraged feminists to renounce the holy duties of motherhood and, as it were, prey upon their as yet unconceived babies, manifests itself henceforth in Lucy in the form of a determined blood lust for children. As she dies, she slides back into a state of primal bestiality, and soon children begin having their throats torn open on Hampstead Heath.

To stop Lucy's viraginous evil, her depredation of future generations, her attempts to drag civilization back into the evil past, the four "husbands" who, in yielding to her blood lust, have contributed to the creation of this monstrous beast, must come together and, in a concerted effort, send her off forever into the arms of a permanent death. At night—under the light of the moon, naturally—the four see Lucy again. "The sweetness was turned to adamantine, heartless cruelty, and the purity to voluptuous wantonness." Becoming aware of her former lovers, "she drew back with an angry snarl, such as a cat gives when taken unawares," and "with a voluptuous smile" she "flung to the ground, callous as a devil, the child that up to now she had clutched strenuously to her breast, growling over it as a dog growls over a bone." Confronted with Van Helsing's crucifix, Lucy's "beautiful color became livid, the eyes seemed to throw out sparks of hell-fire, the brows were wrinkled as though the folds of the flesh were the coils of Medusa's snakes" (217–18).

Arthur, the only man she should have married, the only one whose gentle, fructifying gift of semen she should have wished for if she had been a decent British girl, becomes, in the presence of the other three men, the executioner of the vampire of degenerative polyandry. In punishment and revenge, he "rapes" the female beast whose evil desires had unlawfully siphoned off the manhood of all four. "He looked like the figure of Thor as his untrembling arm rose and fell, driving deeper and deeper the mercy-bearing stake, whilst the blood from the pierced heart welled and spurted up around it." Having forever "broken" Lucy's heart, her symbolic maidenhead, in his vigorously monogamous assertion of masculine right over woman's bestial inclinations, Arthur feels better right away. "A glad, strange light broke over his face and dispelled altogether the gloom of horror that lay upon it." Lucy, too, has now been brought under control:

There, in the coffin lay no longer the foul Thing that we had so dreaded and grown to hate that the work of her destruction was yielded as a privilege to the one best entitled to it, but Lucy as we had seen her in her life, with her face of unequalled sweetness and purity. True that there were there, as we had seen them in life, the traces of care and pain and waste; but these were all dear to us, for they marked her truth to what we knew. One and all we felt that the holy calm that lay like sunshine over the wasted face and form was only an earthly token and symbol of the calm that was to reign for ever (222).

Thus, by means of a little show of monogamous masculine force, Lucy, the polyandrous virago, has been transformed into that ideal creature of feminine virtue of the mid-nineteenth century: the dead woman. Civilization and evolution have, in the nick of time, triumphed over the vampire of degeneration. And this is all man's doing, for the "half-criminal" hidden, according to Lombroso, in every normal woman's bosom had certainly wrecked Lucy's claim to man's respect before her surgical restitution to true femininity. But Dracula has another neck to bite: The far more virtuous jugular of Mina is still to be subjected to the temptation of bestial reversion. Mina's case is quite different, however, for she is truly the ideal modern woman: a virtuous footstool ready to do man's bidding in the world of scientific accomplishment and intellectual evolution.

It is important to keep in mind that the world of Stoker's novel was not the vaguely fantastic world of old-fashioned experience it has since come to seem in twentieth-century cinematic adaptations. On the contrary, Stoker's world was teeming with evidence of technological progress and scientific achievement. His characters always used the latest available scientific and technical equipment. Dr. Seward, for instance, uses a phonograph to record his diary. Van Helsing, as we have seen all too well, practices blood transfusions and travels back and forth between England and Holland at the drop of a hat. Last but certainly not least, Mina is a whiz at the typewriter and she takes great shorthand. Not only destined to be Jonathan Harker's determinedly monogamous wife, she is also his very convenient, practical, and willing personal secretary. She is thoroughly modern but also happy to be able to make fun of the New Woman, whose hilarious demands for what are masculine prerogatives she satirizes in her journal as she thinks about Lucy: "Some of the 'New Women' writers will someday start an idea that men and women should be allowed to see each other asleep before proposing or accepting. But I suppose the New Woman won't condescend in future to accept; she will do the proposing herself. And a nice job she will make of it, too! There's some consolation in that" (100).

Bustling about industriously and making herself modestly useful, Mina has no use for viraginous nonsense. She is the perfect nurse and, as every male in the novel hastens to

X, 3. Franz Flaum (1867–1917), "Vampire," sculpture (ca. 1904)

X, 4. Edvard Munch
(1863–1944),"Vampire,"
woodcut and lithograph
(1895/1902)

emphasize, a "wonderful woman," an updated household nun turned personal secretary in a changing world. But, being a woman, a daughter of Eve, and hence having by her own admission "some of the taste of the original apple that remains still in our mouths" (189), she, too, is subject to the brutalizing, atavistic bite of the animal past, a potential victim for Dracula, who enters the room—and her body—through her woman's heart to try and awaken her bestial nature:

> With his left hand he held both Mrs. Harker's hands, keeping them away with her arms at full tension; his right hand gripped her by the back of the neck, forcing her face down on his bosom. Her white nightdress was smeared with blood, and a thin stream trickled down the man's bare breast which was shown by his torn-open dress. The attitude of the two had a terrible resemblance to a child forcing a kitten's nose into a saucer of milk to compel it to drink (288).

There is in Stoker's account of this scene a striking resemblance to a case recounted in Krafft-Ebing's *Psychopathia Sexualis,* with which—given the evidence of the novelist's interest in the theories of criminal psychology current in his day—he may in fact have been familiar. "A married man presented himself with numerous scars of cuts on his arms. He told their origin as follows: when he wished to approach his wife, who was young and somewhat 'nervous,' he first had to make a cut in his arm. Then she would suck the wound and during the act become violently excited sexually." Having thus presented the case, Krafft-Ebing commented that it recalled "the widespread legend of the vampires, the origin of which may perhaps be referred to such sadistic facts" (85).

It was considered a scientific fact by many turn-of-the-century intellectuals that for woman to taste blood was to taste the milk of desire, and that such a taste might turn an innocent, inexperienced woman into an insatiable nymphomaniac. To exorcise that hunger, to return Mina to her appointed role as man's efficient private secretary, Dracula must be hunted out, followed back into the distant past, destroyed at his point of origin. Just as Renfield, Dracula's degenerate male votary, tried to reach the condition of bestiality by swallowing, one after another, the creatures on the Darwinian food chain, thereby symbolically descending along the ladder of devolution, so Dracula must be forced to retreat from Mina and the hearts of all other British women by first being chased out of England and then back into the evil East from which he emanates. In the end, this creature of moonlight can only be executed in the daylight of modern science by technology and the righteous wrath of a virtuously monogamous husband.

Thus Stoker's *Dracula* is a very carefully constructed cautionary tale directed to men of the modern temper, warning them not to yield to the bloodlust of the feminist, the New Woman embodied by Lucy. For, to adapt words Flaubert used in a different context, this New Woman was herself nothing but "a vampire who satisfied the handsome young men in order to devour their flesh—because nothing is better for phantoms of this kind than the blood of lovers" (*The Temptation of Saint Anthony*, 197). She was the personification of bestiality, forever crawling, like the vampire in a sculpture by Franz Flaum [X, 3], toward her victims. She was the silent drinker of man's essence, graphically portrayed by Edvard Munch [X, 4]. She was the bat-winged woman leading a massed, blinded humanity to the abyss of degeneration, as in a painting of 1897 by Henri Martin. In a characteristic turn-of-the-century adaptation of Poe by Frantisek Kupka, she was the peacock-feathered, winged chimera strewing death and destruction, the conquering worm of death itself.

Attracted by the apparent sense of power imputed to the female vampire by turn-of-the-century culture, women of the period often cultivated the anorexic look of that predator. Art inevitably followed where fashion had led the way. Shortly after the turn of the century, for instance, Lotte Pritzel, a Munich artist, began to produce dolls which clearly took their inspiration from the cult of the vampire [X, 5]. In an article for *The Arts* of April 1923, Helen Appleton Read described these dolls: "They are modeled in colored wax, and are usually kept under glass or in specially arranged niches. They stand about a foot and a half high. Delicate orchids in human form, outgrowths of a neurotic mind." They are, Read continued, "dressed in the richest or most delicate of materials, gold laces and gauzes or cloth of gold brocades. They wear real jewels on their skeleton-like fingers and toes, their hair is sometimes made of fine gold wires. The effect is incredibly alluring." Read saw in Pritzel's work "the same delicate and unhealthy preciosity which we find in the vision of Beardsley, who was a consumptive," and she recognized the latter's "aristocratic erotic fancies of Salome" in the dolls of the Munich artist (279–80).

Read's remark that Pritzel's dolls were "incredibly alluring" provides a striking indication of the bizarre manner in which the male ideal of the dead woman, as well as men's fantasies of woman as vampire, had come to influence women's conception of themselves. Pritzel's dolls were clearly fashioned after the real-life anorexic emaciation of Gabriele D'Annunzio's favorite companion, Ida Rubinstein, whose corpselike body was seen floating as on vampire's wings in Romaine Brooks' painting "Le Trajet" [see II, 28]. It would appear from works such as this, and from surviving photographs, that it was Rubinstein's goal to become as much like the period's archetypal vampire creature as she possibly could. Octave Uzanne, commenting in *The International Studio* in 1897 on the women in Georges de Feure's paintings, gave a striking delineation of these fashionable daughters of Dracula, who, as Hollywood's "vamps," were soon to make their appearance on the silver screen as well. These creatures, Uzanne noted—probably barely able to keep his pen straight in the heat of his masochistic excitement—have the mark of the predator upon them. De Feure's vampire has

a woman's form, regal, triumphant, satanic. Deliciously childlike in its virginal simplicity, her

X, 5. Lotte Pritzel (1887–1952), decorative dolls, wax and cloth (ca. 1912)

X, 6. Philip Burne-Jones (1862–1926), "The Vampire" (ca. 1897)

X, 7. Félicien Rops (1833–1898), "The Absinthe Drinker," etching (ca. 1890)

X, 8. Manuel Rosé (1887–1961), "Interior of a Café" (ca. 1914)

youthful body is outlined in gracious curves, soft as a caress; her heavy masses of hair crown her brows as with a diadem; her face bears the stamp of a strange and prodigious beauty; from her mouth, with its blood-red, kiss-provoking lips, and its two rows of pearls, gleams the light of hell; her nostrils quiver in ardent palpitations; while below her pure forehead are two terrible eyeless orbits, staring vacantly on the world, blind to the victims of her fatal body (100).

Female vampires were now everywhere. They were not just the lurid creatures of poets but nasty, stupid, everyday creatures like the ones about whom Rudyard Kipling was impelled to write after he had visited the New Gallery in London and had seen Philip Burne-Jones' 1897 representation of a lustful vampire in a modern nightgown crouching over the body of a young man [X, 6]. This vampire was clearly a gold digger, Kipling pointed out, as hungry for coin as for blood:

Oh the years we waste and the tears we waste
And the work of our head and hand,
Belong to the woman who did not know
(And now we know that she never could know)
And did not understand (''The Vampire,'' 15).

By 1900 the vampire had come to represent woman as the personification of every-thing negative that linked sex, ownership, and money. She symbolized the sterile hunger for seed of the brainless, instinctually polyandrous—even if still virginal—child-woman. She also came to represent the equally sterile lust for gold of woman as the eternal po-lyandrous prostitute. She was the absinth drinker, her fever for the gold of man's essence fed by her addiction, who was seen by Félicien Rops as lurking in the alleyways of Paris [X, 7]. She was the woman cloaked in darkness who beckoned man to his death, portrayed by Manuel Rosé in his ''Interior of a Cafe'' [X, 8], exhibited in San Francisco at the Panama-Pacific Exposition of 1915. Many men, burdened with thoughts of their awesome responsibility in the evolution of mind, saw in the very presence of a woman—any woman— a hint of ''The awful daring of a moment's surrender / Which an age of prudence can never retract,'' as T. S. Eliot had put it in *The Wasteland*. Like Eliot, these men imagined that every time '' A woman drew her long black hair out tight / And fiddled whisper music on those strings,'' they were seeing creatures with unnatural, viraginous tendencies, po-lyandrous gynanders who had renounced the responsibilities of motherhood and had be-come child murderers, seed-hungry, blood-lusting vampires, ''bats with baby faces in the violent light,'' out to drag men back into the night of retrogression, as in false allurement they ''whistled, and beat their wings / And crawled head downward down a blackened wall'' (*The Waste Land*, 48).

This world—full of the ''Murmur of maternal lamentation,'' filled with ''hooded hordes swarming / Over endless plains, stumbling in cracked earth / Ringed by the flat horizon only''—was truly the wasteland of feminine desire. It was the infernal, because infertile, dead womb of woman's greed for the glitter of gold, that cold material essence of spent seed, of congealed blood. It was the barren territory of the unnatural woman, whose pri-mary desire was to sever man's sex-obstructive, idealistic head from that resistant mass of erectile tissue, his body, whose lingering lusts she craved to rekindle. As we shall see, gold, seed, and blood came to form in the male imagination an ultimate triangle of mur-derous magic around which, in heedless nudity, their peacock tails flashing, danced the cruel, childlike priestesses of man's severed head.

CHAPTER XI

Gold and the Virgin Whores of Babylon; Judith and Salome: The Priestesses of Man's Severed Head

Masochism is the opiate of the executioner's assistant. Unable to share in the isolate, incorporative, sadistic satisfaction of the man whose boot holds down the offender's neck, the masochist usurps the role of victim without descending to the actual victim's position of abject helplessness. Expendable, marginalized, a mere tool of the hierarchies of domination, the assistant dreams of having the executioner do his bidding, of making the master become *his* slave. The offender, the person who, in words or action, illuminates the failures of the dominant value system, must always be a danger to the prevailing structures of power. Even when destroyed, s/he lives on in the executioner's mind as a faint threat of potential insubordination, of alternative solutions which threaten the smooth working of the rituals of mastery.

But the masochist assists in the proper continuation of these rituals. He has no personal being, feeding on defeat to turn his impersonal, parasitic existence into a secret mirrorworld of the executioner's values. In this realm of moonlight and looking-glass magic, of fantastic dreams and majestic feats of submission, he tries to make the executioner see him as the threat he very well knows he isn't by manipulating the master into dealing with his pointless servant as if he were a meaningful, threatening victim: for, at least so the executioner's assistant has come to believe, a victim has individual being in the eyes of the master, while a servant has none. Thus, in his fantasies the executioner's assistant becomes master over his boss by making his master be slave to his wishes.

But in reality the master has better things to do than pay attention to the groveling insignificance of his menials. His attention remains focused on important tasks: on the sadistic dismemberment and arrogant incorporation of real enemies and the destruction of those who pose a serious threat. Hence the executioner's assistant needs a surrogate master—someone even more marginalized than he—who will play the part of his boss, do his bidding, and "beat" him. For in this manner the assistant, in his miserable fantasies of mirrored power, sees himself triumph parasitically by joining in the tortured importance of the victim. At the same time, by manipulating the actions of the surrogate he has recruited to debase him, he lives out a fantasy of control. In directing the baleful ministrations of the surrogate master, the menial has, in his dreams, come to direct the actions of his sadistic boss.

Thus the masochist turns his own marginalized position into a mirror image of the arena of power and his surrogate master into the fantasy image of the executioner as slave to his neglected menial. But all the while the executioner's assistant remains marginalized, continues to live a meaningless existence, does not share in the executioner's wealth, and yet will always do his master's bidding, partly so that he may continue to dream of taking the executioner's place. It is this which makes him the executioner's perfect assistant, the workhorse without whom the imperial sadist could not maintain his power. The assistant is the parasite who gives his master life. In 1900 he had already bought a house, and lived quite comfortably among the artists and writers, accountants, soldiers, middle-level executives, and shopkeepers of the enlightened middle class. The principal surrogate he used to fill his master's role and give meaning to his insignificance as a worker in the quarries of power was his wife, whom he forced to impersonate the devil so that he might play martyr while he watched the executioner of dualistic thought depredate the forces of humane coexistence.

During the second half of the nineteenth century, the economic structure of the United States and of most European nations had changed drastically from the relatively free wheeling, open-market system which had earlier allowed a good many members of the mercantile middle classes, located on various levels of the ladder of economic success, to regard themselves as movers and shakers, primary actors, rather than bit players in the great drama of the world's progress. The formidable growth of personal fortunes, usually the result of expanding local monopolies, rapidly established pools of venture capital which, through acquisition and forcible incorporation, turned many of the previously independent merchants into paid employees of massive commercial conglomerates, of the trade monopolies and trusts which had come to dominate the turn-of-the-century marketplace. In a remarkably short period of time most of the former movers and shakers found themselves relegated to marginal status in the epic of economic growth. Many discovered, to their surprise, that they were stranded in history, as it were, with a good deal of money but with very little personal motivation or actual power. They began to play the stock market and to invest in shadowy schemes. If they did not continue as employees, as mid-level managers, they frequently became rentiers, receiving a fluctuating income from sources which were often hard to identify precisely. The world in which virtually everyone knew his own net worth was rapidly giving way to a volatile economic environment of prolonged depressions and heady periods of economic growth in which one could be ruined or rendered rich virtually overnight. The middle class therefore became the class of the moved, not that of the movers—and this, of all things, just when the progress-obstructive arrogance of the aristocracy had finally come for a decisive fall. Thus the former movers became the class of the shaken, for even with the aristocracy out of the way, a mysterious, indeterminate group of ''others'' continued to shake the tree of gain.

At the same time, the displaced aristocracy felt its power slip away, absorbed by what it saw simply as a group of uncouth representatives of the bourgeoisie. To the members of the aristocracy it seemed as if hordes of the great, half-washed middle classes were usurping their right to rule. They were, of course, being upstaged by the same relatively small group of capitalists with both aristocratic or more proletarian backgrounds which was also creating confusion among the middle classes. Only those with a truly ruthless hunger for acquisition could play leading roles, sit at the head of the table during the great feast of imperialist acquisition which was taking place everywhere and from which, on the material level, both the aristocracy and the middle class were certainly benefiting considerably. Thus, even many of the most comfortably situated men in late nineteenth-century culture felt a vague, poorly defined sense of marginalization. If the robber barons and the already faceless trusts now seemed to have become the true movers and shakers—were the new executioners—the cloudy-browed middle aristocracy and middle bourgeoisie formed but an uneasy band of executioner's assistants. Aware that they were no longer executioners themselves, they looked around for someone to take the blame. And, as always, woman was conveniently available.

When, in 1853, William Holman Hunt's conscience-striken kept woman had raised her youthful body from her lover's lap, presumably to join the ranks of the household nuns, she had (though Hunt was never to know it) only raised her body. Inwardly she had bristled at her broken wings, at the loss of freedom imposed by "moral" man upon that previously happily incautious sparrow, her mind. Forced to babble like a child, prevented from earning her own income, she had taken her husband's money and spent it—often doing so while vindictively acting out the child's role assigned to her, determined to have no greater concern for moderation than a child. "Conscience" had lifted her from the lap of physical prostitution and demanded that now she prostitute her mind so that she might become the new consumer and fill the empty prison of her world with trinkets. And as she did what she was told, her spotlessly monogamous, almost virginal body indeed often acquired the mercantile mind of the dreaded whore. By 1871, Tennie Claflin was pointing to the manner in which women's enforced dependence on men had created a marriage market stocked with women "for sale to the highest bidder." In her book *Constitutional Equality* she insisted that

> a large part of the marriages which are contracted are nothing more nor less than bargains and sales, into which consideration the questions of love and adaptation do not enter. What is more common that to hear women remark, "She has made her market," or, "She has done well?" and what, withal, is more decidedly vulgar?
>
> The truth of the matter is, that "young ladies" are set up, advertised and sold to the highest cash bidder, and where a mutual attraction does not exist, a strict analysis finds no difference between it and the other association of the sexes denominated prostitution (89).

In less direct terms Abba Goold Woolson was to say very much the same thing two years later in *Woman in American Society:* "A young man must labor when he would succeed in the business of his life; a young woman must charm. She must become a siren, and lure lovers and admirers to her side, since she is forbidden to cross the insulating waves to them." And, asked Woolson, does anyone really wonder, then,

> that husband-hunting is the main occupation of our young women? Does any one sneer at them as silly and reckless, because they are ready to stake every thing which they should value for after-life—health, useful acquirements, and culture—in this eager pursuit? or brand them as mercenary, when they turn deaf ears to indigent lovers and smile only upon those who lay full purses at their feet? We make merry over managing mammas, and hold up to ridicule their enticing snares and adroit manoeuvres as if these were fair game for virtuous satirists. But why, indeed, should they not manage? As the world rates marriage, and woman's relation to it, it would seem to be their duty. If it is incumbent upon the father to secure a profitable

business for his sons, it would appear equally incumbent upon the mother to provide for the future well-being of her daughters (65).

Many middle-class women, impressed with the lesson that they were of value only as consumer goods, came to see it as their primary purpose in life to enhance their own status in marriage by surrounding themselves with other expensive consumer goods. In the rapidly expanding economies of Europe and America, moreover, an ever-growing group of consumers was needed to make the fruits of the employees' industry return to those who controlled the means of production. Thus woman, having been consumed in the marriage market, then having become consumptive as a wife through lack of respect, exercise, and freedom, took her revenge by becoming a voracious consumer. Said Woolson: "The majority of women seem to consider themselves as sent into the world for the sole purpose of displaying dry goods; and it is only when acting the part of an animated milliner's block that they feel they are performing their appointed mission" (103).

Claflin spoke with similar disdain of the woman of fashion. "In a pecuniary point of view they know no boundary to their caprices, no limit to their extravagance," she said. With characteristic bluntness she continued:

> What does the woman of fashion do for the world? She begins and ends by deceiving it in part, and herself wholly. Walk up Broadway and count the windows wherein are exposed for sale huge, vile bunches of hair, tortured into all conceivable, unnatural shapes, to transform the natural beauty of the head to a hideous, affected thing. The amount expended on these outrages upon common sense, alone, would educate and render comfortable every child of distress and poverty. What right have you, Woman of Fashion, to thus consume wealth, while children on the next street are crying for bread? Your laces and diamonds, and other superfluous articles of ornamentation which you filch from the public welfare, seeking thereby to hide your deformities or to add to your attractions, would mitigate all the distress that stalks among us, with pale, wan cheek, tearful eye and bleeding feet. The general economy of the universe will hold you responsible for all these inequalities.
>
> How many fortunes have you squandered, homes made desolate, and husbands driven to distraction in the pursuit of your insatiable desire for dress? and how many, when the purses of your husbands or fathers have failed to furnish what you require, and would have, have sold yourselves to others to obtain it?

Clearly, when feminists of such very different tempers as Woolson and Claflin felt it necessary to rail against the rampant consumerism of the women of the 1870s, there is little reason to doubt that many women had indeed become veritable buying machines. Thus, in a roundabout, largely unconscious fashion, nineteenth-century middle-class women had transformed the characteristic trappings of their own marginalization from the productive life of their time into the raw materials for a direct attack on the men who had placed them in the gilded cage of conspicuous consumption. Having little else to do with their lives than be trivial and decorative, they transformed the realm of trivia and decoration into a torture garden for the men who had collectively set out to turn them into the trained seals of a consumer society.

The latter half of the nineteenth century saw the massive spread of prostitution in urban centers. During no period, before or since, was the sight of prostitutes so common, so much taken for granted. Prostitutes were an integral element of social life, their presence a logical result of the normal workings of the law of supply and demand. Moreover, once the men of the middle classes had "elevated" their wives to the position of spotless, quasivirginal household nuns, once they had made them into delicate possessions which needed special handling, they discovered that they had fashioned in their own minds bleak monsters of sexual frustration. They found themselves battling the dual restraints imposed by their wives' frequent invalidism—which discouraged sexual contact—and their own intimidation before their chosen mates' impressively displayed acculturated "sexlessness." They therefore tended to slip rather easily into encounters with the less perfectly acculturated women of the working class, who seemed to be ubiquitously available. These women, forced to choose between endless, deadening hours of back-breaking manual labor at less than subsistence wages or what was under these conditions the slightly more tolerable alternative of prostitution, quite understandably came to populate the streets of the major cities.

In dealing with rural and working-class women, Michelet for once descended from his high perch to describe the plight of the French woman born in proverty. If she married, she could expect to do as much or more than her husband in contributing to the upkeep of the family. "She has the hardest part. He primes the vine at his ease—she scrapes and digs. He has respites—she none. He has festive occasions and friends. He goes alone to the tavern—she goes for a moment to church, and there falls asleep. If he returns at night intoxicated, she is beaten, and often, which is worst, when she is pregnant. Then she endures for a year her double suffering, in heat and cold, chilled by the wind, drenched by the rain, daily" (*Woman,* 26).

The image Michelet draws is a far cry from the idealized representations of French rural life palmed off upon an eager and admiring bourgeois public as realistic documents of virtuous labor by Jean-Francois Millet. These paintings, which filled their viewer's hearts with assurances concerning the peasants' tolerance for back-breaking labor, served as an opiate against a recognition of the realities of working-class misery for millions on the fringes of economic power.

During this period engravings and photographic reproductions of such works as "The Angelus," "The Gleaners," and "The Sower" hung on the walls of untold numbers of petty-bourgeois households. With their mock-religious suggestions of a mystical relation between the peasant and the soil, these works warmed the hearts of those who were eager for proof that they did not have to care about the poor since God guarded the fields. It is a significant expression of our own extraordinary eagerness to duplicate the sentiments of this earlier period's most popular economic thinkers that Millet's misleading cultural confections have once again become absurdly overvalued. Nostalgic for a time when the poor toiled gladly and knew their place, the wealthy once again scramble to own these works, which record the pious simplicities of a time that never was.

In Britain and the other European countries the realities of working-class life were hardly different from what they were in France. The phenomenal increase in open prostitution which resulted is one of the more thoroughly studied aspects of late nineteenth-century culture and its details therefore need not concern us here. However, inevitably the effect of its prevalence, and virtually every male's acquaintance with aspects of the phenomenon, led many, under the expert guidance of the bio-sexists, to the convenient conclusion that something deep in woman's nature was at fault. Bernard Talmey could hence assert that it was foolish to believe, with the "philanthropists," that in general prostitutes tended to be driven to their degrading trade by idleness or necessity. Instead, "one should realize that not a few choose this life to satisfy their nymphomaniac desires." (*Woman,*

115). Otto Weininger, leaning on the weighty experience of his twenty-three years and braced by a thorough reading of Lombroso, declared unhesitatingly that

> prostitution cannot be considered as a state into which men have seduced women. The man may occasionally be to blame, as, for instance, when a servant is discharged and finds herself deserted. But where there is no inclination for a certain course, the course will not be adopted. Prostitution is foreign to the male element, although the lives of men are often more laborious and unpleasant than those of women, and male prostitutes (such as found among waiters, barbers and so on) are always advanced sexually intermediate forms. The disposition for and inclination to prostitution is as organic in woman as is the capacity for motherhood (*Sex and Character*, 217).

Prostitution, Weininger concluded, was the natural outlet for any woman whose inherently polyandrous nature was too strong to be tamed by her acculturation for the purpose of assuming the task of motherhood.

Prostitution having thus conveniently been made the sole responsibility of the prostitute, it was easy for the middle-class male to see himself as the helpless victim of these tempting sirens and vampires of the streets, these lusty creatures of the working class who did not seem to have any of the middle-class woman's reticence about sex. Armand Silvestre, in his *Le Nu au Salon—Champ de Mars* of 1893, quite effectively expressed the pressure of pent-up desires and dark longing felt by many middle-class youths of the late nineteenth century. He did so, appropriately enough, in a commentary on a painting based on Zola's *Nana*. Young men, he said, found their lower instincts aroused by the vigorous beauty of seemingly free, young working-class women who, given their lowly economic status, had not been driven into the asexual invalidism demanded of decent middle-class wives and daughters. Thinking back to the years of his youth, just after 1870, Silvestre remarked:

> How many times did we not sit, my friend Desboutin, who had just returned from Italy, and I, on the terrace of a café which, having been for many years a gathering place for artists and writers, still flourishes on the avenue de Clichy. Here, just after the war, a large group of us— Zola, Manet, Duranty, Fantin-Latour, Degas, and others—used to pass the time of day. Indeed, how many times did we not sit there and, in ecstasy, watch the wonderful women who, during the warm summer evenings, their workday past, with a flower touching their lips, and often with a lover on their arm, would return home along that street, bareheaded and with a beautiful glow of liberation in their eyes! It seemed truly a festival of those released from labor, and never, except perhaps at the times when the tobacco workers at Toulouse returned from work, did I ever see a more delicious procession of gorgeous women in the full unfolding of their youth and health.

Thus the indolent sons of the bourgeoisie watched the hard-working women of the factories and seemed to see a promise of sexual freedom and spontaneity which they sought again in the arms of prostitutes. Inevitably they were disappointed, and, just as inevitably, they blamed the women, coming to see in them, as Zola did in *Nana*, golden flies, predatory sirens, carriers of atavistic evils. Nana, as described by the journalist Fauchery,

> had grown up in the slums, in the gutters of Paris; and now, tall and beautiful, and as well made as a plant nurtured on a dung heap, she was avenging the paupers and outcasts of whom she was the product. With her the rottenness that was allowed to ferment among the lower classes was rising to the surface and rotting the aristocracy. She had become a force of nature,

XI, 1. Albert Matignon (b. 1869), "Morphine" (ca. 1905)

> a ferment of destruction, unwittingly corrupting and disorganizing Paris between her snow-white thighs, and curdling it just as women, every month curdle milk (221).

A confused mixture of sexual desire and guilt, a vague sense of class difference and exploitation, and a desire to hold onto privileges gained made the prostitute seem to these men, as the brothers Goncourt had remarked in their journal, the means whereby the proletariat revenged itself upon the rich.

Victor Barrucand, writing in *La Revue Blanche* in 1895, described how one night he had observed "a traitorous army" of prostitutes and had had a vision of "some sort of revenge of the weak upon the strong, of the revenge of the sacrificial female upon the egoistic male," for

> with a simple undulation of her rump, she was able to trouble man's brain; and with her slowly insinuating ability to fascinate, she picked apart fortunes, the arts, creeds. Venus-Pandemos triumphed over idealistic aspirations; she ridiculed chastity, the family, the fatherland, the future life, drama and the world of dreams. It was the revenge of brute desire, breaking the lyres and the guitars of an aged world's orphic singers whom she had forced into prostration before sex itself (350).

Wedekind's Lulu "was created for every abuse, / To allure and to poison and seduce / To murder without leaving any trace" (*The Lulu Plays*, 10).

The image of the young working-class woman with a rose between her lips rapidly transformed itself into the male's nightmare confrontation with the flower of evil, the woman who wanted money, who did not yield to him in gratitude but in anger, and who, all too often, returned to him a dreadful disease in exchange for his attentions. For in her miserable environment pulmonary illness and venereal infections festered and became the middle-class male's reward for his transgressions. In addition, the world of the prostitute was a world of women hungering for forgetfulness, eager to still their physical and mental pain at any cost. It was, in consequence, a world of addicts, of *morphinomanes.* Fasci- nated by this further evidence of woman's natural depravity, the painters rushed in to depict these women's perverse indulgence.

In the years after 1900, no major exhibition was complete without its images of opium-smoking women or morphine addicts. These works represented the thematic dead end of the collapsing woman and the nymph with the broken back. In 1913 the sculptor Zonza-Briano placed three *morphinomanes* together in a lump of stone, recording these women's despairing inanition with the dispassionate precision of a bureaucrat. In *Jugend* woman was habitually portrayed as the personification of hashish and belladonna. In 1911 Albert Matignon showed her to be half-virgin and half-vampire while in the throes of an opium dream among the paraphernalia of her addiction. Eugène Grasset recorded her in the process of injecting herself in a lithograph which was chilling in its clinical accuracy. In 1905 Matignon had already shown morphine entering the world of fashionable dissipa- tion and had suggested a link between the *morphinomane* and the vampire [XI, 1].

In the *Assiette au Beurre,* a Parisian satirical magazine, the caricaturist Vogel showed death as a procuress offering a nude young woman to a comfortably seated man of means. The American painter Albert Sterner combined hints of tubercular adolescent purity and vampiric emaciation in a young girl with febrile eyes to suggest that the onset of puberty was the kiss of death for a woman whereby she was recruited into the infernal legions of angelic destruction [XI, 2]. Although Sterner's lithograph was ostensibly meant to play

XI, 2. Albert Sterner (1863–1946), "The Angel of Death," lithograph (ca. 1914)

XI, 3. Kimon Loghi (b. 1871), "Post-Mortem Laureatus" (ca. 1896)

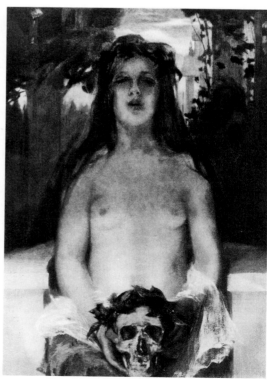

upon the pathetic terminal illness of an innocent girl, Eros and Thanatos mingled in an all too obvious fin-de-siècle fashion to suggest to his viewers the tragic working-class children whose "virginity" was sold over and over again to middle-class males in search of purity. In the same vein, the Rumanian artist Kimon Loghi exhibited at the Universal Exposition of 1900 a much-discussed painting of a barely pubescent girl with a skull in her lap [XI, 3]. Ostensibly his intention was to comment on the sad fact that praise tends to come to the man of genius only after his death, but as Louis Raemaekers was to show when he painted a grinning demi/mondaine holding a skull before her loins for a poster warning against the spread of syphilis, that symbol could be given a rather more direct interpretation. To Léon Frédéric a painting ostensibly warning crowned heads of "The Nothingness of Worldly Glory" became an excuse to personify time as a voluptuous woman with the face of a disdainful virago and still another skull in her hand [XI, 4].

Félicien Rops showed a demimondaine with the livid eyes and sharp teeth of a vampire who carried death's own sickle to symbolize syphilitic death [XI, 5], while Huysmans' Des Esseintes, leader of the strugglers against nature, had a nightmare in which syphilis was personified in the form of a working-class woman who "wore a servant's white apron." She appears first as a "bull-dog woman" and then turns into "a pallid naked woman, green silk stockings moulding the legs," only to change again into a palpitating mass of organic matter, the horrendous vulval "flower" of nature, "the hideous open wound" of the earth:

> Then he noticed the terrifying irritation of the bosoms and of the mouth, discovered on the skin of the body stains of bistre and copper, and recoiled in horror; but the woman's eye fascinated him, and he crept slowly, reluctantly towards her, trying to drive his heels into the ground to stay his advance, dropping to the earth, only to rise again to go to her. He was all but touching her when black Amorphophalli sprang up on every side, and made darts at her belly that was rising and falling like a sea. He put them away from him, pushed them back feeling an infinite loathing to see these hot, moist, firm stems coiling between his fingers. Then, in a moment, the odious plants disappeared, and two arms were seeking to wind themselves about him. An agony of terror set his heart beating wildly, for the eyes, the dreadful eyes of the woman, had become pale, cold blue, terrible to look at. He made a superhuman effort to free himself from her embraces, but with an irresistible gesture she seized and held him, and haggard with horror, he saw the savage Nidularium blossom under her meagre thighs, with its sword blades gaping in blood-red hollows (*Against the Grain,* 93).

No wonder that in the visual vocabulary of the artists of the turn of the century the time-honored representation of death as an old man carrying a scythe was, as in Sterner's lithograph, replaced by the image of death as a woman, sometimes with the wings of an angel but more often with those of a vampire bat—or no wings at all. Almost always she had malice in her heart. Fritz Erler, who in his old age was to become, like Fidus, one of the honored old masters of Third Reich culture, portrayed her as the plague stalking a quiet German town, her breasts jutting forward every bit as aggressively as those of Delville's "Idol of Perversity," to signal that she represented the viraginous, hardened, degenerate feminist set loose upon the world to prey upon the Aryan forces of evolution and eugenics. On the other side of the Atlantic, Daniel Chester French sent a monumental frieze to the Chicago World's Columbian Exposition of 1893 in which woman, portrayed as death, stays the hand of the sculptor. The Italian magazine *Scienza per Tutti* took advantage of the development of X-rays to show that only a gruesome skeleton could be found under a woman's skin. A full-color poster for the magazine, printed in 1909, showed

XI, 4. Léon Frédéric (1856–1940), "The Nothingness of Worldly Glory" (1893)

XI, 5. Félicien Rops (1833–1898), "Mors Syphilitica" (or "Death on the Pavement") etching (ca. 1892)

an elegant lady in a low-cut gown reclining on a sofa. The upper-left quarter of the poster was set up to suggest an X-ray "window." In its greenish glare, half of this beautiful lady's face and half her chest revealed part of a grinning skull and a set of barren ribs. This example of science for all must have made quite an impression on the youths of Italy as they walked past their local corner magazine stands.

Even for Thomas Gotch, that otherwise staunch purveyor of angelic virgins enthroned, woman was "Death the Bride," with a knowing smile, stalking through a field of poppies in a painting exhibited at the Royal Academy in 1895. The Spanish painter Ignacio Zuolaga y Zabaleta needed no symbolism at all. Instead, in painting after painting he exaggerated his feminine subjects' use of makeup to turn their faces into death masks, grotesque, eroticized caricatures of the craggy, carved sorrow of poverty-stricken, prematurely aging prostitutes. However, what is most amazing about these immensely popular portraits by Zuolaga is the fact that they were not images of "Death the Bride," not expressionist portraits of prostitutes at all. Instead, they were meant to be undisguised *ritratos* of the painter's family, acquaintances, and of society ladies who were apparently not averse to having the painter bring out the vampire in them [XI, 6].

An understated yet chillingly explicit painting by Alexandre Falguière shows a young

XI, 6. Ignacio Zuolaga y Zabaleta (1870–1945), ''My Cousin Candida'' (1907)

woman of the people standing on a dark corner, her strong arms folded, her face a cold, arrogant mask. At her feet lies a lady's fan. In her clenched fist she holds a dagger, a stiletto, with a blade so slender that it is almost invisible. The painter has portrayed her as, very calm and at ease, she stands waiting patiently for a kindly, well-bred middle-class gentleman with love on his mind to come along and retrieve the fan for her. At that moment, we now understand, she will stab him in the back to rob him of his money and his life. Clearly, the children of Gautier's Clarimonde—that tempting vampire-prostitute who could make even virtuous men of the cloth behave like simpering tots—were everywhere, lusting after seed and money. Fin-de-siècle men were increasingly reaching the conclusion that one could deal with these man-eating creatures only by following the example of the steadfastly celibate Abbé Sérapion, who by sprinkling holy water and cursing the ''Demon! Impure courtesan! Drinker of blood and gold!'' had finally succeeded in turning Clarimonde into permanent dust (''Clarimonde,'' 148).

With the First World War waiting in the wings, woman also came to be seen as symbolic of the spirit of war. Henri Rousseau showed her—white gown in shreds, a sword in one hand and a torch in the other, the black mane of her hair matching that of her steed—as she rode roughshod over the prostrate naked corpses of men. Jean-Léon Gérôme had given impetus to this form of representation with his sculpture ''Bellona,'' for which he had dragged an obscure figure of classical myth back into the modern light of day, portraying her with her medusa's eyes flashing, her devouring mouth wide open, a sword in her hand, and a gigantic hissing cobra at her feet. War was the means to empire, empire provided access to gold, and many had decided that woman was behind the search for gold. Schopenhauer had put it simply: ''In their hearts,'' he had insisted at mid-century, "women think that it is men's business to earn money and theirs to spend it" ("Of Women," 297).

In 1910, Lovis Corinth left no doubt about what he considered the central weapon in the belligerent god Mars' arsenal of aggression. He placed a voluptuous nude woman at the center of his painting of a group of still innocent young boys who were in the process of examining the martial man's protective armor. Compared to an offensive weapon of such magnitude as this temptress, the painter implied, materials such as swords, helmets, cuirasses, and shields were merely child's play [XI, 7].

The International Studio of June 1915 carried a photograph of a sculpture by Anthony de Francisci called "War" which showed a nude woman with a body so masculine that one might have assumed her to be a man but for her pronounced breasts and the equally pronounced absence of male genitalia (the absence of which Freud had decided to be such a source of envy in all women, and hence presumably even more so in viragoes such as this). Holding in one hand, as if it were a bomb, a skull out of which crawls a snake about to wind itself around her arm, this personification of war reaches up to the sky with her other hand to signify that she is about to drag man down from his lofty intellectual perch and throw him down into the muddy battlefield of material desire which is her appointed domain. At about the same time the German graphic artist and painter Erich Erler placed himself in the service of wartime propaganda in his personification of Great Britain as a nude virago who proved to be "The Beast of the Apocalypse" incarnate [XI, 8]. Erler's etching is a striking indication of the manner in which the turn of the century's virulent antifeminine propaganda was being harvested in the minds of early twentieth-century males. The image contains a plethora of suggestions of woman as maneater and child murderer; a creature half-bestial herself in her pursuit of bestial pleasures and all-destructive in her apocalyptic lack of redemptive feminine passivity.

The goods-consuming middle-class wife who spent her husband's hard-earned money and the life-consuming prostitute who took what he had left in exchange for sharply pun-

XI, 7. Lovis Corinth (1858–1925), "The Weapons of Mars" (1910)

XI, 8. Erich Erler (b. 1870), "The Beast of the Apocalypse," etching (ca. 1915)

ished dreams of Eros thus blended in the fin-de-siècle male's fantasies to form the primal woman, incessantly voracious in her hunger for gold. Finding himself buffeted about in a changing economic environment, and having been marginalized in the struggle for power among the executioners of the world, he sought to discover a cause for the ever-increasing hazards to his economic well-being and for his sense of harrassment in trying to keep up with the breathtaking feats of ostentatious consumption of his friends and colleagues. It was easy to blame woman as the root cause of the problem. In his heart he knew that he had never wanted to climb on this roller-coaster of nerve-wracking enslavement to the great god mammon; his soul, after all, yearned for much greater things.

It was, as everyone knew, Eve who had set the rat race going, and it had been Pandora—the woman made of earth, beautiful and infinitely charming, but endlessly deceptive—who in her greedy curiosity had opened the box containing all the world's ills. Under the strain of his myriad responsibilities as breadwinner, and weighed down by the onus of his awesome spiritual tasks, the late nineteenth-century male began to dream furtively of those paradisiac days before woman had forced the evolutionary process into motion, before progress had become the driving force of life. Given the pressure of women's extravagant expenditures on items of dress and decoration, it became a habit among wits to emphasize that by taking a fig leaf off a branch Eve had inaugurated dressmaking.

Thus she had demonstrated her acquisitive nature right from the start. Evolutionists were as struck by the inherent truth of this fable as those who held fast to the biblical theory of creation. Owen Meredith, in the rhyming record of late nineteenth-century clichés of morality and social truth he called *After Paradise* (1887), has the fox watch Eve clothe Adam in sheep's wool, and sneer, "How wonderful is Woman's whim! / See, Adam's wife hath made a sheep of him!" And in a section of his poem entitled "The Legend of Eve's Jewels," he elaborates on Eve's reasons for being so interested in clothes and gold. It was the serpent himself who had shown her how to use the fleece of lambs to cover her limbs and cloak her body in mystery. "From that day forth Eve eyed with tenderness / The Serpent, to whose craft she owed her dress."

Thus, Meredith stressed, it was the lubricious primal reptile of desire, Eve's serpentine body, that had set the rat race in motion. Worried about the fact that the youthful beauty of her face is beginning to fade, the first lady of *After Paradise* becomes ecstatic as "the Serpent roll'd / His ruby-colour'd rings and coils of gold / Around the form of Eve." She's no fool and realizes straight off that she's gotten the hang of a good thing: "The sense of some new sexual power / Unknown to all her being till that hour / Within it kindled a superb surprise. Back with half-open'd lips and half-shut eyes, / She lean'd to its rich load her jewell'd head." Thus Eve assumes the pose of the nymph with the broken back the moment she realizes the infinite concealing power of adornment, and the serpent, her bestial alter-ego, remarks contentedly:

> By the bright blaze of thine adornment, see
> What in the years to come thy sex shall be!
> Mere female animal, much weaker than
> The male its master, not the Queen of Man,
> Scarce even his mate, that sex was born; but more
> Than it was born shall it become. Such store
> Doth in it lurk of secret subtilty,
> Such seed of complex life, as by-and-by
> Shall grow into full Woman; and, when grown,
> The Woman shall avenge, tho' she disown,
> The Female, her forgotten ancestress.
> Mother of both, my glittering caress
> Now wakes beneath thy bosom's kindled snow
> Whole worlds of Womanhood in embryo!
> A penal law controls Man's fallen state.
> It's name is Progress: and, to stimulate
> That progress to its destin'd goal, Decay,
> Woman, with growing power, shall all the way
> Its course accompany—from happiness
> And ignorance to knowledge and distress.

With swelling pride, the tumescent serpent now insists that, in her desire to keep him occupied, woman shall yearn ever more intensely for jewels, adornment, gold. Adam's labor power and his inventive ability will be pushed forever onward in the service of Eve's desire: "So shall the Feminine Force that set him on / Still keep him going till his course be done—" and "on the day / He lost that Paradise he ne'er had won, / Here was his progress, thanks to thee, begun." Behold, then:

The Future Woman; form'd to civilize,
Corrupt, and ruin, raise, and overthrow
Cycles of social types that all shall owe
To her creative and destructive sway
Their beauty's blossom, and their strength's decay.
Behold, then, in thyself the primal source
Of Human Progress, and its latest force!

Thus, if money was the root of all evil, it was so primarily because woman's sexual desire, expressed through her vanity and love of adornment, was the source of all desire for money. Woman's desire for gold was the root of all progress and all evil. José-Maria de Hérédia declared so quite as directly as Meredith when he described how the conquistadors thronged to the Pacific, "where in a haze of gold / The promised El Dorado beckoned to wealth untold, / Mingling, in monstrous visions, treasure and fierce desire / For the Amazonian virgins of the tropic zones of fire" (*The Trophies,* 207). But if gaining the gold and the virgin became confused in many late nineteenth-century men's minds, so did the acts of spending their money and spending their seed. Coventry Patmore—in a passage from his *Angel in the House* that was quoted with great enthusiasm by Ruskin in *Sesame and Lilies*—had already explained why it was so important to keep the household nun walled up properly:

Ah, wasteful woman, she who may
    On her sweet self set her own price,
Knowing man cannot choose but pay,
    How has she cheapen'd paradise;
How given for nought her priceless gift,
    How spoil'd the bread and spill'd the wine,
Which, spent with due, respective thrift,
    Had made brutes men, and men divine (79).

Clearly, woman's sexual hunger and her hunger for gold were one and the same. That was why the woman who did not see her primary task as the production of new life, of children, was in effect "spilling the wine" of future generations, pursuing gold for its own sake, cheapening the paradise of her warm womb and making it into a cold Pandora's box of economic evils. In an issue of *La Revue Blanche* published in 1896, Jules Bois, discussing "The Battle of the Sexes," bewailed the fact that woman "no longer wants to content herself with being the fated breeding ground of generations" (364), and described the womb of the vengefully infertile woman as "a fountain of life from which flows, every month, death itself, in a flood of debris, a wreckage which repeats itself continuously—a tide of blood which gushes forth in memory of shame and cruelty" (365).

Eduard von Hartmann, a disciple of Schopenhauer and Darwin and a truly blindfolded pioneer in the exploration of the psychology of the unconscious, saw the husbands of the second half of the nineteenth century as caught in a titanic struggle against the "cruel, heartless, and selfish thoughtlessness with which a woman tries to encumber her husband with all her burdens." In the collection of his essays which was published in an English translation in 1895 under the title *The Sexes Compared,* Hartmann saw the embattled middle-class husband faltering under the strain of having to cope with his wife's "hysteria and melancholy, which constantly threaten to develop into madness if her will is not satisfied, and her depression dissipated by diversions. The husband has to strain every nerve to obtain the money necessary to satisfy her desires'' (42).

It was extremely important, Lombroso and Ferrero were quick to point out, to keep woman occupied within the family environment and to remove all sources of sexual excitement, for

> the normal woman is deficient in moral sense, and possessed of slight criminal tendencies, such as vindictiveness, jealously, envy, malignity, which are usually neutralised by less sensibility and less intensity of passion. Let a woman, normal in all else, be slightly more excitable than usual, or let a perfectly normal woman be exposed to grave provocations, and these criminal tendencies which are physiologically latent will take the upper hand (*The Female Offender*, 263).

Hartmann, however, saw in a husband's slavish attempts to satisfy his wife's craving for ornament and entertainment a perhaps even more dangerous outcome:

> If he should want to use this money, which is sufficient to gratify his wife's craving, for the purpose of supporting his family, she would consider it an encroachment on her rights. Every attempt on the part of the husband to increase his family must appear criminal to the wife, should his income be already inadequate to appease all her wants. In this case, her natural selfishness combines with economic considerations to defeat the goal of marriage (42).

Indeed:

> It never enters her mind that her husband's occupation, which is undertaken for the support of the family, entails a far severer martyrdom on him than she suffers by fulfilling all her natural duties, and that it shortens his life in a far greater degree (46).

The consequence of all this, in Hartmann's opinion, was economic waste. In addition, woman's hunger for goods had dramatic negative implications for the future evolution of humanity, as was demonstrated by "the smaller number of children in the families of the higher, compared with those of the lower classes" (49).

Frank Norris took all these concerns very seriously when he came to write *McTeague*, in which he dramatized the horrible effects of woman's hunger for gold in the relationship between McTeague and Trina. In Norris' exploration of the world of the half-civilized, of those who live in the gray area between the "civilized" world of the middle class and the jungle of "brute labor," the male—in this case McTeague—finds himself precariously but relatively comfortably nestled on the edge of civilization. Then, however, woman awakens his sexuality. In her first encounter with McTeague, Trina, sedated and hence the image of true late nineteenth-century feminine languor, reclines "unconscious" and "very pretty" in the dentist's chair. At that very moment, Norris lets us know, McTeague's "animal nature," the urge to rape, "stirred and woke" (22). Norris observes "A woman had entered his small world and instantly there was discord. The disturbing element had appeared" (39). If animal attraction was what brought man and woman together, it was that same attraction which destroyed. "The instant Trina gave up, the instant she allowed him to kiss her, he thought less of her." Sex was, in Norris' eyes, a part of material exchange: The man offered to the woman his physical force, and the woman absorbed that force. Yet in doing so she lost her capacity to inspire the male's quest for spiritual transcendence. "It belonged to the changeless order of things—the man desiring the woman only for what she withholds; the woman worshipping the man for what she yields up to him. With each concession gained the man's desire cools; with every surrender made the woman's adoration increases'' (62–63).

For Norris it was clear that women hungered for a power they did not possess, and if male potency was what women desired, then the material symbol of that male potency—gold—became equally desirable and, in a sense, obtainable to them in a way in which ''maleness'' itself never could be. Hence women's inordinate desire for gold. But the desire for gold displaced women's true function in civilization: motherhood and the passive nurturing of the generations of the future. Relieved of their nurturing role, however, women could only degenerate. Thus, their very yearning for what they had lost—their virginal purity, innocence, and beauty, and the sexual power bestowed on them by virtue of these qualities—was symbolized by their desire for gold, which thereupon became the source of their reversion to bestial instincts. The more Trina yields her physical self to McTeague, the more desperately she identifies with the five thousand dollars she has won in a lottery. ''She clung to the sum with a tenacity that was surprising; it had become for her a thing miraculous, a god-from-the-machine'' (114).

Trying to hoard what she has already accumulated, Trina now becomes the source of McTeague's downfall, draining his physical energies but refusing to share her hoard of gold. Childless, degenerate, hungry for everything physical, Trina's emotions ''had narrowed with the narrowing of her daily life. They reduced themselves at last to but two, her passion for her money and her perverted love for her husband when he was brutal'' (227). Finally Trina degenerates to the level of bestial unconsciousness, having destroyed her husband's energies by vampirizing his physical strength and hoarding it in the form of her gold. ''Her love of money for the money's sake brooded in her heart, driving out by degrees every other natural affection. She grew thin and meagre; her flesh clove tight to her small skeleton; her small pale mouth and little uplifted chin grew to have a certain feline eagerness of expression; her long, narrow eyes glistened continually, as if they caught and held the glint of metal'' (259). Trina has become one of Dracula's daughters.

It is at this point in the trajectory of her degeneration that she completely loses the capacity to distinguish between masculine energy—male seminal power—and the sterile ego-nutritive qualities of her gold. ''Not a day passed that Trina did not have it out where she could see and touch it. One evening she had even spread all the gold pieces between the sheets, and had then gone to bed, stripping herself, and had slept all night upon the money, taking a strange and ecstatic pleasure in the touch of the smooth flat pieces the length of her entire body'' (264). When, at this turn of events, McTeague breaks into Trina's miserable living quarters and kills his estranged wife, taking her gold, Norris has succeeded in making it seem as if her death were a just punishment for this woman's monetary marital infidelity, and that McTeague, in taking her gold, is symbolically repossessing his spent manhood—although it is clear that his effort comes too late and that his wife has succeeded in dragging him down with her into the Death Valley of atavistic bestiality.

If this ritualistic exposition of woman's desire for gold weren't enough, Norris reinforced the theme with a subplot in which he demonstrated that he knew all about the fashionable scientific theories of the times which attributed the degenerate nature of ''inferior'' races to their general effeminacy and their inherently low level of evolutionary achievement. Norris therefore took pains to show their position to be on a level with that of women among the lower forms of human development. Like Trina, two other characters in *McTeague* demonstrate an inordinate lust for gold. One of them is Zerkov, a Polish

Jew, "a dry, shrivelled old man of sixty odd," with "the thin, eager, cat-like lips of the covetous; eyes that had grown keen as those of a lynx from long searching amidst muck and debris; and claw-like, prehensile fingers—the fingers of a man who accumulates, but never disburses," who sees in gold "the virgin metal, the pure unalloyed ore, his dream, his consuming desire," and who regards "each piece of money as if it had been the blood in his veins" (32–33). The other is Maria Macapa, of Latin extraction, "a strange woman of a mixed race" who hungers after the memory of a "service of gold plate" which once belonged to her family before it disintegrated as a result of miscegenation. Their story prefigures what is to happen to Trina and McTeague. For, united in their perverted hunger for gold, they marry. Unable to think of anything but gold, and linked merely in their degeneracy, they can create only "a child, a wretched, sickly child, with not even strength enough nor wits enough to cry . . . a strange, hybrid little being, come and gone within a fortnight's time, yet combining in its puny little body the blood of the Hebrew, the Pole, and the Spaniard" (176). Even so, the very act of giving birth, notwithstanding the terminally degenerate characteristics of the creature the two produce, is enough to snap Maria out of the dementia of her lust for gold and make her forget about the fabled "service of gold plate" whose history she used to narrate to her gold-lusting husband. Zerkov, a man who was never manly, squealing for gratification, and finding Maria, who has now become the mother-woman, no longer responsive to his sterile needs, thereupon murders her for gold that never existed. This done, Norris, with bathetic irony, makes McTeague and Trina move into the apartment vacated by their predecessors on the road to degeneration.

The equation Norris had made between woman's hunger for gold and her sterile hunger for masculine energy, for seed—her viraginous hunger for unfructifying sex, which, in Nicholas Cooke's words, caused woman to "cease to be the gentle mother, and become the Amazonian brawler"—was to be found everywhere in the work of the painters of the period. The latter were especially fond of representations of the classical Danae myth, in which Zeus, that endlessly resourceful philanderer, managed to impregnate the daughter of Acrisius of Argos by letting himself drop upon her loins in the form of a shower of gold—and this at a time when she had been isolated in a tower by her father to prevent just such an impregnation. The story gave the painters of the 1890s a chance to make a moral statement about woman's predatory nature, while also establishing a fashionable equation between woman's hunger for gold and her hunger for seed. Moreover, it gave them a fine excuse to exploit the visual theme of a woman in the throes of physical ecstasy.

In 1891 Carolus-Duran painted Danae luxuriously leaning back while a shower of gold coins was beginning to drop out of the sky. In the same year Chantron showed Danae eagerly pulling aside a curtain to make sure that the coins would find their mark. In 1896 Antoine-Auguste Thivet depicted a particularly hungry-looking Danae as she opened her legs to a well-directed shower of gold [XI, 9]. Across the Rhine Carl Strathmann depicted a characteristically pudgy, Germanic Danae, far removed from her tower, marching among a riot of fertile flowers while a veritable cloudburst of gold coins all around her paid tribute to Zeus' extraordinary potency [XI, 10]. The visitors to the Berlin Secession Exhibition of 1909 must have been deeply impressed. Edmund Dulac painted a series of designs for tapestries, among which were a Circe and a reclining Danae, arms behind her head, a smile of extreme contentment on her face, and a substantial shower of gold coins tumbling down steadily toward her loins.

In 1895 Max Slevogt painted Danae as a prostitute lying casually on her bed, arms

XI, 9. Antoine-Auguste Thivet (1822–after 1900), "Danae" (1896)

XI, 10. Carl Strathmann (1866–1939), "Danaë" (ca. 1905)

behind her head and nude except for a flimsy bit of gauze which barely served to cover her pubis. Gold coins fall like rain from the ceiling of her working-class room. A shabbily dressed older woman—perhaps her mother—holds out the folds of her skirt to form an ample receptacle for the falling coins. Not much classical symbolism here! In France Paul Désiré Trouillebert showed Danae in a state of restful satisfaction, sleeping off her lustful encounter, her hand still placed with rakelike tensility among the coins which symbolized her conquest. In 1908, in a now famous image, Gustav Klimt pushed the theme to the edge of excessive specificity in a decorative—though not very decorous—version of Danae's lubricious pleasure, in which he showed Danae, her face flushed with excitement, with her knees drawn up to her head as if she had positioned herself for intercourse. Thus Klimt made Danae turn herself into a circular form in a work filled with suggestions of circularity, while a mountain stream of gold coins rushed across her loins. In the same year Fritz Erler displayed a mural in Munich, simply entitled "Gold," which showed a hieratically posed half-naked woman, her raised hands spewing forth a rain of gold coin, while kings, bishops, beggars and soldiers gathered around her. In the foreground the painter placed another well-constructed and decidedly naked young woman who had pushed her breasts forward with her hands to catch the gold and let it nestle between them.

By the turn of the century, then, it had become commonplace to insist that woman, in her hunger for gold, was responsible for the manner in which the economic environment seemed to be changing. She had come to be seen as the secret force which had taken the reins of economic selfhood out of the hands of many whose fathers had still appeared comfortably in control of their own financial futures. Suppositions of this sort made her into an ideal surrogate executioner to the many masochistic executioner's assistants among the men of the turn-of-the-century middle class. Woman became the victimizer of choice of the period's self-pityingly marginalized male. By identifying her as the culprit, he could forego the search for other causes, and by using her as an executioner's surrogate, he could indulge his pleasure in manipulating the supposed manipulator. In the aggressive woman's flashing sword, directed at his head by the masochist's own volition, the executioner's attention was being called to the towering importance of his much-neglected assistant's still evolving but direly endangered intellect.

In his sculpture "The Eternal Idol" [XI, 11] Rodin placed masochistic man in his preferred relationship to woman: A helpless hero, he kneels before the perverse creature,

XI, 11. Auguste Rodin (1840–1917), "The Eternal Idol," sculpture (1889)

his arms as if tied behind his back, unable to breach the surface of woman's inviolate, virginal, childlike body except with the vulnerable caresses of his lips. Rodin's woman, the idol of perversity personified, leans away from man, preferring her icy isolation on her pedestal. His sculpture is virtually a direct illustration of a passage in Leopold von Sacher-Masoch's *Venus in Furs,* in which Severin, the plangent hero of that maudlin production of the late nineteenth-century middle-level male mind, describes how he fell in love with the marble textures of ''a stone statue of Venus'' and now pays it homage at all hours: ''Often I visit that cold, cruel mistress of mine by night and kneel before her, my forehead or my lips pressed to the cold pediment on which her feet are standing—and my prayers ascend to her'' (10).

Venus in Furs, first published in 1870, when the children of the household nuns of the 1850s were beginning to reach maturity, is the story of one among them named Severin, who often remembers his ''mother whom I had loved so deeply, and whom I had to watch as she was slowly devoured by a terrible illness'' (98). Severin also points out that, as a Platonist growing up among women less perfect than his mother, ''to me on the verge of adolescence, the love of woman seemed something particularly base and ugly'' (27). By his own admission, Severin is indeed of middle-level competence in all respects: ''I am nothing but a dilettante, a dilettante in painting, in poetry, in music and in several of the other so-called unprofitable arts which, however, secure for their masters these days the income of a cabinet minister or even of a petty princeling. Above all, I am a dilettante in life'' (8). Securely locked into the most extreme form of dualistic thinking, he believes that man ''has only one choice: to be the tyrant or the slave of woman. No sooner does he give way than his neck is under the yoke, and then the whip will begin to fall'' (7). Severin seeks transcendent satisfaction. Knowing full well that he is not a master, he wants to be a slave, the most abject, most indispensable, because most maltreated, slave around.

Women, however, do not understand his inexplicable need to be recognized as a meaningful entity through the ritual of maltreatment. Wanda, the living counterpart to his stone Venus, remarks rather sensibly, ''You seem to regard love, and particularly woman, as something hostile, something to guard yourself against, even unsuccessfully—as if its power were a kind of pleasant torment, a piquant cruelty. A truly modern attitude'' (15). Given the ''insensitivity'' of her observation, Wanda is obviously a perverse, mindless female who does not understand the transcendent yearning of the male intellect for gratuitous punishment. Or, rather, she has lived herself perfectly into the role of the mid-nineteenth-century helplessly cringing female. She is at heart a masochist herself and therefore is not interested in playing the sadist. ''I can indeed imagine belonging to one man for my entire life,'' she remarks, ''but he would have to be a real man, a man who would dominate me, subjugate me by his own innate strength'' (23). In other words, she too craves the attentions of an executioner—which Severin obviously is not—but since Severin needs the attentions of an executioner even more than she, they come to understand each other.

With the foot-stomping impatience of a spoiled child, Severin, the quintessential effeminate man of the time, tells Wanda, ''I want to be abused and betrayed by the woman I love, and the more cruelly the better,'' because ''one can only truly love that which stands above one'' (25). She at last complies and whips him, but with a definite reluctance. She does it only to be of service to him and as part of an attempt to mold herself into the woman of his dreams. She knows very well what is going on. ''For me,'' she says, ''this brutal game goes against the grain. If I were really the kind of woman who whips her slaves you would be horrified'' (38). Yet Wanda's reluctance makes her a perfect ''medium.'' For, notwithstanding all his protestations, Severin wants to be in control.

Wanda is to be, as it were, his own suppressed, bestial nature rising to the surface. That is why she must wear furs and thereby acquire the appearance of a cat. ''To me,'' Severin explains, ''woman represented the very personification of nature . . . she was cruel, like Nature herself who throws aside whatever has served her purpose as soon as she needs it no longer'' (32). Thus, by making Wanda become what he considers woman to be, Severin can still maintain a sense of being in control of his world, of being the manipulator. If Wanda is to be a vampire who bites him—as she does quite literally—she is to be one because Severin can yield his ''intellect'' to her without having to denounce himself for succumbing to ''the languor of satiety.'' Instead he can blame her as the personification of ''brute nature'' on the prowl. For—and here Sacher-Masoch shows clearly that he had read Carl Vogt with avidity—''despite the march of civilization, woman remains the same as when she came from the shaping hand of nature, she has the nature of a *savage,*—faithful or fruitless, magnanimous or cruel, according to the impulse which sways her at the moment'' (43).

Sacher-Masoch felt his world progressively turning into a congregation of aliens. It was a world in which his hero, Severin, must measure his own insignificance by the subsumption of his Aryan presence into a world of Semites and peasants, ''of Mazovians in homespun linen and greasy-haired Jews'' (58). Severin yearns to be recognized in an act of supreme humiliation, not by the woman he manipulates into acts of aggression— indeed, not by woman at all, that creature whose ''character is the want of character'' and who is inferior to him ''in brains and bodily strength''—but by a real male, a superman, a godlike master, a true representative of the class of executioners whose mousy menial he is. Since woman—Wanda—is the surrogate who, as she herself acknowledges, ''needs a master,'' she can become man's conduit to the superman. Woman, then, for Severin is only a means to true mastery, the symbol of mastery and not mastery itself.

In *Venus in Furs* Severin succeeds in catching the executioner's attention when Wanda melts away in abject masochistic submission before a glorious Greek, a true Apollo, with ''icy gaze'' and ''savage masculinity,'' a man with all the beauty but none of the ''weakness'' of a woman. Using Wanda to make this superman notice him, Severin undergoes his ultimate, most delicious humiliation, a consummate act of personal submission which shows that even in his marginalized position in the world he is still needed by the master executioner, and that the Lord must still pay attention to this menial. To prove that supermen are still men, and that even mousy males have a privileged place in the masculine future, the ''handsome Greek,'' the executioner himself, ''in his riding-boots, close-fitting white breeches and short velvet jacket,'' the man who personifies the theory of the survival of the fittest, whips Severin, his whining servant. ''And,'' Severin admits, to no one's surprise, ''the most shameful thing of all was that to begin with I still felt a certain, mad, supersensual stimulation under the lash of Apollo's whip and the cruel laughter of my Venus—but Apollo whipped the poetry out of me, with blow after blow, until at last I simply clenched my teeth in helpless rage and called down curses upon my voluptuous fancies, on woman, and on love'' (108).

Thus the ritual of recognition, of renunciation, of the transcendence of woman becomes a ritual of passage necessary for the construction of an exclusively masculine world of masters and slaves, of executioners and their trusty helpers. The rejection of the surrogate, of woman, the enemy, the ''She-Dionysus,'' in favor of a symbolic unification of male master and effeminate slave in unquestioning homage to the principle of male aggression "cures" Severin of his dependence on woman and makes him recognize "that to *labour* and *do my duty* was as comforting as a draught of fresh water" (emphasis in original). The masochism, then, of the late nineteenth-century male, and his manipulation of the image of woman as an all-destroying, rampaging animal was an expression of his attempt to come to terms with the implications of his own marginalization, his removal from the true seats of power in his society. It was not at all a backhanded compliment to woman's power over him; it was rather the creation of a surrogate master who could be sacrificed—indeed, destroyed if necessary—once the true masculine, the true "Aryan" master-slave bond of collaboration in man's depredation of the "inferior species" of being had established itself.

Woman, who at mid-century had of necessity been the first to explore the realm of masochism, now found herself pushed into the role of the surrogate sadist, so that the male could vent his pent-up frustrations in an orgy of masochistic self-indulgence. With her apparent hunger for gold, her outward purity and inward lust, her seeming self-sufficiency and blood thirsty virginity, she was the perfect foil to the pervasive masochism of the artists and intellectuals—the cultural middlemen—of the turn of the century. Spending the male's money, woman symbolically wasted his seed, and in wasting his seed she caused him to lose the most precious source of nourishment of his transcendent intellect. What was worse, she could spend his money while remaining physically a virgin. Even a daughter could thus participate in the unmanning of her own father. In 1839 Sarah Ellis had already drawn a graphic picture of woman as the domestic financial vampire. "We have seen pictures," she said, "of birds of prey hovering about their dying victim; but I doubt whether a still more repulsive and melancholy picture might not be made, of a man of business, in the decline of life, when he naturally asks for repose, spurred and goaded into fresh exertions, by the artificial wants and insatiable demands of his wife and daughters" (*The Women of England*, 201).

Virgin vampires, adolescents lusting after seed, unconscious whores who drained the

XI, 12. Max Liebermann (1847–1935), "Samson and Delilah" (ca. 1902)

XI, 13. Alexander Oppler (b. 1869), "Samson and Delilah" (ca. 1908)

veins of man's intellect, who were out to atrophy his head—what better surrogates could there be to take the role of the executioner in man's masochistic fancies? Severin certainly was not deaf to their siren call. "Reading in the Book of Judith," he admitted, "I envied the grim hero Holofernes because of the queenly woman who cut off his head with a sword, I envied him his beautiful sanguinary end" (12).

Symbolic castration, woman's lust for man's severed head, the seat of the brain, that "great clot of seminal fluid" Ezra Pound would still be talking about in the 1920s, was obviously the supreme act of the male's physical submission to woman's predatory desire. Turn-of-the-century artists searched far and wide to come up with instructive examples of such emasculating feminine perfidy. Some, like the British painter John Collier, tried to turn a fierce-looking medusa-eyed, but very mauldin and stagey Clytemnestra into a scourge of masculine potency, her enormous double-bladed axe still bleeding the spent Agamemnon's energies onto the pale marble at her feet.

The story of Samson and Delilah, always an excellent exemplum in the hands of those who wished to highlight the perfidy of woman, also became a perfect subject for painters who thought it necessary to expose woman's head-hunting capacity. True, Delilah had only gone after Samson's hair, but, given the poor muscle man's fate, she might as well have lopped off his head at the same time and thus have spared the foolish womanizer his humiliating consciousness of his emasculation. The painters showed Delilah crawling like a panther to her prey, appreciatively pawing Samson's mane in anticipation of her success, or otherwise simply as a regal figure enthroned, the ideal image of a dominatrix. To Max Liebermann she was clearly a modern incarnation of Everywoman, triumphantly holding up her victim's hair the way an indian might have help up the scalp of a conquered enemy. Meanwhile, naked and unmanned, Samson lies across the temptress' lap. The position in which Liebermann chose to paint the former strong man's trimmed cranium, pushed down by Delilah's imperious hand, deliberately amplified the suggestion that this victim of woman's lust for seed had indeed lost his head as well as his hair [XI, 12]. The sculptor Alexander Oppler sought to demonstrate the manner in which Delilah might have used her lap as a sacrificial altar. In his version of the theme Delilah appears poised to bear down upon her exhausted lover's neck with a cleaver; she seems to be evaluating her prey the way a butcher might size up a turkey for decapitation [XI, 13]. The mutual inspiration of artists and writers during the late nineteenth century is best expressed by Sacher-Masoch, who has Severin—at the point when he is to be whipped for the first time by his glorious Greek—contemplate a mural "on the ceiling where Samson, lying at Delilah's feet, was waiting for the Philistines to put out his eyes. At that moment the picture seemed a symbol, an eternal parable of passion and lust, of the love of man for woman. 'Each of us in the end is a Samson,' I thought, 'and in the end, willingly or not, is betrayed by the woman he loves, whether she wears a coat of cloth or of sables.' " (*Venus in Furs,* 107).

Sacher-Masoch's writings were eagerly sought after by intellectuals, especially in France, where his complete works were published in translation, and where, in 1883, he received high official recognition in the form of the Cross of the Legion of Honor. Emile Zola and Victor Hugo were among his admirers, as was Camille Saint-Saëns, who in 1877 had seen his own opera *Samson and Delilah* open in Weimar. The operatic stage, in fact, became one of the major stalking grounds of the headhuntresses, as well as of temptresses of every comparable sort. Oscar Wilde's *Salome* virtually made Richard Strauss' career when he used it as the libretto for his opera. Strauss also teamed up with Hugo von Hofmannsthal to warn the world of the madness of women in *Elektra* and to argue, in *Die Frau ohne Schatten,* that women who remain childless demonstrate their nothingness by having no shadow. In fact, with reference to the works of Wagner and Strauss alone an encyclopedic register of the period's idols of perversity might be established without any difficulty. Alban Berg brought morbid song to Wedekind in his *Lulu,* while Antonin Dvořák dragged *Rusalka* onto the operatic stage. Paul Hindemith managed to hear music in Oskar Kokoschka's thesis that a murderer could give hope to women who wished to consolidate their immersion in nothingless *(Mörder, Hoffnung der Frauen),* while in France *Thaïs* was set dancing by Massenet. Carlo Gozzi's fable about Turandot, a cruel oriental princess who could think of nothing better to do than to order that the men who loved her be decapitated, was the subject of operas by Ferruccio Busoni and by Giacomo Puccini, who chronicled the doings of the exotic headhuntress at the very end of his career. Gozzi's fable also found its way into the visual arts in such graphic works as Joseph von Divéky's demonstration of what it meant to love a woman [XI, 14].

Certain themes, however, were especially attractive to the artists of the period. Sacher-Masoch had already singled out the story of Judith and Holofernes, and the painters rushed in to show why. They tried to document every stage of her encounter with man, the enemy, from the moment she raised her sword in the full splendor of her perfidous nudity, as Franz von Stuck documented, through her triumphant, self-satisfied contemplation of the male's severed head. Carl Strathmann depicted Judith as a plumply enflowered vamp

XI, 14. Joseph (von) Divéky (1887–1951), "Turandot" (ca. 1916)

XI, 15. Benjamin Constant (1845–1902), "Judith" (1885)

with sharp, hungry teeth and a malicious glint in her eyes. Many others showed her as she gleefully clawed at the severed head, hungry for the vital fluids it contained. In the Bible Judith had been a paragon of self-sacrificial martyrdom for a noble cause. The late nineteenth-century painters, however, unmasked her as a lustful predator, an anorexic tigress. The sculptor Hermann Halm revealed that Judith had placed her victim's head, still bleeding, in her lap, the better to reap its precious benefits, and other sculptors and painters showed how she had taken man's head and had stomped on it maliciously with her dainty foot.

In 1885 Benjamin Constant painted an imperial, orientalized Judith sporting a sword of phenomenal proportions—not the sort of thing one could easily sneak into an enemy camp but certainly quite symbolic of his young woman's viraginous tendencies [XI, 15]. When an engraving of this painting was published in *Les Lettres et les Arts* in 1886, it was accompanied by a poem in which Jean Lahor succinctly brought together most of the period's platitudes about woman's nature:

Judith has chosen to devote her body to her country;
She has prepared her breasts to tempt her dreadful lover,

XI, 16. Carl Schwalbach (b. 1885), "Judith" (ca. 1914)

XI, 17. F. Humphry Woolrych (1864–1941), "Judith" (ca. 1887)

Painted her eyes and brightened their somber scintillation,
And she has perfumed her skin—destined to return much faded.

Pale, she has stepped forward to stage her massacre—
Her large eyes crazed with ecstasy and terror;
And her voice, her dance, her lean, hypnotic body
Have served the dark Assyrian as dread intoxicants.

In the arms of her triumphant master, suddenly
She has cried out, closing her eyes as if she were a child.
Afterward the man, relaxed, descends into a bestial slumber:

Caught as much within a horror of love as of dark death,
Her conscience free, woman has lashed out at man:
Coldly and with slow determination she has sliced off his head.

Carl Schwalbach's highly stylized Judith [XI, 16] showed the influence of modernism on a painter whose imagination was otherwise firmly locked within the fin de siècle. Influenced by the expressionist mentality, Schwalbach chose to emphasize—in fact, to exaggerate—Judith's feminine genetic deficiencies, giving her a forehead so low that it quite obviously did not contain the slightest cavity in which to stow a brain of her own. The painter thus provided his viewers with a convenient explanation why this bestial creature of the lower orders should have been so eager to come into possession of the intellectual cranium of Holofernes, with its high forehead and scholarly mien.

Before the turn of the century, more obviously idealized representations of Judith had been the fashion. At this time the representation of Judith as a skittish, imperious adolescent of the sort portrayed by Constant, had predominated. But the earlier decades had also seen the production of a few images strikingly resistant to the idealist mode. F. Humphry Woolrych, for instance, painted a "Judith" [XI, 17] which remained the startlingly realistic portrait of a studio model who, self-possessed and irritated at the masculine nonsense which was going on around her, was holding her prop sword with the cynical and disdainful expression of a woman who has seen more than enough of the idealizing pretenses of the artists for whom she was posing. In Woolrych's oil sketch (probably executed while he was studying at the Ecole des Beaux-Arts in Paris) we can almost see the model think that she'd know very well what to with the silly stick in her hand, if it weren't that her livelihood depended on such modeling sessions. This little painting thus provides a striking and unexpected antidote to the male fantasies of the period. It was Woolrych's strength as a painter that he had too much respect for the personalities and physical peculiarities of his sitters to be able to produce the idealized, mythologized images of feminine evil so much in vogue at the time that he painted his "Judith." Woolrych could paint wispy sylphs in the woods with the best of his contemporaries, but when he executed portraits he painted ordinary people. His nudes, too, tended to be portraits of rather ordinary women who just happened to be naked. It is not as if there were no threat in the eyes of Woolrych's "Judith," but it is the threat of laconic common sense and not that imagined threat of transcendent evil which so stirred the masochistic male of the fin de siècle.

For women the story of Judith inevitably took on the sort of ambiguous attraction which any image of mock-mastery tends to have among those whom it is meant to calumniate. Many a turn-of-the-century society party undoubtedly offered tableaux vivants such as those described by Edith Wharton in *The House of Mirth* as taking place at Bellomont, in which "scenes were taken from old pictures" and enacted by the women present for the entertainment of the assembled guests. Occasionally a mischievous young lady must have felt a gratifying thrill of illicit excitement as she portrayed Judith in the manner in which Frances Benjamin Johnston photographed the Countess Marguerite Cassini in 1904, posing as Judith, hand viraginously on hip and eyes appropriately glowering, while impressively displaying a long sword.

The story of Judith was popular among the intellectuals of the late nineteenth century, but the exploits of her biblical companion Salome became the true centerpiece of male masochistic fantasies. What better source for the fruitful conjunction of the period's numerous libidinous fetishes than this virginal adolescent with a viraginous mother, a penchant for exotic dances, and a hunger for man's holy head? Like so many of the fin de siècle's central fantasies of perverse sexuality, the story of Salome was given its initial impetus by Gustave Moreau, who in 1876 added the dastardly dancer to his repertoire of perverse women, which by this time already included animal lovers such as Leda, Europa, and Pasiphaë, as well as various sphinxes and chimeras. In addition, Moreau had produced a gaggle of sketches, drawings, and several finished paintings of a young Thracian girl carrying the head of Orpheus [XI, 18].

This last theme proved especially attractive, for it permitted the tantalizing juxtaposition of young virgins and the sad evidence of an idealistic and poetic male's rudely intercepted attempt at cranial evolution. The Orpheus myth had, for all practical purposes, been the later nineteenth century's perfect entry into the realm of the fantasy of the severed head. Armand Silvestre made this link obvious when he commented on a sculpture of Judith and Holofernes by Tony Noël in his *Le Nu au Salon; Champ de Mars* for 1984. The "haughty disdain with which the murderess pushes the bleeding head of her victim away with her foot" reminded him forcefully of the ferocity of "one of those maenads who, in a similar fashion, used their white feet to batter the decapitated head of the divine Orpheus."

As the most famous poet of classical myth, Orpheus was a perfect role model for the artists and intellectuals of the later nineteenth century. Could he not tame the beasts of the fields simply through the power of song? He was the ideal symbol of mind over matter, of the evolving male brain's triumph over the animal self. In the later nineteenth century an elaborate narrative developed concerning the symbolic significance of Orpheus' tragic demise, loosely based upon scattered suggestions taken from classical sources. It was held that after losing Eurydice forever to Hades, Orpheus had conceived a relentless hatred for women and their perfidious hold over men's hearts. He was thought to have been particularly harsh in his condemnation of the lifestyle of the maenads and their bestial rituals. Nothing, of course, was more galling to a maenad than to be pestered about her erotic proclivities. Tired of the poet's arguments in favor of mind over matter, these wild women had pounced on him, beaten him, and quite literally made mincemeat out of him—an incident affectingly documented by the painter Emile Lévy in his 1866 rendition of "The Death of Orpheus," which shows the impetuous young ladies howling and hacking away at the gentle poet, who is just about to lose his head. Enraged at the fact that Orpheus restlessly continued to sing about man's ideal future even after his head had been severed from his body, the maenads scooped the offending cranium onto the poet's lyre and threw it into the sea. When it finally floated back to land, it was supposed to have been found by gentle, domesticated virgin nymphs. Like Moreau's modest Thracian maiden of 1865, these nymphs were thought to have carried the poor victim of unacculturated feminine lust to a well-deserved hero's grave.

By 1876 Moreau was ready to embark on a more detailed representation of the manner in which an intellectual can lose his head over women. That year he contributed to the Paris salon a Salome shown dancing on the very tips of her toes and carrying a white lotus flower balanced on a snakelike stem, her marmoreal limbs encompassed by the shell-and-jewel-encrusted sparkle of her veils—and all this before an appropriately owl-eyed Herod [XI, 19]. With this work Moreau inaugurated the late nineteenth-century's feverish exploration of every possible visual detail expressive of this young lady's hunger for St. John the Baptist's head.

Within months of the exhibition of Moreau's painting at the salon, Flaubert was writing his story "Herodias," perhaps directly inspired by Moreau's creation. Flaubert, as usual, embellished the scanty details of the biblical account in Matthew, which tells of Salome's dance to please Herod and her demand, at her mother's bidding, for the head of the imprisoned St. John. Within the sharp, tactile textures of his hothouse prose, Flaubert made Herodias, Salome's mother, into an ambitious nineteenth-century virago and Herod into an effeminate aesthete. His Jokanaan was a John the Baptist who inveighed against Herodias in the measured tones of an Orpheus showing his idealistic disdain for the materialistic greed of the maenads:

XI, 18. Gustave Moreau (1826–98), "Thracian Maiden with the Head of Orpheus" (1865)

XI, 19. Gustave Moreau (1826–1898), "Salome" (1876)

Ah! It is you Jezebel! You took his heart with the creak of your slipper; you neighed like a mare; you set up your bed on the mountains to perform your sacrifices. But the Lord shall tear off your earrings, your purple robes, your linen veils, the bracelets on your arms, the rings on your toes, and the little golden crescents that tremble on your brow, your silver mirrors, your ostrich-feather fans, the mother-of-pearl pattens that increase your stature, the arrogance of your diamonds, the scents of your hair, the paint on your nails, all the artifices of your carnality (108).

Flaubert's virginal Salome is a blind tool of her calculating mother, who had made certain that her daughter would grow up to be an innocent lure in service to her power-hungry parent. Thus Flaubert makes that daughter into a virgin whore, instinctively mimicking in dance the details of a passion her body had not yet experienced in fact:

The girl depicted the frenzy of a love which demands satisfaction. She danced like the priestesses of the Indies, like the Nubian girls of the cataracts, like the bacchantes of Lydia. She twisted from side to side like a flower shaken by the wind. The jewels in her ears swung in the air, the silk on her back shimmered in the light, and from her arms, her feet, and her clothes there shot out invisible sparks which set the men on fire. A harp sang, and the crowd answered it with cheers. Without bending her knees, she opened her legs and leant over so low that her chin touched the floor. And the nomads inured to abstinence, the Roman soldiers skilled in debauchery, the avaricious publicans, and the old priests soured by controversy all sat there with their nostrils distended, quivering with desire (121).

Blindly, innocently carnal, Flaubert's Salome is not yet personally hungry for man's head. She can't even remember the name of the man whose head her mother has told her to demand. In Flaubert's tale it is still mom who does the scheming. But Herodias was not to survive much longer as the central personage in this drama of decapitation. As in Moreau's painting, Salome was stepping into center stage. In a watercolor called "The Apparition," executed slightly later than his oil painting, Moreau showed a Salome who, even while dancing, was having lustful visions of the Baptist's severed head, which the painter showed suspended in the air above the dancer's eagerly outstretched hand, its precious fluids dripping, in a garish tapestry of blood onto the floor. In this painter's work, as well as in the images of many of his contemporaries, the iconography of Salome as headhunter had clearly come to dominate. The Salomes who, in paintings of the 1870s by Henry Regnault, Alfred Stevens, and Benjamin Constant, had still been shown waiting in anecdotal, Orientalist fashion, with or without chargers in their laps but certainly always without the saint's head, now began to carry around the bloody trophy of their seed-hungry exertions wherever they went.

The appearance in 1884 of Joris-Karl Huysmans' *Against the Grain* was perhaps the crucial factor in bringing together, in the public's mind, the carnal virgin of Flaubert's tale and the hungry headhunter in Moreau's images of Salome's dance. Huysmans' hero, des Esseintes, waxes masochistically ecstatic over Moreau's interpretation of the Salome theme. In Moreau, we are told,

> Des Esseintes saw realized at last the Salomé, weird and superhuman, he had dreamed of. No longer was she merely the dancing-girl who extorts a cry of lust and concupiscence from an old man by the lascivious contortions of her body; who breaks the will, masters the mind of a King by the spectacle of her quivering bosoms, heaving belly and tossing thighs; she was now revealed in a sense as the symbolic incarnation of world-old Vice, the goddess of immortal Hysteria, the Curse of Beauty supreme above all other beauties by the cataleptic spasm that stirs her flesh and steels her muscles,—a monstrous Beast of the Apocalypse, indifferent, irresponsible, insensible, poisoning, like Helen of Troy of the old Classic fables, all who come near her, all who see her, all who touch her (53).

Des Esseintes still thought that Moreau's Salome had been repulsed by the appearance of John the Baptist's head in that artist's watercolor of the dancer's vision, but, in fact, Moreau's young lady was reaching in ecstatic hunger and not "petrified, hypnotized by terror," as des Esseintes would have it. Clearly, Huysmans was not as attuned to the pleasures of pain as some of his contemporaries. Sacher-Masoch would not have liked such an imputation of fear, weakness—indeed, of human feeling—in a woman's heart. Huysmans' failure of interpretation was rapidly pointed out by other painters, Edouard Toudouze among them. This painter made quite a splash at the Salon of 1886 with his own version of a triumphant Salome. Toudouze's Salome [XI, 20] is a pixieish adolescent who likes to snuggle up to the bestial carvings on the arm of the throne on which she sits, staring at us with the faraway eyes of a satisfied dreamer. She's had her fill of excitement, and the sainted head of Jokanaan lies forgotten in its charger at her feet. Toudouze's Salome thus has all the blind sensuality des Esseintes saw in the work of Moreau, but she

XI, 20. Edouard Toudouze (1848–1907), "Salome Triumphant" (ca. 1886)

does not have the sense of ultimate abhorrence he attributed to that painter's temptress. Instead, Toudouze's version shows her to be the childlike creature of Flaubert's tale without that adolescent's innocence. Toudouze's Salome is not so much deliberately evil as playfully wasteful of man's essence. She is both a carnal temptress and a virgin who, with her self-centered passion tarnishes the gold of purity she has collected in her Danae-like hunger for the seed of man's spiritual ambitions.

In the turn-of-the-century imagination, the figure of Salome epitomized the inherent perversity of women: their eternal circularity and their ability to destroy the male's soul even while they remained nominally chaste in body. Salome became the endlessly multiplied image of woman as Concha, the relentlessly perverse near-adolescent, eighteen-year-old temptress of Pierre Louÿs' *Woman and Puppet,* who "smiled with her legs as she spoke with her torso" (165), and who, as a child of the people and as a worker in a cigarette factory, uses her radiantly innocent beauty to tempt the novel's principal narrator, Don Mateo, into offering her his gold as he visits her workplace: "That golden coin thrown to that child was the fatal moment of my play. From then I date my present life, my moral ruin, my downfall and all the changes that you see on my brow" (184).

Mateo, a man of the world, shares with the other principal male figure in Louÿs' novel the belief that men "ask and women give themselves. Why should it be otherwise?" He therefore tries to make Concha understand that her defiant independence represents the childish behavior of a thoughtless creature intent upon inflicting moral suffering upon an eager male. In Concha's statements, however, the voice of the feminists of the time echoes with remarkable clarity. From a twentieth-century point of view, her remarks don't seem those of a petulant child but rather those of a thoughtful, strong, self-possessed young woman. "I belong to myself and I keep myself," she says. "There is nothing more precious to me than my own self, Mateo. No one is rich enough to buy me from myself." And when she says, "I am not contented with what satisfies other women. Not only do I wish all happiness, but I want it for all my life" (210–12), she may sound a little idealistic and ambitious, but only a typical turn-of-the-century male, who expected it as his birthright to have women yield themselves to him at his bidding, could have regarded these statements as signs of vicious perversity in a woman. Livid with the pain of his bruised masculine dignity and his desire, Mateo searches for revenge. According to his account of Concha's behavior, we are to understand that out of pure, inexplicable spite she symbolically "decapitates" him by "uniting" with a local boy "there . . . under my eyes . . . at my feet," protected from the raging Mateo by the grill of a gate to her courtyard (215).

Having set up the conditions for a justified revenge, Mateo, when next he comes within reach of her, attacks Concha violently and brutally. He beats her and chokes her to, by his own admission, within an inch of her life, and he is determined to finish his ministrations by raping her. But in the midst of her brutal debasement, Concha, as pointed out in an earlier chapter, thanks him delightedly for the pain he is inflicting upon her. In condescending to hurt her, Mateo presumably has proven how much he loves her. Moved by the strength of his passion, she says, "You are not going to take me by force. I await you in my arms. Help me rise . . . I told you that I had a surprise for you? Well then, you shall see presently, you shall see. I am still a virgin. The scene of last night was only a comedy, to hurt you" (218).

Mateo, still unbelieving, soon discovers that she is right. Thus the adolescent whore has "betrayed" her attacker once more by perversely proving that she is still a virgin. Mateo, meanwhile, has proved the point of his story: Virgin or whore, or virgin-whore, all women are the same and will attempt to unman you either way. Mateo must discover, in fact, that virginity is the worst form of feminine whoredom, because in her virginity woman maintains her self-sufficiency, and hence her power to "decapitate" the male by making him wait in impotent longing for her compliance to his wishes. Then, when he loses patience, she, in effect, perversely "forces" him to rape her, to "slay" her in order to regain his masculinity. In Mateo's eyes, then, man's assault upon the virgin-whore is a just result of her perversity in not giving in to man's desire. The enticement to rape was the virgin's unconscious demand that man "break her circle." But it was also a lure to make man throw his golden coin into the lap of a child, to make him waste his spiritual being. That enticement is thus revealed to be the virgin's capital crime.

No wonder, then, that Salome became the archetypal image of woman as serpent, as brute nature's virgin dancer, whose only reason for being, like Concha's, lay in "the movements of the arms, the legs, of the supple body and the muscular loins, born indefinitely from a visible source, the very center of her dance, her little brown belly" (208).

Mallarmé's Salome, in "Hérodiade," murmurs contentedly as she gazes fixedly at herself in her mirror:

> The horror of my virginity
> Delights me, and I would envelope me
> In the terror of my tresses, that, by night,
> Inviolate reptile, I might feel the white
> And glimmering radiance of thy frozen fire,
> Thou that art chaste and diest of desire,
> White night of ice and of the cruel snow!

"All about me," murmurs Salome, "lives but in mine own / Image, the idolatrous mirror of my pride."

Arthur Symons, whose own poems rarely matched the force of his partial translation of Mallarmé's dramatic fragment (which was first published in the December 1896 issue of *The Savoy*) tried to emphasize Salome's status as Everywoman in his sequence "The Dance of the Daughters of Herodias," written in 1897: "They dance, the daughters of Herodias, / Everywhere in the world," he intoned, adding, "When they dance, for their delight, / Always a man's head falls because of them," something that occurred "whenever they have danced his soul asleep" (*Poems,* vol. 2, 38–40). In a much later poem he described Salome's virgin bloodlust: "Her perverse, pure eyes malign / See, instead of signs of wine, / Frantic, to her vision, blood" ("Salome", *Lesbia,* 79) In the July 1898 issue of *La Wallonie,* Stuart Merrill dedicated his poem "Ballet" to Gustave Moreau, and also saw the Eternal Feminine in the virgin's dance:

> Bestial locks of hair streaking their hot pink lips,
> Arms barbarously weighted down with bracelets lifted high
> In velvet movements to the moonlit set around them,
> They murmur in deep malice and yet strangely mute:
>
> We are, oh mortal men, the dancers of desire;
> Salomes, and our bodies twisted into lures of pleasure,
> Will edge your hours of love toward depravity arcane.

"The Jewess Salome," Charles Besnard chanted in the Parisian magazine *L'Image* of March 1897, "while dancing, arches her supple loins / And lewdly brings balance to her hips":

Her loosened hair hangs down upon her heels—
And as, half-naked for her dance, she sings,
Desire steals into hearts like a shiver
Stirring the flanks of a stallion in rut.

Besnard had dedicated his poem to Jean Lorrain, who published in *La Revue Blanche* in 1893 under the rubric "Petites Salomés" a description (in turn dedicated to Oscar Wilde) of a performance by La Laus, a dancer whose movements brought to his mind "those proud and supple nudities, rustling with gemstones and flashing with light, which that master-sorcerer Gustave Moreau builds into his infinitely precious watercolors." In the movements of La Laus Lorrain saw "those dancing Salomes, wrapped round with agates and with sparks of electricity, those princesses of lewdness and unconscious cruelty who offer, with the swooning grace of monstrous flowers, the mystery of their sex and their smile to aging kings who have become children once more." La Laus awoke in her audiences—of this Lorrain was certain—"all the debaucheries of ancient Asia, all the ambiguous and bloody mysteries of lost religions, all the crimes committed by and about that sex—all those desperate betrayals lived again in her, Delilah, Salome and Thaïs at once, all the courtesans, all the flute players, all the perditions" (42).

While the theme of Salome as a bestial virgin Jewess, whose dance revived the dead embers of carnal life in even the most chaste of men, was passed around among the writers of the period's most determinedly purple prose, the painters became involved in their own scientific-archeological explorations of the link between gender and race in the realm of degeneration. Friedrich Fuchs, in *Venus* (1905), a monumental two-volume study of the representation of woman in art, commended the French painters of the later nineteenth century for having been among the first to emphasize Salome's Semitic origins. He pointed to these painters' concern for bringing out Salome's "racial nuances" and marvelled gratefully at their "ethnographic thoroughness," which he linked to the Orientalist vogue among French painters (vol. 2, chap. 8, 12). An anonymous writer for *Famous Pictures Reproduced,* a book notable for its equation of a painter's artistic success with the extent of his

XI, 21. Max Slevogt (1868–1932), "Salome's Dance" (1895)

XI, 22. Hugo von Habermann (1849–1929), "Salome" (or "Herodias") (ca. 1896)

XI, 23. Otto Friedrich (1862–1937), "Salome" (ca. 1912)

ability to uncover vast fields of feminine epidermis, showed how well these French painters were bringing their message across to the general public. Commenting on a Salome by Jules Lefebvre, this anonymous celebrant remarked that the master had succeeded in portraying in his painting of the daughter of Herodias, "an essentially Semitic type of the antique period, with the sensuous and soulless beauty of the tigress rather than the woman, bearing the charger which is to receive the head of John the Baptist, and the sword which is to decapitate him, as indifferently as if it were a dish of fruit" (119).

With such august satraps of official art as Lefebvre to guide them into the exploration of the intimate kinship between woman and the "degenerate races," it is not surprising that many of the younger artists who cranked out their obligatory "Salomes" for the salons came to pay much more than mere passing attention to the links between race and gender. Some, like Max Slevogt, explored the degenerative effects of the dance of the seven veils itself. Slevogt, in fact, was eager to demonstrate his personal knowledge of the popular theories which juxtaposed the "degeneracy" of women, Jews, and Africans. In his painting it was specifically left to viciously caricatured members of these "degenerate races" to ogle and leer at the wild agitation of Salome's dance [XI, 21].

Others, like Hugo von Habermann, chose to highlight the bestial degeneracy of the woman herself [XI, 22]. Habermann's widely reproduced painting was variously titled "Herodias" and "Salome," proving that in the minds of many the separate entities of the mother and her daughter had blended into a single image of ferocious vampire bloodlust with a touch of Ophelia thrown in for good measure. Still others followed the example of writers who had emphasized the universality of the Salome theme, as is the case with Jacques-Emile Blanche, who portrayed her as a stylish, sarcastically observant turn-of-the-century society lady. Most, however, chose to portray the evil temptress as she lewdly fondled the blood-stained remains of the saint's head. Otto Friedrich's triumphant anorexic Salome of around 1912, shown dancing, head in hand, on the circular pattern of her feminine inscrutability, is a characteristic example of this genre of representation [XI, 23]. A work by Gustav Klimt painted a decade earlier, now usually given the title "Judith I" but listed in the issue of *Die Kunst* in which it was reproduced as "Salome" [XI, 24], shows that the two great decapitators of the fin de siècle were sometimes as difficult to distinguish as Salome and Herodias. Klimt's Salome/Judith is a heady mixture of vampire lore, high fashion, and the period's obsession with the notion that the headhuntress had desired to obtain hands-on knowledge of John the Baptist's head. Nikolaus Friedrich sculpted a frieze showing a stark naked, deliberately crude-featured Salome poking her livid fingers into the dead man's eyes, a detail first popularized in painting by Lovis Corinth in his gory delineation of an unconvincingly orientalized *Dirne* playing with her capital catch [XI, 25].

Pierre Bonnaud, in his version, went all out to demonstrate his knowledge of the symbolic significance of the situation. He placed his Salome against a circular background, put a swanlike crown of red passion flowers on her head, placed a gold viper round her arm, made her sit on an enormous tiger skin, and had this catlike comforter glaring at the

XI, 24. Gustav Klimt (1862–1918), "Salome" (or "Judith I") (1901)

XI, 25. Lovis Corinth (1858–1925), "Salome Receiving the Head of St. John the Baptist" (ca. 1896)

XI, 26. Fritz Erler (1868–1940), "Dance," mural (1898)

XI, 27. Juana Romani (1869–1924), "Salome" (1898)

viewer with vivid ferocity. To cap things off, he showed the haughty lady lubriciously fingering the sorry remains of man's luckless brain. Meanwhile, back in Breslau, Fritz Erler covered four walls with mural decorations devoted to the Salome theme for the very up-to-date music room of a wealthy art lover's villa. His personification of "Dance" is none other than the period's favorite headhuntress herself, her hair standing upon her head like the flames of a fire, as well it might, considering the imperious manner in which she has planted the doubtlessly still bleeding head of the reluctant saint between her thighs [XI, 26].

Erler, as already noted, went on to become a much-admired favorite of the leaders of the millennial Third Reich—a clear indication that the Reich's objections to degenerate art had more to do with issues of style than with content, at least as long as things were kept monumentally and stylishly classical. It is disconcerting to realize that the thematic content and the narrative impact of what was still a largely representational form of early modernist art—the work of the expressionists, the Fauves, and the post-impressionists—has remained virtually unanalyzed in terms of its ideological conformity with the themes of "official" art. In the wake of the twentieth century's fervent admiration for style over content, any image which at least in form contradicted the late nineteenth-century official art world's preoccupation with immediately recognizable shapes and three-dimensional volume seems to have been welcomed, even if it's content was expressive of the crudest forms of turn-of-the-century misogyny.

In its absolutist rejection of the formal values of late nineteenth-century art, the early twentieth century demonstrated all too effectively both the wastefulness and the basically uncreative nature of rigid dualistic thinking. There was, within the more official modes of representation of the years around 1900, quite as much room for the expression of contestatory sentiment as there was within the context of stylistic innovation. Few artists, however, whether they were inclined toward stylistic innovation or had remained traditional in terms of the formal elements of their art, demonstrated any inclination to be nonconformist in their subject matter or in its interpretation. It is therefore no wonder that, given the extreme pressures placed upon them to prove their worth as artists, only a handful of women attempted to infuse any unorthodox elements into their own explorations of such a determinedly "male" theme as the representation of Salome. To tackle the theme at all was generally seen as a woman's declaration that she wanted to be "one of the boys," that she was impatient with the themes of babies, domesticity, landscape, and still life which were habitually thought to be the only legitimate realm for women artists.

The technically brilliant, and in her day quite popular, Juana Romani was among the few women painters who did attempt a version of the theme of Salome [XI, 27]. The painting she exhibited at the Paris Salon of 1898 was a superbly executed but stylistically and ideologically orthodox representation of a Salome, platter in hand and sword in lap, waiting for Herod's response to her request for John the Baptist's head. Romani's Salome is a characteristically petulant, snake-eyed, very nasty creature, with the Lombroso-approved degenerative features of the woman-criminal. Only the absence of the masochistic melodrama of a prominently featured decapitated head indicates that the work was painted from a woman's point of view. Both the painting's subject matter and its expression are completely orthodox—and "male." No wonder, then, that the French government acquired it for the state.

Most other versions of the Salome theme produced by women during the years around 1900 followed Romani's lead. There was, however, at least one very striking departure from this pattern. In 1890 the American painter Ella Ferris Pell exhibited a "Salome" at the Salon des Artistes Français which, although stylistically orthodox, was radically at variance with the prevailing mode in conceptual terms. A student of William Rimmer at Cooper Union in New York, Pell had been introduced to the idealist symbolism prevalent among the more ambitious European painters at what was for an American a relatively early date. When she moved to Paris during the late 1880s, she worked under the tutelage of three of the most highly regarded masters of official art: Jean-Paul Laurens, the period's most celebrated painter of historical subjects; Gaston Saintpierre, a technically superb painter specializing in Orientalist single-figure compositions; and Fernand Humbert, a highly regarded painter of society portraits.

Pell's years as a student of these painters' orthodox methods of representation clearly came to fruition in her brilliantly executed, tonally exquisite "Salome" [XI, 28] a modish single-figure composition in Saintpierre's Orientalist manner. Like Romani's, her Salome is waiting, charger in hand, for the Baptist's head. Every detail of Pell's painting is representative of the later nineteenth century's most revered qualities of technical mastery in

XI, 28. Ella Ferris Pell (1846–1922), "Salome" (1890)

academic art, highlighted by its subtle exploitation of a single frontal light source, which allowed the painter to emphasize and contrast the textures of flesh and cloth. As a technical achievement, then, this Salome stood quite as much above the average salon painting as Romani's. Yet the French state did not run in to buy this painting. Instead, both at the salon exhibition in 1890 and the following year, when it was exhibited in the United States at the National Academy of Design, the painting was greeted with stony silence on the part of the critics, although it did receive a "second place of honor" at the latter institution. In retrospect, its emphatic neglect is anything but surprising: Pell's Salome was, whether consciously intended or not, a truly revolutionary feminist statement for its period.

In Pell's painting a number of the most characteristic turn-of-the-century attributes of the biblical temptress are absent. She does not glare at us with a look of crazed sexual hunger; she does not have the wan, vampire features of the serpentine dancer; nor does she show herself to be a tubercular adolescent. Instead, she is a woman of flesh and blood, not a mythologized flower of evil. She may be young, but she is healthy and strong. A woman of the people—clearly and without disguise a painter's model—her features have none of the artificial refinement so prized among the women of the ruling class, nor any of the signs of "bestial degeneration" favored by the symbolists. She is as "white and plump" and "big-hipped" as Nana, but she has none of the perverse qualities which made that figment of Zola's imagination "a disturbing woman with all the impulsive madness of her sex, opening the gates of the unknown world of desire," a creature "with the deadly smile of a man-eater," characterized by the "wave of lust . . . flowing from her as from a bitch in heat" (44–46). Such a portrayal would have had admirers by the score yelping at the feet of this Salome.

Instead, Pell's Salome makes a revolutionary statement simply by being nothing but the realistic portait of a young, strong, and radiantly self-possessed woman who looks upon the world around her with confidence, with a touch of arrogance, even—but without any transcendent viciousness. Superhuman evil the men of the turn-of-the-century would have been able to handle; a woman with "a neck on which her reddish hair looked like an animal's fleece," would have sent shivers of masochistic pleasure up the spines of male viewers, for like Nana such a woman could, in the long run, be trampled in the mud or executed. But Pell's Salome, a real life-woman, independent, confident, and assertive, was far more threatening, far more a visual declaration of defiance against the canons of male dominance than any of the celebrated viragoes and vampires created by turn-of-the-century intellectuals could ever have been. Such a woman could not be disposed of in as cavalier a fashion as the evil women in man's mind. Her indomitable reality was this feminist Salome's most formidable weapon, far more dangerous than any imaginary decapitating sword.

It is unclear how much of Pell's representation was the result of conscious deliberation. In 1893 the painting finally did catch the attention of the American critic Edgar Mayhew Bacon. In an article entitled "The making of Masterpieces," for *The Quarterly Illustrator,* Bacon referred to "Salome" as the artist's "finest general achievement," and he quoted her as saying that she had tried to depict a Salome "unable to perceive the spiritual light emanating from it [the decapitated head], a light which illuminates herself, and by which alone she is visible in history." This was certainly as orthodox an interpretation as anybody could have wished for. But Pell may have decided to play it safe and not antagonize her interlocutor now that she was finally given a small modicum of critical attention. Another of her comments about the painting quoted by Bacon, that "the purely physical nature of Salome revolts against the ugliness of the decapitated head'' (310), is a good deal more complex in its implications than the habitual identification of Salome as a blood-lusting animal, hungry for John the Baptist's head. This latter statement to some extent parallels Huysmans' interpretation of Moreau's ''Apparition,'' but it has been stripped by Pell of Huysmans' taste for symbolist excess. In Pell's interpretation Salome's gesture of revulsion becomes an action determined by entirely different considerations than those which were acceptable within the context of the orthodox masculine fantasies concerning the temptress' uncontrollable sexual hunger. Pell's interpretation suggests a genuine distaste, a fundamental radical feminist rejection of the ideological premises of male society, a sentiment the men of her time, who preferred to play with ideas and not confront realities, invariably found intolerable.

Clearly, the leaders of turn-of-the-century culture could not look kindly on such a truly individualistic expression of radical defiance by a woman painter. Although Pell maintained a certain degree of prominence in New York in the 1890s by serving as president of that city's Ladies' Art Association, eking out a passable existence as a landscape and portrait painter, she remained virtually ignored by the critics. Her painterly skills and orthodox mode of expression seem to have had far less of an influence on her career than the negative consequences of her association with the feminist circle of Helen Campbell. If, in its ideological implications, Pell's stylistically orthodox rendition of Salome was considerably more daring than the stylistically advanced but ideologically timid work of such a painter as Mary Cassatt, her work could, precisely because of this, ultimately be dismissed all the more easily by the champions of progressive art. In consequence, Cassatt was hailed as a revolutionary. She died in material comfort and never lost her fame. Pell, on the contrary, who came from a prominent and still illustrious East Coast family, was ignored as an artist, disowned by her relatives, and buried in a pauper's grave.

Who, then, can be so bold as to censure the women artists of the turn of the century for simply desiring to be given the chance to paint—itself novel enough as an idea among women? Most refused to venture into the treacherous realm of overtly controversial ideological and stylistic statements. The wages of genuinely individualistic artistic expression, of that sort of individuality which, like Pell's, succeeds in expressing itself despite the apparent orthodoxy of its formal trappings, are rarely high. In the case of the woman artist of the turn of the century, these wages, no matter what her social background happened to be, were most likely punitive silence, universal disdain, and a pauper's death.

The cultural leaders of the years around 1900 much preferred the depiction of a simple world of dualistic absolutes, of easily identifiable abject household nuns and monstrous devil-women juxtaposed with godlike imperial males and pitiful effeminate victims. In *Psychopathia Sexualis* Krafft-Ebing had justified his coining of the word masochism after Sacher-Masoch by contending that until the publication of *Venus in Furs* ''this perversion . . . was quite unknown to the scientific world as such'' (87). Obviously, when at an earlier date women such as Emily Brontë or Christina Rossetti had tried to turn their marginalization into a source of dramatic longing for the small measure of admiring attention a self-sacrificial death might bestow upon them, their attempts were not deemed to merit the same sort of attention as the scientists came to bestow upon the more violent male expressions of a similar desire which Sacher-Masoch had begun to chronicle so minutely in 1870. Yet to the men of the later nineteenth century the sight of women refusing to continue to play the role of the ever-submissive victim of man's desire for mastery was evidence of a fundamental betrayal of trust. No longer able to lord it securely over the women in their lives, many men had, as we have seen, become rapidly conscious of their own removal from the centers of social and economic power. Obviously the wild woman, the gynander, the viraginous man-woman, was trying to usurp the male's place as the executioner's trusty assistant. Given the mathematics of sexual completion so popular at the time, this meant that the men who consorted with the virago, the feminist, were by their very nature effeminate. The male masochistic scheme was to act out to the fullest the role of suffering which was otherwise rightfully that of woman, and thus submit to the woman, who had become man, until the executioner, the godlike Greek, could come along to team up with the dominated male to put woman back in her place.

In an 1894 issue of *La Revue Blanche,* Henry Albert, discussing Strindberg's fascination with predatory women, sought to unmask the Scandinavian author as precisely such an effeminate creature: "Strindberg's hatred for woman is but a fear of woman. His ostentatious pride in being a man of strength is nothing but the weakness of a beggar hungry for love crawling before woman," Albert insisted. And, he continued, Strindberg "cries at the drop of a hat, he practices crying as if it were a sport. Next to tears, humiliation is his greatest pleasure." Moreover, said Albert, "while he believes he despises woman and tramples her underfoot, he actually places her on a pedestal. . . . The phobia against women which has come to pervade his entire creative production was set in motion by the sort of masochism, expressive of the need to suffer for the object of one's affections, with which we are all too familiar in our current literature." For Albert this attitude was proof of Strindberg's inability to be a true man, to be of the party of the godly Greek, the imperial sadist. "He wishes to calumniate woman, to push her down to the lowest level of our social order, and in wishing to calumniate woman he only ends by calumniating man. They seem so flat and trivial, these women he describes to us as monsters of perversity and domination—women whom a man who is sure for himself would throw down at his feet in a single gesture of disdain!" (495–96).

The principal interest of Albert's remarks lies in the manner in which they illuminate the masochist's project toward the achievement of mastery. Instead of directly taking the executioner's position, that of the lordly master who tramples on the world because it is his, the masochists, the Strindbergs of the late nineteenth century, needed to isolate, identify, and elevate the presumed source of their humiliation so that the destruction of their enemy—woman—would ultimately lift them to a position of parity with the imperial sadist, the godly Greek of worldly power. But for those who, like Albert, saw through the masochist's scenario, the executioner's helper still remained a mere executioner's helper. All the more so, in fact, because he had allowed himself to become the weak-kneed foil to the trivial pursuits of the viraginous woman. A man losing his head to a woman hence did not represent a triumph for man. The offending creature, the female predator, Salome, must be killed outright to redeem the sacrificial male. As a result of her justifiable execution, the male would regain his masculine essence, Samson would regain the god-given strength taken from him by Delilah in her capillaceous depredations.

Lombroso and Ferrero had insisted that "women who kill their lovers from passion have a virile strength of sentiment." In *Venus in Furs* Wanda becomes the ideal foil to Severin because she has all the polyandrous hunger which supposedly characterized the New Woman. The normal, submissive woman's "natural" attitude was one of self-destruction. The Italian phrenological duo thought they knew that "pure, strong passion, when existing in a woman, drives her to suicide rather than to crime." That was why "the true crime of love—if such it can be called—in a woman is suicide" (*The Female Of-*

fender, 276). The death of the male martyr at the hands of woman therefore justified the imposition of a death sentence upon her for crimes of gross insubordination. To execute woman was to redress the balance of nature, to exorcise the beast, and return man to his just position of imperial dominance—at least in the pitiful fantasy world of the evermore marginalized turn-of-the-century middle-class male.

Jules Laforgue, in his confused perusal of the Salome theme in his *Six Moral Tales*, developed an aspect of the virgin whore's bloodlust which had been given early publicity by Heinrich Heine in his poem "Atta Troll." Heine, embellishing upon folk tradition and the Brothers Grimm—and cheerfully admitting that none of the information he used was to be found in the Bible—depicted Herodias, not Salome, as carrying John the Baptist's head with her wherever she went, covering that sorry sacrificial cranium with "ardent kisses" in what was a clear instance of the awful consequences of the archetypal criminal woman's amorous transport. Heine, having already declared with characteristic light-heartedness: "Where women are concerned / It is impossible to tell / Where the soft Angel-child / Fades into the queen of Hell," now asked rhetorically, "Would a woman hunger for the head / of a man she does not love?" (60–61).

In line with Heine's presentation, Laforgue had his Salome, "exorcised of her virginity, by the gift of Jokanaan's head," try to revive it, as it "shone like the head of Orpheus," by kissing his dead eyes with all the petulant impetuosity the poet saw as characteristic of a young girl left unsatisfied after her initiation into the pleasures of love. Unfortunately, said Laforgue, "her electric caresses did not draw anything from the face except grimaces which had no consequences." Next "she stood up, and for ten minutes she exposed her ripeness to the mystic nebulae." But not even the sight of her nakedness can revive the head. "Then she unfastened Orion's troubled gold and grey opal. Like a sacramental wafer, she laid it in the mouth of Jokanaan. Then she kissed his mouth mercifully and hermetically, and she stamped his mouth with her corrosive seal." But the sainted cranium remains impassive, and "with a fractious 'come, now!' she struck the grinning head with both of her small feminine fists."

Disgusted with the head's continuing refusal to become serviceable again, Salome decides to get rid of this useless remnant of John the Baptist's manhood:

> Since she wanted the head to fall into the sea without striking the rocks, she used some force. The relic described a satisfactory phosphorescent parabola. Oh! the noble parabola! But the unlucky little astronomer had badly miscalculated her swing. She fell over the balustrade, and with a cry that was human at last, she toppled from rock to rock. With death rattling in her throat, and in a picturesque anfractuosity which was washed by the waves, far from the sounds of her country's festivities, lacerated into nakedness, with her sidereal diamonds cutting into her flesh, with her skull battered in, paralyzed by vertigo—in a bad way, as you might put it, she died for an hour (244).

Earlier in his account of Salome's exploits Laforgue had declared: "Every honest man believes / In the evolution of his Species" (230), and his description of the virgin-nymphomaniac's ultimate violation by nature thus becomes the revenge of civilized decency upon the crude machinations of brute desire. Throughout his *Six Moral Tales* Laforgue cultivated a tone which alternated between sophomoric flippancy and sophisticated irony, but which also often returned to an all-too-seriously meant sententiousness. It is the latter mode which dominates the poet's conclusion to his tale of Salome, in which he remarks that she was "less the victim of illiterate chance than of having wished to live in the artificial, instead of simply and honestly, like the rest of us'' (245). Like Sacher-Masoch, Laforgue permits himself the ultimate pleasure which accrues to the masochist returned to power through the outrages of his tormentor. He turns John the Baptist's head, symbol of male potency debased by woman, into the source of her punishment, a punishment which he could hence describe with all the sadistic relish of a deposed dominator at last determined to reclaim his manhood. In Laforgue's tale, Salome's imagined aggression against man therefore became full justification for her punitive rape by nature in the service of man's evolution.

Elements of the Salome theme delineated by Moreau, Flaubert, Huysmans, and Laforgue were to come together by 1891, to produce—first in French and shortly afterward in an English translation by Lord Alfred Douglas—Oscar Wilde's famous play, which did more than any other single image or piece of writing to make the headhuntress' name a household word for pernicious sexual perversity. Wilde's *Salome* is a very carefully designed dramatization of the struggle between the bestial hunger of woman and the idealistic yearnings of man. The play works up to a conclusion in which the masculine mind is led, through temptation and submission, to an understanding of the need for woman's immediate physical destruction. In Wilde's symbolic drama a wholesale manipulation of the image of woman as aggressor serves as a cleansing ritual of passage designed to expose her mindless perfidy and insatiable physical need. As such the work climaxes in a categorical renunciation of any communication between male and female, and, in effect, becomes a call to gynecide.

Wilde's Salome is Everywoman. The play opens with remarks by pages and soldiers of Herod's court, who get matters going by equating Salome with the moon, with vampirism, and with death. The moon, which reminds them of ''a princess who has little white doves for feet,'' and of a dancing woman, ''is like a woman rising from a tomb. She is like a dead woman. One might fancy she was looking for dead things'' (195). Salome herself is as pale as death, and ''like the shadow of a white rose in a mirror of silver'' (197). The princess' own perception of the moon is characteristic: ''She is like a little piece of money, a little silver flower. She is cold and chaste. I am sure she is a virgin'' (200). The moon is a reflection of Salome, who reflects the moon. Together they form the icy circle of woman's emotional self-containment and material greed. They represent Flaubert's ''virgins of Babylon who prostitute themselves to the Goddess''—a goddess symbolized in *The Temptation of St. Anthony* as ''a block of stone representing the sexual organ of a woman'' (236–37). Wilde's Salome is, in Jokanaan's words, ''a basilisk'' born ''from the seed of the serpent,'' a reptile able to kill a man simply by looking at him (202).

Wilde's play pitches sight against sound. Both may be primary senses, but for Wilde the battle between sight and sound represented the struggle between materialism and idealism, between the feminine and the masculine. Salome, Everywoman, the moon is ''seen,'' perceived solely in terms of her physical beauty, her material presence. One *looks* at Salome, at woman who, in turn, as Jokanaan says of Herodias, always gives ''herself up unto the lust of her eyes'' (205). The young Syrian, who will become a suicidal victim to the princess' beauty, is told, ''You are always looking at her. You look at her too much. It is dangerous to look at people in such fashion. Something terrible may happen'' (197). When Salome confronts Jokanaan, the prophet exclaims, ''Who is this woman who is looking at me? I will not have her look at me. Wherefore doth she look at me with her golden eyes, under her gilded eyelids." And he asserts emphatically, "It is not to her that I would speak" (206).

The voice is the realm of thought, of ideas. Jokanaan does not live in the realm of the eyes. Salome cries in anger, "It is his eyes above all that are terrible. They are like black holes burned by torches in a tapestry of Tyre" (206). The prophet is all voice, all intellect; hence he cannot be tempted. but to the virgin whore he is a set of sense organs and physical textures: "I am amorous of thy body, Jokanaan! . . . Suffer me to touch thy body." Rejected, she exclaims, "It is thy hair that I am enamoured of, Jokanaan . . . Suffer me to touch thy hair." Finally, in what must have seemed to Wilde a triumph of ironic statement about woman's tendency to turn symbol into fact and thought into matter, she decides that "it is thy mouth that I desire, Jokanaan . . . Suffer me to kiss thy mouth" (208). But safe in his world of sound, the prophet rejects the hungry Salome.

Less firm in his idealism than Jokanaan, Herod is a man torn between the worlds of sight and sound. He is capable of transcendence and aware of the dangers of sight. When the young Syrian kills himself over Salome, Herod remarks, "I remember that I saw that he looked languorously at Salomé. Truly, I thought he looked too much at her." Whereupon Herodias remarks sardonically, "There are others who look too much at her" (213).

But Herod, unlike the women about him, is also capable of auditory transcendence. "I hear in the air something that is like the beating of wings, like the beating of vast wings." The world of the spirit is calling him. but when he asks Herodias, "Do you not hear it?" she responds with the deaf, self-reflective coldness of woman: "I hear nothing" (213). Soon Herod is pulled into the realms of sight and physical desire. Tempted by material pleasure, he begs, "Salomé come and eat fruits with me. I love to see in a fruit the mark of thy little teeth" (214). Falling victim to the temptation of sight, Herod seals the doom of John the Baptist, the voice of moral transcendence. As the disembodied voice of Jokanaan sings his imprecations against her, Salome, "the daughter of Babylon with her golden eyes and her gilded eyelids," teams up with her mother, Herodias, who also "cannot suffer the sound of his voice," in order to silence the prophet forever. Herod makes his promise to give Salome whatever she wants if she will but dance for him, and as he does so he sees that the vampire-moon, the virgin predator, in anticipation of what is to come, "has become red. She has become red as blood" (226).

Laconically Wilde now notes that "Salomé dances the dance of the seven veils." Woman's perfidious, bestial lust for blood, for seed, for man's head surfaces. Salome, *not* at her mother's bidding but as the moon, the very personification of all feminine lust, asks for "the head of Jokanaan in a silver charger." At once Herod realizes the nature of his error: "I have looked at thee over-much. Nay, but I will look at thee no more. One should not look at anything. Neither at things, nor at people should one look. Only in mirrors is it well to look, for mirrors do but show us masks" (229). When Salome insists on being given the head of the prophet, Herod must comply. The princess listens for sound, but the execution of a man of the spirit produces only silence. "Why does he not cry out, this man? Ah! if any man sought to kill me, I would cry out, I would struggle, I would not suffer . . . No, I hear nothing. There is a silence, a terrible silence." The spirit of man does not speak to the lustful body of woman. Vindictively Salome remarks, "I tell thee, there are not dead men enough" (233–34).

Once she has been given the head of the prophet, Salome's feminine destiny is brought to fulfillment:

I am athirst for thy beauty; I am hungry for thy body; and neither wine nor apples can appease my desire. What shall I do now, Jokanaan? Neither the floods nor the great waters can quench my passion. I was a princess, and thou didst scorn me. I was a virgin, and thou didst take my virginity from me. I was chaste, and thou didst fill my veins with fire . . . Ah! ah! wherefore didst thou not look at me? If thou hadst looked at me thou hadst loved me. (236).

The spectacle of Salome's bestial passions makes Herod shiver. Wiser now, he comes to a final resolve. Henceforth, he declares, "I will not look at things, I will not suffer things to look at me" (236). But the outrages of feminine desire continue. In a passage in which Wilde directly equates semen and the blood which feeds man's brain, Salome, Woman, the vampire hungry for blood, tastes the bitter seed of man, depredates the spirit of holy manhood: "Ah! I have kissed thy mouth, Jokanaan, I have kissed thy mouth. There was a bitter taste on my lips. Was it the taste of blood? . . . Nay; but perchance it was the taste of love . . . They say that love hath a bitter taste . . . But what matter? what matter? I have kissed thy mouth" (237).

Seeing the ravages wrought upon man's spirit by the materialistic thirst of woman, by Salome's golden eyes, Herod, the executioner, now comes to that ultimate point of personal transcendence for which the male masochists of the late nineteenth century were striving in their fantasies of humiliation, the point at which woman, the surrogate, the cruel vampire, can be exorcised. John the Baptist has succeeded at last, through the sacrifice of his physical selfhood, in bringing Herod to a recognition of his ultimate responsibility to the realm of the word, to the transcendent realm of the masculine spirit. Herod, the plaything of women, can now finally become Herod the Great, master of men, the Lord High Executioner. When Herod takes the step which will exorcise the demon forever; when he turns to Salome and commands his soldiers to "Kill that woman!" we witness turn-of-the-century culture completing its long, fantastic, ritualistic indictment of woman for crimes she never planned, and for outrages she only committed in the skittish, nerve-wracked minds of economically ever more marginalized men.

Salome's hunger for the Baptist's head thus proved to be a mere pretext for these men's need to find the source of all the wrongs they thought were being done to them. Salome, the evil woman, became their favorite scapegoat, became the creature whose doings might explain why the millennium, that glorious world of the mind's transcendence over matter, did not seem to loom as near as their impatient souls desired. All that was needed to bring peace back to earth, these men fantasized, was to offer them the job of executioner, let them be permitted to take the sword from Salome's hands and, in a cleansing massacre, drive woman from them in the final climacterium of their transcendence of the flesh.

Thus, in an acute, lurid, antifeminine symbolism and a fashionable, loudly proclaimed espousal of evolutionist and eugenicist theories, men everywhere, of every possible political persuasion, declared their emancipation from the viraginous, decapitating sword of woman's regressive, degenerative concern for the real. By the first decade of the twentieth century, antifeminine attitudes, often accompanied by a wholesale espousal of misogyny, had become the rule rather than the exception in both Europe and the United States. Ordinary acts of love were seen as no more than the submission of the soul to the materialistic enticements symbolized by woman's hunger for gold, which in turn was dramatized by her hunger for man's head. To emphasize this link, the stories of Salome and Judith

XI, 29. Thomas Theodor Heine (1867–1948), "The Execution" (1892)

XI, 30. Ludwig von Hofmann (1861–1945), "The Valley of Innocence" (1897)

were scarcely sufficient. The turn-of-the-century artists therefore branched out into a welter of representations of other sword-wielding women.

Among the most remarkable of these variants is a truly flamboyant creation dating from 1892 by the German painter and caricaturist Thomas Theodor Heine [XI, 29]. The work is a rendition of a man's nightmarish encounter with a viraginous dominatrix, and shows him being led by the goat of lubricity over a rickety, narrow walkway into the vulval castle of carnal desire. The nooselike harness of roses around his neck is held by a sword-toting woman who, once they have passed through the fatal portal, is certain to turn upon him, decapitate him, and throw his emasculated remains among the lividly red-beaked, flesh-eating black swans. The latter, swarming in untold masses upon the waters of universal, undifferentiated femininity, feed off the spent bodies of woman's male victims. The painting, which was widely reproduced—in 1920 *Jugend* even printed a double-page color photograph of it—is a masterpiece of decorative and compositional elegance. Ironically, given the extreme antifeminine implications of its symbolism, it is rather less crude and blatant in its statement than Heine's work generally tended to be.

In that respect, a drawing not by Heine but by Ludwig von Hoffman, one of his colleagues among the *Jugend* crew, entitled "The Valley of Innocence" is more characteristic. A plumply enticing, towering young woman with the petulant look of a child plays with the body of a tiny young man as if he were nothing but a yo-yo. She's testing his head. Next to her is a knife, before her a small pile of decapitated male bodies and, waiting their turn, an army of fools [XI, 30].

In 1908, Albert von Keller painted an image of a naked young woman—Salome and Judith rolled into one—casually holding a long, sharp sword. On a bed next to her, in the tempest-swept, half-razed tent pitched on the battlefield of the turn-of-the-century's war against woman, lies the decapitated nude body of a man. The man's head, already half-forgotten, lies abandoned in the dust at the woman's feet. The dark-haired woman's beak-like, Semitic face, low forehead, and livid lips are meant to bespeak her degeneracy. The decapitated man's hand still touches the laurel wreath he was reaching for when she attacked him. She has stepped on it disdainfully. The artist gave this crudely symbolic painting a simple title. He called it "Love." It's purpose was not moral but aggressive. The audience which took this widely reproduced work to be a masterpiece of philosophic statement was not invited to contemplate this image in a thoughtful fashion. Instead, having been presented with the ravages of woman's materialistic power over man, viewers

XI, 31. Albert von Keller (1844–1920), "Love" (1908)

were expected to turn to the guards surrounding the fallen hero's tent and, with Wilde's Herod, sound the call of gynecide: "Kill that woman!"

Women, however, were everywhere, and they were ultimately not as mysterious, as unfathomable, as alien as the artists and intellectuals were trying to make them appear. The women of the turn of the century had to suffer untold humiliation from the baroque, self-justifying fantasies men had built up around them. But dinner had to be cooked and faltering egos had to be coddled. Gynecide was indeed an extravagant fantasy, but, as the world was to discover all too soon, genocide was not. Salome and Judith were both Jewish women, as the intellectuals of the turn of the century did not tire of pointing out. As such, they combined the crimes of women with those of a "degenerate race." If women's prosaic, everyday presence made it impossible for most men to maintain a constant sense of enmity toward them, the Jew was still there, guilty of the same crimes as woman. In his shadowy presence he was Dracula come out of the East; the mysterious, effeminately artistic, but heartless Svengali; and Zerkov, the greedy "Man with the Rake, groping hourly in the muck-heap of the city for gold, for gold, for gold." Woman, the feminine gender, as a surrogate for the executioner, as the modern Venus in her bestial furs, could ultimately only be conquered by being "transcended" in a slow, laborious process. As man soared toward the ideal, there was no reason not to keep her conveniently available to cook meals, take care of the kids, manage the servants, or submit to an occasional bit of masculine backsliding into the realm of physical pleasure. Salome, however, was the denatured woman, the Semitic degenerate, an expendable enemy. She could therefore justifiably be killed for her crimes.

The images of the viraginous woman and the effeminate Jew—both equally eager to depredate the gold, the pure seed of the Aryan male—began to merge. The deadly racist and sexist evolutionary dreams of turn-of-the-century culture fed the masochistic middle-class fantasy in which the godlike Greek, the Führer, the lordly executioner, leader of men, symbol of masculine power, at last moved by his assistant's marginalization, would kill the vampire, set his trusty servant free, and bring on the millennium of pure blood, evolving genes, and men who were men. If it was difficult to execute one's wife—not to say inconvenient—there was always the effeminate Jew. Fantasies of gynecide thus opened the door to the realities of genocide. The twentieth century was to prove that it had understood the murderous implications of the imperial symbolism of evolutionary inequality all too well.

Bibliography

I. **SOURCES QUOTED**

II. **SERIALS**
    A. Exhibition Catalogues
    B. Periodicals

III. **PRINCIPAL TEXTS CONSULTED**

N.B. The list of sources and credits following the bibliography, which includes complete references to all picture sources, contains important bibliographical information not duplicated here.

I. SOURCES QUOTED

Adam, Paul. "Des Enfants." *La Revue Blanche,* 9 (1895), 350–53.

Adams, Brooks. "The New Struggle for Life Among Nations." *McClure's Magazine,* 12 (April 1899), 558–64.

Albert, Henri. "Auguste Strindberg." *La Revue Blanche,* 7 (1894), 481–98.

Alger, William Rounseville. *The Friendships of Women.* Boston: Roberts Bros., 1868.

Aurier, G.-Albert. "Renoir." *Mercure de France* 3, no. 20 (August 1891), 103–6.

Bacon, Edgar Mayhew. "The Making of Masterpieces." *The Quarterly Illustrator,* 1, no. 4 (1893), 297–312.

Balzac, Honoré de. *The Wild Ass's Skin.* Trans. Herbert J. Hunt. New York: Penguin Books, 1977.

Barrès, Maurice. *Du Sang, de la Volupté et de la Mort* [1894/1903]. Paris: Plon, 1959.

Barrucand, Victor. "Des Filles." *La Revue Blanche,* 8 (1895), 347–52.

Baudelaire, Charles. *Oeuvres Complètes.* Ed. Y.-G. le Dantec and Claude Pichois. Paris: Bibliothèque de la Pléiade, 1961.

Beerbohm, Max. "The Pervasion of Rouge." In *The Works of Max Beerbohm,* 5th edn., with a bibliography by John Lane. London: John Lane, 1923.

Belot, Adolphe. *Mademoiselle Giraud, My Wife.* Trans. "A.D." Chicago: Laird & Lee, 1892.

Bennett, Arnold. *Our Women; Chapters on the Sex-Discord.* London: Cassell & Co., 1920.

Bernard, Charles. "Salomé." *L'Image,* no. 4 (March 1897), 119.

Blackburn, Henry. *Grosvenor Notes, 1885: A Complete Catalogue with Notes.* London, 1885.

——— [On Solomon J. Solomon's "Sacred and Profane Love"]. *Academy Notes* (1889), p. xix.

Bois, Jules. "La Guerre des Sexes." *La Revue Blanche,* 11 (1896), 363–68.

Bouyer, Raymond. "The Pastel Drawings of Aman-Jean." *The International Studio,* 31 (1907), 285–90.

Brontë, Emily. "The Prisoner, A Fragment." Reprinted in *The Norton Anthology of English Literature,* 4th edn. Ed. Meyer H. Abrams et al. New York: W. W. Norton, 1979, pp. 1314–6. For ms. versions of this poem, see C. W. Hatfield, *The Complete Poems of Emily Brontë,* (New York: Columbia Univ. Press, 1941), pp. 236–42.

Bushnell, Horace. *Women's Suffrage: The Reform Against Nature.* New York, 1869.

Butler, Charles. *The American Lady* [1836]. Philadelphia: Hogan & Thompson, 1842.

Campbell, Harry. *Differences in the Nervous Organization of Man and Woman.* London, H. K. Lewis, 1891.

Carpenter, Edward. *Love's Coming of Age: A Series of Papers on the Relations of the Sexes.* Manchester: Labour Press, 1896.

———. *Woman, and Her Place in a Free Society.* Manchester: Labour Press, 1894.

Chopin, Kate. *The Storm and Other Stories, with The Awakening,* Ed. Per Seyersted. Old Westbury, N.Y.: The Feminist Press, 1974.

Claflin, Tennie C. [Tennessee]. *Constitutional Equality, a Right of Woman.* New York, 1871.

Comte, Auguste. *System of Positive Polity, or Treatise on Sociology, Instituting the Religion of Humanity.* 4 vols [1851–54]. Trans. Richard Congreve and Frederick Harrison. London: John Henry Bridges, 1875–77.

[Cooke, Nicholas Francis]. *Satan in Society* [1870] (by "A Physician"). Cincinnati: C. F. Vent, 1876.

D'Annunzio, Gabriele. *The Child of Pleasure [Il Piacere].* Trans. Georgina Harding. London, 1898.

Darwin, Charles. *The Descent of Man* [1871]. 2nd edn. London, 1874.

———. *On the Origin of Species.* A Facsimile of the First Edition, with an introduction by Ernst Mayr. Cambridge, Mass.: Harvard Univ. Press, 1964.

Dickens, Charles. *Hard Times.* Ed. George Ford and Sylvère Monod. New York: Norton, 1966.

Donnelly, Ignatius. *Caesar's Column: A Story of the Twentieth Century* [1890]. Cambridge, Mass.: Harvard Univ. Press, 1960.

"Le Doyen de Nos Peintres: Ernest Hebert." *Je Sais Tout,* 3, no. 5 (1907), 671–78.

Dreiser, Theodore. *Sister Carrie* Unexpurgated edn. New York: Penguin/American Library, 1981.

Drummond, Henry. *The Ascent of Man.* London: Hodder and Stoughton, 1901.

du Maurier, George. *Trilby* [1894]. London: Everyman's Library, 1969.

Eliot, George [Marian Evans]. "[Margaret Fuller and Mary Wollstonecraft]." *The Leader* (October 1855). Reprinted in *The Norton Anthology of English Literature.* 4th edn. Ed. Meyer H. Abrams et al. New York: W. W. Norton, 1979, pp. 1655–61.

Eliot, T. S. "The Waste Land." In *The Complete Poems and Plays, 1909–1950.* New York: Harcourt, Brace, 1959.

Ellis, Havelock. *Man and Woman: A Study of Human Secondary Sexual Characteristics.* London, 1894.

———. *Studies in the Psychology of Sex.* 2 vols. (vol. I, pts. 1–4; vol. II, pts. 1–3). New York: Random House, 1936.

Ellis, Sarah Stickney. *The Women of England, Their Social Duties and Domestic Habits.* New York: D. Appleton & Co., 1839.

Famous Art Reproduced, Noted Modern Paintings and Sculptures of All Schools and from All Nations, with Descriptive and Critical Text. Chicago: Rand McNally, 1900.

Famous Paintings of the World. Introd. General Lew. Wallace. New Haven, Conn.: Butler & Alger, 1897.

Famous Pictures Reproduced from Renowned Paintings by the World's Greatest Artists. Chicago: John R. Stanton Co., 1917.

Flaubert, Gustave. "Herodias" and "A Simple Heart." In *Three Tales.* Trans. Robert Baldick. New York: Penguin, 1961.

———. *Salammbô.* Trans. A. J. Krailsheimer. New York: Penguin, 1977.

———. *The Temptation of Saint Anthony.* Trans. D. F. Harmigan. London: H. S. Nichols, 1895.

Fleming, George. "On a Certain Deficiency in Women." *The Universal Review,* 1 (May-August, 1888), 398–406.

Ford, Ford Madox. See Hueffer, Ford Madox.

Forel, August. *The Sexual Question; A Scientific, Psychological, Hygienic and Sociological Study* [1906]. Revised ed. Trans. C. F. Marshall. New York: Physicians and Surgeons Book Co., 1925.

France, Anatole. *Thaïs.* Trans. Basia Gulati. Chicago: Univ. of Chicago Press, 1976.

Freud, Sigmund. *Civilization and Its Discontents* [1930]. Trans. James Strachey. New York: Norton, 1962.

Fuchs, Friedrich. *Venus: Die Apotheose des Weibes.* 2 vols. Berlin; Willy Kraus, 1905.

Furniss, Harry. *The Confessions of a Caricaturist.* 2 vols. London: T. Fisher Unwin, 1901.

Gaskell, Catherine Milnes. "Women of Today." *The Nineteenth Century* 26, no. 153 (Nov. 1889), 776–84.

Gaskell, Elizabeth. *Mary Barton.* Ed. Stephen Gill, Harmondsworth: Penguin, 1970.

Gautier, Théophile. "Clarimonde" ["La Morte Amoureuse"]. In *One of Cleopatra's Nights and Other Fantastic Romances.* Trans. Lafcadio Hearn. New York: Brentano's, 1899.

Ghent, William J. *Our Benevolent Feudalism.* New York: Macmillan, 1902.

Gide, André. *Corydon.* With a comment on the second dialogue by Frank Beach. New York: Farrar, Straus, 1950.

Gilman, Charlotte Perkins. *The Yellow Wallpaper.* Old Westbury, The Feminist Press, 1973.

Gorki, Maxim. *The Artamonov Business* [1925]. Trans. Alec Brown. New York: Grosset & Dunlap, 1968.

Gourmont, Remy de. "A Woman's Dream." *Remy de Gourmont: Selections from All His Works.* 2 vols. Trans. Richard Aldington. New York: Covici-Friede, 1929, vol. 1, pp. 114–17.

Great Pictures in Private Galleries. 2 vols. London: Cassell & Co., 1905.

Haggard, H. Rider. *She.* New York: Hart Publishing Co., 1976. Facsimile of original 1887 ed.

Hamel, Maurice. *Salon de 1890.* Edition de Luxe. Paris, 1890.

Hamerton, Philip Gilbert. *Painting in France After the Decline of Classicism.* Boston, 1895.

Hardy, Thomas. *Jude the Obscure* [1895]. London: Macmillan, 1957.

Hart, James D. *The Popular Book: A History of America's Literary Taste* [1950]. Berkeley, Calif.: Univ. of California Press, 1963.

Hartmann, Edward von. *The Sexes Compared; and Other Essays.* Sel. and trans. A. Kenner. London: Swain Sonnenschein, 1895.

Hefting, Victorine, et al. *J. Th. Toorop, De jaren 1885 tot 1910.* Otterlo: Rijksmuseum Kröller-Müller, 1978.

Heine, Heinrich. "Atta Troll." In Heine, *Werke in Vier Banden.* Die Bibliothek Deutscher Klassiker, 39. Ed. Klaus Briegleb. Munich: Carl Hanser, 1982, vol. 4, pp. 9–88.

Hérédia, José-Mariá de. *"The Trophies," with Other Sonnets.* Trans. John Myers O'Hara and John Hervoy. New York: John Day, 1929.

Hopkins, Gerard Manley. "A vision of the Mermaids." In *The Poems of Gerard Manley Hopkins.* 4th rev. and enl. edn. London: Oxford Univ. Press, 1970.

Howells, William Dean. *A Hazard of New Fortunes.* New York: Signet Classics, 1965.

Hueffer, Ford Madox [Ford Madox Ford]. "Women and Men." *The Little Review,* 4 (1917–18); no. 9 (Jan. 1918), 17–31; no. 11 (March 1918), 36–51; no. 12 (April 1918), 54–65.

Huxley, T. H. "On the Natural Inequality of Men." *The Nineteenth Century,* 27, no. 155 (January 1890), 1–23.

Huysmans, Joris-Karl. *Against the Grain.* Trans. John Howard. Introd. Havelock Ellis. New York: Dover, 1969.

James, Alice. *The Diary of Alice James.* Ed. Leon Edel. New York: Penguin Books, 1982.

James, Henry. *The Bostonians.* New York: Modern Library, 1956.

———. "Communistic Societies." In *Literary Reviews and Essays on American, English and French Literature.* Ed. Albert Mordell. New Haven, Conn.: College and University Press, 1957.

Jersey, Countess of [J. E. Jersey]. "Ourselves and Our Foremothers." *The Nineteenth Century,* 27, no. 155 (Jan. 1890), 56–64.

Kipling, Rudyard. "The Vampire." In *The Works of Rudyard Kipling.* New York: Walter J. Black, (n.d.), p. 15.

Klee, Paul. *The Diaries of Paul Klee, 1898–1918.* Ed. with an Introd. by Felix Klee. Berkeley, Calif.: University of California Press, 1964.

Krafft-Ebing, Richard von. *Psychopathia Sexualis.* Trans. Franklin S. Klaf. New York: Bell Publishing Co., 1965.

Kraus, Karl. *Selected Aphorisms.* Ed. and trans. by Harry Zohn. Montreal: Engendra Press, 1976.

Laforgue, Jules. *Oeuvres Complètes.* 4 vols. (vols. I and II: *Poésies;* vol. III: *Moralités Légendaires;* vol. IV: *Mélanges Posthumes*). Paris: Mercure de France, n.d.

———. *Six Moral Tales.* Ed. and trans. by Frances Newman. New York: Horace Liveright, 1928.

Lahor, Jean. "Judith." *Les Lettres et les Arts,* 1, no. 1 (1886), 144.

Larkin, Oliver W. *Art and Life in America.* New York: Rinehart & Co., 1949.

LeConte, Joseph. *Evolution: Its Nature, Its Evidences, and Its Relation to Religious Thought* [1888/1891]. 2nd rev. edn. New York: D. Appleton, 1897.

Lee, Vernon [Violet Paget]. "Dionea." In *Arts and Letters* [English edn. of *Les Lettres et les Arts*], 4 (December 1888), 241–68.

———. "A Frivolous Conversion." In *Vanitas; Polite Stories.* 2nd edn. London: John Lane, 1911.

Le Fanu, Joseph Sheridan. *Carmilla.* In *Best Ghost Stories of Sheridan le Fanu.* Ed. E. F. Bleiler. New York: Dover, 1964.

Lemaître, Jules. "L'Amour selon Michelet." *La Revue de Paris,* 5, pt. 5 (Sept.-Oct. 1898), 732–43.

Linton, Mrs. E. Lynn. "The Wild Women as Social Insurgents." *The Nineteenth Century,* 30, no. 176 (Oct. 1891), 596–605.

——— *Modern Women.* New York, 1889.

Livermore, Mary, on gynecologists: see George L. Austin, M.D. *Perils of American Women, or a Doctor's Talk with Maiden, Wife and Mother.* With a recommendatory letter from Mrs. Mary A. Livermore (Boston, 1883). Reprinted in *Root of Bitterness: Documents of the Social History of American Women.* Ed. Nancy F. Cott. New York: 1972, pp. 292–3.

Lombroso, Caesar, and William Ferrero. *The Female Offender.* Introd. by W. Douglas Morrison. New York: Appleton, 1899.

Lorrain, Jean. "Petites Salomés: La Laus." *La Revue Blanche,* 4 (1893), 41–42.

Louÿs, Pierre. *Woman and Puppet.* In *The Collected Works of Pierre Louÿs.* New York: Liveright 1932.

Ludovici, Anthony M. *Enemies of Women: The Origins in Outline of Anglo-Saxon Feminism.* London: Carroll and Nicholson, 1948.

———. *Woman: A Vindication.* New York: Alfred A. Knopf, 1923.

Maeterlinck, Maurice. *The Treasure of the Humble*. Trans. Alfred Sutro. New York, Dodd, Mead, n.d.

Mayreder, Rosa. *A Survey of the Woman Problem [Zur Kritik der Weiblichkeit]*. Trans. Herman Scheffauer. New York: G. H. Doran, 1913.

Meredith, George. "Modern Love." In *Poems*. New York, Scribner, 1923.

Meredith, Owen [Robert, Lord Lytton]. *After Paradise, or Legends of Exile*. Boston, 1887.

———. "The Earl's Return." "Lucile." In *The Poetical Works of Owen Meredith*. Boston, n.d. [ca. 1900].

Mérimée, Prosper. *Lokis*. In *The Venus of Ille, and Other Stories*. Trans. Jean Kimber. London: Oxford University Press, 1966.

Merrill, Stuart. "Ballet." *La Wallonie* (July 1898). Reprinted in *The Little Review*, 5, no. 6 (October 1918), 54.

Michelet, Jules. *Woman [La Femme]*. Trans. J. W. Palmer. New York: Carleton, 1860.

Mirbeau, Octave. Review of *Lilith* by Remy de Gourmont. *Mercure de France*, (1900).

Möbius, Paul J. *Ueber den Physiologischen Schwachsinn des Weibes*. 9th enl. edn. Halle: Carl Marholm, 1908.

Monkhouse, Cosmo. *British Contemporary Artists*. New York, 1899.

Nadal, Victor. *Le Nu Au Salon*. Paris, 1903.

Natanson, Thadée. "XVᵉ Salon des Femmes Peintres et Sculpteurs." *La Revue Blanche*, 10 (1896), 186–87.

Neumann, Erich. *The Great Mother: An Analysis of the Archetype*. 2nd edn. Trans. Ralph Manheim. Princeton, N.J.: 1963.

Nordau, Max. *Degeneration*. 3rd ed. New York: D. Appleton & Co., 1895.

———. *Paradoxes*. Chicago: L. Schick, 1886.

Norris, Frank. *McTeague* [1899]. Ed. Carvel Collins. New York: Rinehart, 1950.

One Hundred Crowned Masterpieces of Modern Painting, with descriptions by J. E. Reed and other writers. 2 vols. Philadelphia, n.d. [ca. 1890].

Pater, Walter. *Marius the Epicurean*. 2 vols. London: Macmillan, 1885.

Patmore, Coventry. *The Poems of Coventry Patmore*. Ed. with an introd. by Frederick Page. London: Oxford University Press, 1949.

Péladan, Joséphin. *La Décadence Latine*. 14 vols., 1884–96. Vol. 9: *La Gynandre*. Paris, 1892.

Phillips, Claude. "The Salons." *The Magazine of Art* (1894), 898–329.

Pound, Ezra. *Pavanes and Divagations*. New York: New Directions, 1958.

Proudhon, P.-J. *La Pornocratie, ou les Femmes dans les Temps Modernes* [1858]. Paris: A. Lacroix, 1875.

Rachilde [Marguerite Eymery-Vallette]. "The Blood Drinker" [La Buveuse de Sang]. *Contes et Nouvelles, Suivi du Théâtre* Paris: Mercure de France, 1900, pp. 71–80.

Read, Helen Appleton. "Dolls." *The Arts*, 3 (April 1923), 278–81.

Régnier, Henri de. "La Femme de Marbre." *La Revue de Paris*, 7, pt. 1 (Jan.-Feb. 1900), 225–41.

Richardson, Samuel. *Pamela*. New York: Norton, 1958.

Robertson, Walford Graham. *Time Was*. London: Hamish Hamilton, 1931.

Robinson, William J. *Married Life and Happiness, or Love and Comfort in Marriage* [1922]. New York: Eugenics Publishing Co. 1933.

Rockefeller, John D. "Sunday School Address." Quoted in *Our Benevolent Feudalism*, by William J. Ghent. New York: 1902, p. 29.

Romanes, George J. "Mental Differences between Men and Women." *The Nineteenth Century*, 21, no. 123 (May 1887), 654–72.

Rosenhagen, Hans. *A. von Keller*. Bielefeld: Velhagen & Klasing, 1912.

Rossetti, Christina. "After Death." In *Poems*. New Enl. Edn. London: Macmillan, 1894.

Rossetti, Dante Gabriel. ''Aspecta Medusa''; ''The Day-Dream''; ''Eden Bower''; ''My Sister's Sleep''; and ''The House of Life.'' In *Poems*. Ed. Oswald Doughty. London: J. M. Dent, Everyman's Library, 1961.

Ruskin, John [on William Lindsay Windus]. *Pre-Raphaelitism: Lectures on Architecture and Painting*. London: Everyman's Library, n.d., p. 328.

———— *Sesame and Lilies; The Two Paths; The King of the Golden River*. London: J. M. Dent, Everyman's Library, 1970.

Sacher-Masoch, Leopold von. *Venus in Furs* [1870]. Trans. John Glassco. Burnaby, B.C., Canada: Blackfish Press, 1977.

Saltus, Edgar. *Historia Amoris: A History of Love Ancient and Modern*. New York: Brentano's, 1906.

Samain, Albert. ''Une.'' *Oeuvres*. 2 vols. Vol I: *Au Jardin de l'Infante*. Vol II: *Le Chariot d'Or; La Symphonie Héroique; Aux Flancs du Vase*. Paris: Mercure de France, 1913, I, pp. 97–98.

Sangster, Margaret E. *Winsome Womanhood; Familiar Tales on Life and Conduct*. New York, 1900.

Schnitzler, Arthur. ''Mother and Son.'' Trans. Agnes Jacques. In *Vienna 1900: Games with Love and Death*. New York: Penguin, 1976.

Schopenhauer, Arthur. ''Of Women'' and ''The Metaphysics of the Love of the Sexes.'' In *The Will to Live: Selected Writings*. Ed. Richard Taylor. New York: Frederick Ungar, 1967.

Seyffert, Oskar. *Dictionary of Classical Antiquities*. Rev. and ed. with additions by Henry Nettleship and J. E. Sandys. 3rd ed. New York: World Publishing Co., Meridian Books, 1956.

Shinn, Earl [''Edward Strahan'']. *The Art Treasures of America*. 3 vols. Philadelphia: Geo. Barrie, 1879–80.

Silvestre, Armand. *Le Nu au Salon: Champ de Mars*. Paris, 1889, 1892, 1893, 1984.

————. *Le Nu au Salon des Champs Elysées*. Paris, 1889, 1890, 1891, 1896, 1897.

Soissons, S. C. de [Charles Emmanuel de Savoie, Comte de Carignan]. *Boston Artists: A Parisian Critic's Notes*. Boston, 1894.

Southey, Robert. *The Life of Wesley and the Rise and Progress of Methodism*. 2 vols. New York, 1820.

Spencer, Herbert. *First Principles* [1862]. 6th edn. New York: D. Appleton & Co, 1901.

————. *Social Statics* [1850]. New York: Robert Schalkenbach Foundation, 1954.

Spielmann, M. H. Notes to Arthur Hacker's ''Circe.'' *Royal Academy Pictures*. London, 1893.

Sprague, Charles J. ''The Darwinian Theory.'' *The Atlantic Monthly*, 17 (Oct. 1866), 415–25.

Stoker, Bram. *Dracula* [1897]. New York: Signet Classics, 1965.

Strindberg, August. *By the Open Sea*. Trans. Ellie Schleussner. New York: Haskell House, 1973.

————. ''On the Inferiority of Woman'' [''De l'Infériorité de la Femme'']. Trans. Georges Loiseau. *La Revue Blanche* (1895) 14–20.

Stringer, Arthur. ''Metempsychosis.'' *The Smart Set*, 2, no. 1 (July 1900), 90.

Strong, Theodore. ''The Witch.'' *The Smart Set*, 2, no. 2 (August 1900), 49–50.

Sumner, William Graham. ''Sociology.'' In *War, and other Essays*. Ed. Albert Galloway Keller. New Haven, Conn.: Yale University Press, 1919.

Swinburne, Algernon Charles. *Chastelard: A Tragedy*. London, Chatto & Windus, 1894.

————. ''Laus Veneris'' and ''Atalanta in Calydon'' *Collected Poetical Works*. 2 vols. London: William Heinemann, 1924.

Symons, Arthur. ''Herodiade: From the French of Stéphane Mallarmé.'' *The Savoy*, no. 8 (Dec. 1896), 67–8.

————. *The Collected Works of Arthur Symons*. Vol. 2: *Poems*. London: Martin Secker, 1929.

————. *Knave of Hearts: Poems 1894–1908*. New York: John Lane, 1913.

————. ''Stella Maligna''; ''The Vampire.'' In *Lesbia, and Other Poems*. New York: E. P. Dutton, 1920.

————. *Selected Poetry and Prose.* Introd. by R. V. Holdsworth. Cheadle Thelme: Carcanet Press, 1974.

————. *Silhouettes.* 2nd edn., rev. and enl. London: Leonard Smithers, 1896.

Talmey, Bernard S. *Woman; A Treatise on the Normal and Pathological Emotions of Feminine Love* [1904]. 6th enl. and rev. edn. New York: Practitioners Publishing Co., 1910.

Tardieu, Emile. "Psychologie du Faible." *La Revue Blanche,* 9 (1895), 154–55.

Taylor, J. Russell. "The Posing of Vivette." Illus. by Albert B. Wenzell. *Scribner's Monthly,* 22 (1897), 693–700.

Tennyson, Alfred. "To a Lady Sleeping"; "By an Evolutionist"; "The Lady of Shalott"; and "Elaine." In *Poems and Plays.* London: Oxford University Press, 1975.

Tharp, Ezra. "Thomas Wilmer Dewing." *Art and Progress* 5 (March 1914), 155–61.

Tolstoy, Leo. *What Is Art? and Essays on Art.* Trans. Aylmer Maude. Tolstoy Centenary Edition, vol. 18. London: Oxford University Press, 1929.

————. *What Then Must We Do?* Trans. Aylmer Maude. Tolstoy Centenary Edition, vol. 14. London: Oxford University Press, 1936.

Twain, Mark [Samuel Langhorn Clemens]. *Eve's Diary.* New York: Harper & Brothers, 1906.

Uzanne, Octave. "On the Drawings of M. Georges de Feure." *The International Studio,* 3, no. 10 (Dec. 1897), 95–102.

Vauxcelles, Louis. *The Salons of 1907* [English Edn. of *Les Salons, Edition de luxe 1907*]. Paris, 1907.

Veblen, Thorstein. *The Theory of the Leisure Class* [1899]. Introd. by John Kenneth Galbraith. Boston: Houghton Mifflin, 1973.

[Vedder, Elihu]. Exhibition review. *The International Studio,* 10 (1900), ii.

Villiers de l'Isle-Adam. "The Eleventh-Hour Guest." In *Cruel Tales.* Trans. Robert Baldick. London: 1963, pp. 77–101.

Vogt, Carl. *Lectures on Man: His Place in Creation, and in the History of the Earth.* Ed. James Hunt. London: Longman, Green, 1864.

[Wagner Debate]. "Bayreuth et l'Homosexualité: Réponse a M. Henry Gauthier-Villars." *La Revue Blanche,* 10 (April 1, 1896), 304–6.

Wagner, Richard. Libretto. *Lohengrin.* By Richard Wagner. Cond. Raphael Kubelik. Deutsche Grammophon 2720036.

Walton, W., et al. *The Chefs-D'Oeuvre of the Exposition Universelle* [1900]. 10 vols. Philadelphia: Geo. Barrie, 1901.

Wedekind, Frank. *The Lulu Plays and Other Sex Tragedies.* Trans. Stephen Spender. London: John Calder, 1972.

Weininger, Otto. *Sex and Character.* London: William Heinemann n.d. [ca. 1906].

Wharton, Edith. *The Descent of Man, and Other Stories.* London: Macmillan, 1904.

————. *The House of Mirth.* New York: Rinehart, 1962.

Wilde, Oscar. "The Garden of Eros"; "The Sphinx." In *The Complete Writings of Oscar Wilde.* New York: 1907–9, vol. 5. "Salomé" [in French], vol. 2.

————. *The Picture of Dorian Gray and Selected Stories.* New York: Signet Classics, 1962.

————. *Poems.* 15th edn. London: Methuen, 1921.

————. *Salome.* In *Aesthetics and Decadents of the 1890's.* Ed. Karl Beckson. New York: Vintage Books, 1966, pp. 195–237.

Wister, Owen. *The Virginian* [1902]. New York: Pocket Books, 1956.

Wolf, Eva Nagel. "Elenore Abbott, Illustrator." *The International Studio,* 67, no. 266 (March 1919), xxvii.

Wollstonecraft, Mary. *A Vindication of the Rights of Woman* [1792]. Ed. Carol H. Poston. Norton Critical Editions. New York: Norton, 1975.

Woolson, Abba Goold. *Woman in American Society.* Boston, 1873.

Yeats, William Butler. "Leda and the Swan." In *The Collected Poems*. New York: Macmillan, 1956.

Zola, Emile. *Nana*. Trans. George Holden. Harmondsworth: Penguin Books, 1972.

———*The Sin of Father Mouret*. Trans. Sandy Petrey. Englewood Cliffs, N.J.: Prentice-Hall, 1969.

II. SERIALS

N.B.   The turn of the century was the great age of the painter-illustrated magazine, very different in this respect from our own age of photo-illustrated periodicals. Most of the periodicals listed here were profusely illustrated with the work of contemporary artists. The exceptions (those with few or no illustrations of this sort) are preceded by an asterisk. some of them were short-lived, while others continue to appear—if in a much altered form—even today. Generally speaking, for the purposes of this book I have scrutinized only those items which were published within the period 1880–1920.

Exhibition Catalogues

Academy Notes (annual). Ed. Henry Blackburn. London: Chatto and Windus, 1875–99.

Catalogues, Annual International Exhibition, Carnegie Institute. Pittsburgh, 1896–1914.

Catalogues Illustrés de Peinture et Sculpture (Exposition des Beaux-Arts, Champs Elysées) [Salon des Artistes Français] (annual). Paris, 1878–1914.

Catalogues Illustrés de la Société Nationale des Beaux-Arts (Champ de Mars) (annual). Paris, 1890–1914.

Esposizione Internationale d'Arte della Città di Venezia. Catalogo (biennial). Venice, 1895–1920.

Grosvenor Notes (annual). Ed. Henry Blackburn. London, 1877–99.

Le Nu au Salon (annual). With commentaries by Victor Nadal. Paris, 1902–6.

Le Nu au Salon des Champs Elysées (annual). With commentaries by Armand Silvestre. Paris, 1888–97.

Le Nu au Salon—Champ de Mars (annual). With commentaries by Armand Sylvestre. Paris, 1889–97.

Offizieler Katalog der internationalen Kunstausstellung im Glaspalast zu München (annual). Munich, 1899–1920.

Royal Academy Pictures (annual). London, 1887–1914.

Le Salon, Edition de Luxe (annual). With commentaries. Paris, 1881–1907.

Periodicals

Art and Progress. New York. (Title changed to *American Magazine of Art* in 1915.)

L'Art Flammand et Hollandais. Brussels.

The Artist. London (with "American Survey" section in New York edn.).

Les Arts. Paris.

Les Arts et Les Lettres. Paris. (English edition called *Arts and Letters* and published in New York.)

Brush and Pencil. Chicago.

The Century. New York.

Cosmopolitan Magazine. New York.

Deutsche Kunst und Dekoration. Darmstadt.

Emporium. Bergamo.

Gazette des Beaux-Arts. Paris.

Harper's Monthly. New York.

Harper's Weekly. New York.
L'Illustration. Paris.
L'Image. Paris.
The International Studio. New York [i.e., *The Studio*, with American supplement, from 1897.]
Jugend. Munich.
Die Kunst für Alle. Munich. (Title changed to *Die Kunst* in 1899.)
McClure's Magazine. New York.
The Magazine of Art. London.
Mercure de France. Paris.
*The Nineteenth Century. London.
Pearson's Magazine. London.
*La Plume. Paris.
The Quarterly Illustrator. New York.
*La Revue Blanche. Paris.
The Savoy. London.
The Scottish Art Review. Glasgow.
Scribner's Monthly. New York.
*The Smart Set. New York.
The Strand. London.
The Studio. London.
The Universal Review. London.
The Yellow Book. London.

III. PRINCIPAL TEXTS CONSULTED

N.B.   Innumerable novelists and poets wrote an untold number of works whose focus and imagery would warrant their inclusion in this bibliography. To list them all would be impossible; to list a few would serve little purpose. This, therefore, is a listing only of *nonfiction* sources.

Adam, Paul. "Critique des Moeurs." *La Revue Blanche,* 7 (1894), 337–44.
———. "Des Mères Futures." *La Revue Blanche,* 10 (1896), 390–93.
Adams, Henry. *The Education of Henry Adams* [1906]. Cambridge, Mass., 1961.
Aldington, Richard. "Decadence and Dynamism." *The Egoist,* April 1, 1915, 56–57.
Allen, Grant. "The Girl of the Future." *The Universal Review,* 7, no. 25 (May-August 1890), 49–64.
Allston, Margaret. *Her Boston Experiences: A Picture of Modern Boston Society and People.* Boston, 1900.
Adriaens-Pannier, A., et al. *150 Ans d'Art Belge dans les Collections des Musées Royaux des Beaux-Arts de Belgique.* Brussels: Musées Royaux, 1980.
Anderson, David L.; Maas, Georgia S.; and Savoye, Diane-Marie. *Symbolism: A Bibliography.* New York, 1975.
Anderson, Margaret Steele. *The Study of Modern Painting.* New York, 1914.
Anderson, Ross. *Abbott Handerson Thayer.* Syracuse, N.Y.: Everson Museum, 1982.
Andreas-Salomé, Lou. *The Freud Journal of Lou Andreas-Salomé.* Trans. Stanley A. Leavy. London, 1965.
Andree, Rolf, et al. *Arnold Böcklin: Die Gemälde.* Basel, 1977.
Auerbach, Nina. *Woman and the Demon: The Life of a Victorian Myth.* Cambridge, Mass., 1982.
Aurier, G.-Albert. "Le Coeur de la Femme." *Mercure de France,* 8 (May 1893), 26–27.
Bade, Patrick. *Femme Fatale: Images of Evil and Fascinating Women.* New York, 1979.
Bairati, Eleonora, et al. *La Belle Epoque: Fifteen Euphoric Years of European History.* New York, 1978.

Barker-Benfield, G. J. *The Horrors of the Half-Known Life: Male Attitudes Toward Women and Sexuality in Nineteenth-Century America.* New York, 1977.

Barnes, Earl. *Woman in Modern Society.* New York, 1912.

Barthes, Roland. *Michelet par Lui-Même.* Paris, 1954.

Bashkirtseff, Marie. *The Journal of a Young Artist, 1860–1884.* Rev. edn. Trans. Mary J. Serrano. New York, 1919.

Bayros, Franz von. *The Amorous Drawings of the Marquis von Bayros.* New York, 1968.

Beauvais, Jean de. "Prostitution, Criminalité, Uranisme," [review of Lombroso and Ferrero, *La Femme Criminelle*]. *La Revue Blanche,* 11, pt. 2 (1896), 277–79.

Bebel, August. *Woman Under Socialism.* Trans. Daniel de Leon. New York, 1904.

Beck, Hilary. *Victorian Engravings.* London: Victoria and Albert Museum, 1973.

Beckson, Karl, ed. *Aesthetes and Decadents of the 1890's.* New York: Vintage, 1966.

Beer, Thomas. *The Mauve Decade* [1926]. New York: Vintage Books, 1961.

Belden, Franklin Edson, et al. *Masterpieces of Art: Painting and Sculpture from the World's Great Galleries and Famous Private Collections.* Akron, Ohio, 1907.

Bell, Malcolm. *Sir Edward Burne-Jones: A Record and Review.* 3rd edn. London, 1894.

Bell, Nancy [Nancy d'Anvers]. *Representative Painters of the XIXth Century.* London, 1899.

Belot, Adolphe. *Mademoiselle Giraud, Ma Femme.* 59th edn. Paris, 1880.

Bendz, Ernst. "A Propos de la *Salomé* d'Oscar Wilde." *Englische Studien,* 51 (1917–18), 48–70.

Bénédite, Léonce. *The Luxembourg Museum: Its Paintings.* Paris, 1913.

———— *La Peinture au XIXe Siecle.* Paris, 1909.

Bernier, Georges, ed. *Sarah Bernhardt and Her Times.* New York: Wildenstein, 1984.

Binion, Rudolph. *Frau Lou: Nietzsche's Wayward Disciple.* Princeton, N.J. 1968.

Blunden, Maria, and Blunden, Godfrey. *Impressionists and Impressionism.* New York, 1980.

Boime, Albert. *The Academy and French Painting in the Nineteenth Century.* London, 1971.

————. *Thomas Couture and the Eclectic Vision.* New Haven, 1980.

Borowitz, Helen Osterman. "Visions of Salome." *Criticism,* 14, no. 1 (Winter 1972), 12–21.

Bowler, Peter J. *The Eclipse of Darwinism: Anti-Darwinian Evolution Theories in the Decades Around 1900.* Baltimore, Md., 1983.

————. *Evolution: The History of an Idea.* Berkeley, Calif., 1984.

Bowness, Alan, et al. *The Pre-Raphaelites.* London: Tate Gallery, 1984.

Bowness, Alan; Ebbinge-Wubben, J. D.; Hofstatter, Hans, et al. *Het Symbolisme in Europa.* Rotterdam: Museum Boymans-van Beuningen, 1975.

Bowra, C. M. *The Heritage of Symbolism* [1943]. New York, 1961.

Boyle, Eleanor Vere (Gordon) ["E.V.B."]. *The Peacock's Pleasaunce.* London, 1908.

Bragdon, Claude. *Delphic Woman.* New York, 1945.

Bredt, E. W. "Die Bilder der Salome." *Die Kunst,* 4 (1902–3), 249–55.

Breeskin, Adelyn D. *Mary Cassatt: A Catalogue Raisonné of the Oils, Pastels, Watercolors, and Drawings.* Washington, D.C., 1970.

————. "Thief of Souls." *Romaine Brooks.* Washington, D.C.: National Collection of Fine Arts, 1971.

Breicha, Otto. *Gustav Klimt: Die goldene Pforte, Werk—Wesen—Wirkung.* Salzburg, 1978.

Briggs, Asa, ed. *The Nineteenth Century: The Contradictions of Progress.* London, 1970.

Brinkley, Robert C. *Realism and Nationalism, 1852–1871.* New York, 1935.

Brinton, Christian. *Catalogue of Paintings by Ignacio Zuolaga.* New York: Hispanic Society of America, 1909.

Brome, Vincent. *Freud and His Early Circle.* New York, 1968.

Broude, Norma, and Garrard, Mary D., eds. *Feminism and Art History: Questioning the Litany.* New York, 1982.

Burger, Fritz. *Einführung in die Moderne Kunst.* Berlin, 1917.

Burke, Mary Alice Heekin, and Fink, Lois Marie. *Elizabeth Nourse, 1859–1938: A Salon Career.* Washington, D.C.: National Museum of American Art, 1983.

Burne, Glenn S. *Remy de Gourmont: His Ideas and Influence in England and America.* Carbondale, Ill., 1963.

Burne-Jones, Sir Edward. *Sir Edward Burne-Jones.* Introd. by Malcolm Bell. London, n.d.

Burollet, Thérèse, Walker, Mark Steve, et al. *William Bouguereau, 1825–1905.* Montreal: Montreal Museum of Art, 1984.

Busst, A. J. L. "The Image of the Androgyne in the Nineteenth Century." In *Romantic Mythologies.* Ed. Ian Fletcher. London, 1967, 1–95.

Cachin, Françoise, and Moffett, Charles S. *Manet, 1832–1883.* New York: Metropolitan Museum of Art, 1983.

Caird, Mona. "A Defense of the So-Called 'Wild Women.'" *The Nineteenth Century,* 31, no. 183 (May 1892), 811–29.

Callen, Anthea. *Women Artists of the Arts and Crafts Movement, 1870–1914.* New York, 1979.

Carluccio, Luigi. *Il Sacro e il Profano Nell'arte dei Simbolisti.* Torino: Galleria Civica d'Arte Moderna, 1969.

Carpenter, Edward. *The Intermediate Sex: A Study of Some Transitional Types of Men and Women.* New York, 1912.

———. *Intermediate Types among Primitive Folk: A Study in Social Evolution.* 2nd edn. London, 1919.

Carrington, Fitzroy, ed. *Pictures of Romance and Wonder by Sir Edward Burne-Jones, Bart.* New York, 1902.

Carroll, Charles, et al. *The Salon: A Collection of the Choicest Paintings Recently Executed by Distinguished European Artists.* 2 vols. New York, 1881.

Cassou, Jean; Langui, Emile; and Pevsner, Nicholas. *Gateway to the Twentieth Century: Art and Culture in a Changing World.* New York; 1962.

Casteras, Susan P. *The Substance or the Shadow: Images of Victorian Womanhood.* New Haven: Yale Center for British Art, 1982.

Celebonović, Aleksa. *Some Call It Kitsch: Masterpieces of Bourgeois Realism.* New York, n.d.

Child, Theodore. *Art and Criticism, Monographs and Studies.* New York, 1892.

Christie, J. D. "A Working Man's Reply to Professor Huxley." *The Nineteenth Century,* 27, no. 157 (March 1890), 476–83.

Christie, Jane Johnstone. *The Advance of Woman.* Philadelphia, 1912.

Clark, T. J. *The Absolute Bourgeois: Artists and Politics in France, 1848–1851.* Greenwich, Conn., 1973.

———. *Image of the People: Gustave Courbet and the Second French Republic, 1848–1851.* Greenwich, Conn., 1973.

———. *The Painting of Modern Life: Paris in the Art of Manet and His Followers.* New York, 1984.

Clay, Jean. *Modern Art, 1890–1918.* New York, 1978.

Clement, Clara Erskine. *Women in the Fine Arts.* Boston, 1904.

———, and Hutton, Laurence. *Artists of the Nineteenth Century and Their Works.* 2 vols. Boston, 1894.

Coeuroy, André. *Wagner et l'Esprit Romantique.* Paris, 1965.

Coles, William A. *Alfred Stevens.* Ann Arbor, Mich.: University of Michigan Museum, 1977.

Colette, Sidonie Gabrielle. *Le Pur et l'Impur.* Paris, 1971.

Commager, Henry Steele. *The American Mind.* New Haven, Conn. 1950.

Cook, Clarence. *Art and Artists of Our Time.* 6 vols. New York, 1888.

Coolus, Romain. "Notes Dramatiques: Auguste Strindberg." *La Revue Blanche,* 8 (1895), 88–91.

Cowper, Katie. "The Decline of Reserve among Women." *The Nineteenth Century,* 27, no. 155 (Jan. 1890), 65–71.

Crane, Walter, *An Artist's Reminiscences.* New York, 1907.

Crespelle, J.-P. *Les Maîtres de la Belle Epoque.* Paris: 1966.

Curtis, George William. *From the Easy Chair.* New York, 1892.

Curry, David Park. *James McNeill Whistler at the Freer Gallery of Art.* Washington, D.C.: Freer Gallery, 1984.

Daffner, Hugo. *Salomé: Ihre Gestalt in Geschichte und Kunst.* Munich, 1912.

Dantec, Félix le. "La Virulence du Sexe." *La Revue Blanche,* 30 (Jan.-April 1903), 230–37.

Darwin, Charles. *Selected Writings.* Ed. Philip Appleman. Norton Critical Editions. New York, 1970.

Décaudin, Michel. "Un Mythe 'Fin de Siècle': Salomé." *Comparative Literature Studies,* 4, nos. 1–2 (1967), 109–17.

Degerl ["Von Einem Arzte"]. *Kranke Seelen, Brief und Belehrung an Vera, die Märtyrerin.* Leipzig, 1903.

Delevoy, Robert L. *Symbolists and Symbolism.* Trans. Barbara Bray, Elizabeth Wrightson, and Bernard C. Swift. New York, 1978.

———; Croes, Catherine de; Ollinger-Zinque, Giselle. *Fernand Khnopff.* Brussels, 1979.

Densmore, Emmet, M. D. *Sex Equality: A Solution of the Woman Problem.* 2nd edn. New York, 1907.

Dooijes, Dirk, and Brattinga, Pieter. *A History of the Dutch Poster, 1890–1960.* Amsterdam, 1968.

Doort, M. C. "Beitrag zur Kenntniss der Frauenseele." *Jugend,* 22 (1904), 426.

Douglas, Ann. *The Feminization of American Culture.* New York, 1977.

Douglas, Norman. *Sirenland and Fountains in the Sand* [1911–12]. London, 1957.

Dowling, Linda. "The Decadent and the New Woman in the 1890's." *Nineteenth Century Fiction,* 33 (March 1979), 434–53.

Duca, Lo. *Histoire de l'Erotisme.* Paris, 1969.

Duckers, Alexander, *Max Klinger.* Berlin, 1976.

Dujardin, Edouard. "Réponse de la Bergière au Berger." *La Revue Blanche,* (1892), 65–77.

Dumur, Louis. "De l'Instinct Sexuel, et du Mariage." *Mercure de France,* 1 no. 2 (March 1890), 65–71.

———. "Petites Aphorismes sur les Femmes." *Mercure de France, 4, no. 28 (April 1892), 329–35.*

———. *"De la Vénalité de l'Amour Chez la Femme," Mercure de France,* 2, no. 13 (Jan. 1891), 39–43.

Dunlap, Knight. *Personal Beauty and Racial Betterment.* St. Louis, Mo., 1920.

Dunlop, Ian. *The Shock of the New: Seven Historic Exhibitions of Modern Art.* New York, 1972.

Eberhard, Dr. E. F. W. *Feminismus und Kultur Untergang: Die erotischen Grundlagen der Frauenemanzipation.* 2nd rev. edn. Vienna, 1927.

Ehrenreich, Barbara, and English, Deidre. *Complaints and Disorders: The Sexual Politics of Sickness. Old Westbury, N.Y., 1973.*

———For Her Own Good: 150 Years of the Experts' Advice to Women.* Garden City, N.Y., Anchor Books, 1979.

———. *Witches, Midwives, and Nurses: A History of Women Healers.* 2nd edn. Old Westbury, N.Y., 1973.

Ellis, Havelock. *The Task of Social Hygiene.* London, 1912.

Eldredge, Charles C. *American Imagination and Symbolist Painting.* New York: Grey Art Gallery, 1979.

Ellmann, Richard. *The Identity of Yeats.* London, 1954.

Engen, Rodney K. *Victorian Engravings*. London, 1975.

F., L. R. [L.R.F.]. *Esther, ou l'Education Paternelle; Poème en Six Chants Dédié aux Demoiselles à Marier*. Paris, 1839.

Faderman, Lillian. "Female Same-Sex Relationships in Novels by Longfellow, Holmes and James." *The New England Quarterly*, (September 1978), 309–32.

———. *Surpassing the Love of Men: Romantic Friendship and Love Between Women from the Renaissance to the Present*. New York, 1981.

———, and Eriksson, Brigitte. *Lesbian Feminism in Turn-of-the-Century Germany*. Iowa City, Iowa, 1980.

Farwell, Beatrice, et al. *The Cult of Images: Baudelaire and the 19th-Century Media Explosion*. Santa Barbara: Univ. of California at Santa Barbara Art Museum, 1977.

Faulkner, Harold U. *Politics, Reform and Expansion, 1890–1900*. New York, 1959.

Fay, Antoinette, et al. *Autour de Lévy-Dhurmer: Visionnaires et Intimistes en 1900*. Paris: Editions des Musées Nationaux, 1973.

Fee, Elizabeth. "Nineteenth-Century Craniology: The Study of the Female Skull." *Bulletin of the History of Medicine*, 53, no. 3 (Fall 1979), 415–33.

Fetterley, Judith. *The Resisting Reader: A Feminist Approach to American Fiction*. Bloomington, Ind., 1978.

Fink, Lois. "Children as Innocence from Cole to Cassatt." *Nineteenth Century*, 3, no. 4 (Winter 1977), 71–75.

Finot, Jean. *Problems of the Sexes*. Trans. Mary J. Safford. New York: Putnam's, 1913.

Fish, Arthur. *Henrietta Rae*. London, 1905.

Fiske, John. *The Destiny of Man Viewed in the Light of His Origin*. Cambridge, Mass., 1884.

Fitzgerald, Penelope. *Edward Burne-Jones*. London, 1975.

Flexner, Eleanor. *Century of Struggle: The Woman's Rights Movement in the United States* [1959]. New York, 1974.

Frecot, Janos; Geist, Johann Friedrich; and Kerbs, Diethart. *Fidus, 1868–1948: Zur ästhetischen Praxis bürgerlicher Fluchtbewegungen*. Munich, 1972.

Free, Renee, et al. *Victorian Olympians*. Sydney: Art Gallery of New South Wales, 1975.

Freud, Sigmund. *Beyond the Pleasure Principle*. Trans. James Strachey. New York, 1928.

——— *Dora: An Analysis of a Case of Hysteria* [1905]. Ed. Philip Rieff. New York, 1963.

———. *New Introductory Lectures on Psychoanalysis* [1933]. Trans. James Strachey. New York, 1965.

———. *Three Contributions to the Theory of Sex* [1905]. Trans. A. A. Brill. New York, 1962.

Fuchs, Eduard. *Geschichte der erotischen Kunst*. 3 vols. Munich, 1908.

Gamble, Eliza Burt. *The Evolution of Woman: An Inquiry into the Dogma of Her Inferiority to Man*. New York, 1894.

Gaubert, Ernest. *Rachilde*. Paris, 1907.

Gaunt, William *The Aesthetic Adventure*. New York, 1967.

Gauthier-Villars, Henry. "Bayreuth et l'Homosexualité." *La Revue Blanche*, 10, pt. 1, (March 15, 1896), 252–55.

Gerdts, William H. *The Great American Nude; A History in Art*. New York, 1974.

———. *Revealed Masters: 19th-Century American Art*. New York: American Federation of Arts, 1974.

Gilbert, Sandra M., and Gubar, Susan. *The Madwoman in the Attic: The Woman Writer and the Nineteenth-Century Literary Imagination*. New Haven, 1979.

Gilman, Charlotte Perkins. *The Living of Charlotte Perkins Gilman; An Autobiography*. New York, 1975.

———. *Women and Economics* [1898]. Ed. Carl Degler. New York, 1966.

Gilman, Richard. *Decadence: The Strange Life of an Epithet*. New York, 1979.

Glenn, Constance W., and Rice, Leland. *Frances Benjamin Johnston: Women of Class and Station.* Long Beach, Calif.: Art Museums and Galleries, California State University, 1979.

Gold, Arthur, and Fizdale, Robert. *Misia: The Life of Misia Sert.* New York, 1980.

Goldwater, Robert. *Symbolism.* New York, 1979.

Gould, Stephen Jay. *The Mismeasure of Man.* New York, 1981.

Gourmont, Remy de. *Lilith; suivi de Théodat.* Paris, 1906.

————. *The Natural Philosophy of Love.* Trans. with a Postscript by Ezra Pound. New York, 1931.

————. *Promenades Littéraires.* 2nd edn. Paris, 1906.

————. Review of *"Que Faire de Nos Filles?"* by B. H. Gausseron. *Mercure de France* 3 no. 20 (August 1891), 125.

Gramont, Elisabeth de. *Romaine Brooks: Portraits, Tableaux, Dessins.* Paris, 1952.

Griffin, Susan. *Pornography and Silence: Culture's Revenge Against Nature.* New York, 1981.

Guercio, Antonio del. *La Pittura dell'Ottocento.* Torino, 1982.

Gunn, Peter. *Vernon Lee; Violet Paget, 1856–1935.* London, 1964.

Haack, Friedrich. *Die Kunst des XIX. Jahrhunderts und der Gegenwart.* II. Teil: *Die Moderne Kunstbewegung.* 6th rev. edn. Eslingen a.N., 1925.

Haas, Willy. *Die Belle Epoque.* Munich, 1967.

Haeckel, Ernst. *The Riddle of the Universe at the Close of the Nineteenth Century.* Trans. Joseph McCabe. New York, 1902.

[Hake, Alfred Egmont]. *Regeneration: A Reply to Max Nordau.* Introd. by Nicholas Murray Butler. New York, 1896.

Hale, Oron J. *The Great Illusion, 1900–1914.* New York, 1971.

Hall, Delight. *Catalogue of the Alice Pike Barney Memorial Lending Collection.* Washington, D.C.: National Collection of Fine Arts, 1965.

Haller, John S., and Haller, Robin M. *The Physician and Sexuality in Victorian America.* Urbana, Ill., 1974.

Hammer Galleries. *The Goddess and the Slave: Women in Nineteenth Century Art.* New York, 1977.

Harbison, Robert. *Deliberate Regression.* New York, 1980.

Harding, James. *Artistes Pompiers: French Academic Art in the 19th Century.* New York, 1979.

Harris, Ann Sutherland, and Nochlin, Linda. *Women Artists, 1550–1950.* Los Angeles: Los Angeles County Museum, 1976.

Harris, Joseph. *The Tallest Tower: Eiffel and the Belle Epoque.* Boston, 1975.

Harrison, Fraser. *The Dark Angel: Aspects of Victorian Sexuality.* New York, 1977.

Harrison, Martin, and Waters, Bill. *Burne-Jones.* London, 1973.

Hartman, Mary S. *Victorian Murderesses.* New York, 1977.

Hartmann, Sadakichi. *A History of American Art* [1901]. Rev. edn. 2 vols. New York, 1932.

Hayes, Carlton J. H. *A Generation of Materialism, 1871–1900.* New York, 1941.

Hellerstein, Erna Olafson; Hume, Leslie Parker; and Offen, Karen M. *Victorian Women: A Documentary Account of Women's Lives in Nineteenth-Century England, France, and the United States.* Stanford, Calif. 1981.

Hemmings, F. W. J. *Culture and Society in France, 1848–1898: Dissidents and Philistines.* New York, 1971.

————. *The Life and Times of Emile Zola.* New York, 1977.

Henderson, Marina. *D. G. Rossetti.* London, 1973.

Hérédia, José-Maria de. *Les Trophées.* 107th edn. Paris, n.d.

————. *Les Trophées.* Ed. W. N. Ince. London, 1979.

Hertzberg, Arthur. *The French Enlightenment and the Jews: The Origins of Modern Anti-Semitism.* New York, 1968.

Heyman, Therese Thau. *Anne Brigman, Pictorial Photographer/Pagan/Member of the Photo-Secession.* Oakland, Calif., Oakland Museum, 1974.

Hiley, Michael. *Victorian Working Women: Portraits from Life.* Boston, 1980.

Hill, Christopher. *Reformation to Industrial Revolution.* The Pelican Economic History of Britain, Vol. II. Harmondsworth, 1969.

Hilton, Timothy. *The Pre-Raphaelites.* New York, 1970.

Hinz, Berthold. *Art in the Third Reich.* Trans. Robert and Rita Kimber. New York, 1979.

Hirth, Georg. "Das Gehirn unsre lieben Schwestern." *Jugend,* 1, no. 3 (1896), 49–50.

———. "Faustrecht oder—Gretchenrecht?" *Jugend,* 6, no. 32 (1901), 527–28.

———. "Wir Feministen!" *Jugend,* 7, no. 22 (1902), 363–64.

———, ed. *Dreitausend Kunstblätter der Münchner "Jugend" (1896–1909).* Munich, 1909.

Hobsbawm, E. J. *The Age of Capital, 1848–1875. New York: Mentor Books, 1979.*

———. *Industry and Empire: The Making of Modern English Society.* Vol. II: *1750 to the Present Day.* New York, 1968.

Hobson, Anthony. *The Art and Life of J. W. Waterhouse, R.A., 1849–1917,* London, 1980.

Hoenderdos, P. *Ary Scheffer, Sir Lawrence Alma-Tadema, Charles Rochussen, of de Vergankelijkheid van de Roem.* Rotterdam: Rotterdamse Kunststichting, 1974.

Hofmann, Werner. *The Earthly Paradise: Art in the Nineteenth Century.* New York, 1961.

———. *Das irdische Paradies: Motive und Ideen des 19. Jahrhunderts.* Munich, 1974.

———. *Nana: Mythos und Wirklichkeit.* Cologne, 1973.

———, ed. *Experiment Weltuntergang; Wien um 1900.* Munich, 1981.

Hofmann, Werner, Busch, Gunter; Burollet, Therese; et al. *Symboles et Réalités: La Peinture Allemande, 1848–1905.* Paris: Musée du Petit Palais, 1984.

Hofstadter, Richard. *Social Darwinism in American thought* [1944]. Boston, Beacon Paperbacks, 1955.

Hofstätter, Hans H. *Symbolismus und die Kunst der Jahrhundertwende.* Cologne, 1965.

———*Geschichte der Europäischen Jugendstilmalerei.* Cologne, 1965.

———. *Art Nouveau Prints, Illustrations, and Posters* [1968]. With Contributions by W. Jaworska and S. Hofstätter. New York, 1984.

Holten, Ragnar von. *L'Art Fantastique de Gustave Moreau.* Pairs, 1960.

Horney, Karen. *The Adolescent Diaries of Karen Horney.* New York, 1980.

Hough, Graham. *The Last Romantics.* London, 1947.

Howe, Jeffrey, et al. *Fernand Khnopff and the Belgian Avant-Garde.* New York: Barry Friedman Gallery, 1983.

Huddleston, Sisley, *Paris Salons, Cafés, Studios.* New York, 1928.

Hudson, Derek. *Lewis Carroll; An Illustrated Biography.* 2nd edn. London, 1976.

———*Munby, Man of Two Worlds.* London, 1972.

Huneker, James Gibbons. *Ivory, Apes and Peacocks* [1915]. New York, 1957.

Hunt, John Dixon. *The Pre-Raphaelite Imagination, 1848–1900.* London, 1968.

Huot, Sylviane. *Le "Mythe d'Hérodiade" Chez Mallarmé: Genèse et Evolution.* Paris, 1977.

Huysmans, Joris-Karl. *Certains.* Paris, 1889.

Ince, W. N. *Hérédia.* London, 1979.

Ironside, R. "Burne-Jones and Gustave Moreau." *Horizon,* 1, no. 6 (June 1940), 406–24.

Jackson, Holbrook. *The Eighteen Nineties: A Review of Art and Ideas at the Close of the Nineteenth Century* [1913]. New York: Capricorn Books, 1966.

Janik, Allan, and Toulmin, Stephen. *Wittgenstein's Vienna.* New York, 1973.

James, Henry. *The Notebooks.* Ed. F. O. Matthiessen and Kenneth B. Murdock. New York, 1947.

Jerome, Helen. *The Secret of Woman.* New York, 1923.

Johnson, Diane Chalmers. *American Art Nouveau.* New York, 1979.

Johnson, Wendell Stacy. *Living in Sin: The Victorian Sexual Revolution.* Chicago, 1979.

Jones, Harvey L. *Mathews: Masterpieces of the California Decorative Style*. Santa Barbara, Calif., 1980.

Jones, Howard Mumford. *The Age of Energy: Varieties of American Experience, 1865–1915*. New York, 1971.

Jullian, Phillippe. *D'Annunzio*. Trans. Stephen Hardman. New York, 1973.

———*Dreamers of Decadence: Symbolist Painters of the 1890's*. Trans. Robert Baldick. New York, 1971.

———*The Orientalists: European Painters of Eastern Scenes*. Oxford, 1977.

———*The Symbolists*. London, 1973.

———; Bowness, Alan; and Lacambre, Geneviève. *French Symbolist Painters: Moreau, Puvis de Chavannes, Redon and Their Followers*. London: Arts Council of Great Britain, 1972.

Jussim, Estelle, *Slave to Beauty: The Eccentric Life and Controversial Career of F. Holland Day, Photographer, Publisher, Aesthete*. Boston, 1981.

Kahane, Martine. "La Peinture Wagnérienne." *Beaux-Arts Magazine,* 9, (Jan. 1984), 54–59.

Kallir, Jane. *Austria's Expressionism*. New York: Galerie St. Etienne, 1981.

Kauffer, E. McKnight, ed. *The Art of the Poster: Its Origin, Evolution and Purpose*. London, 1924.

Kelder, Diane. *The Great Book of French Impressionism*. New York, 1980.

Kenealy, Arabella. *Feminism and Sex-Extinction*. New York, 1920.

Kermode, Frank. *Romantic Image*. London, 1957.

Kery, Patricia Frantz. *Great Magazine Covers of the World*. New York, 1982.

Key, Ellen. *Love and Marriage*. Trans. Arthur G. Chater. New York, 1911.

———*The Woman Movement*. Trans. Mamah Bouton Borthwick. New York, 1912.

Kimbrough, Sarah Dodge. *Drawn from Life*. Jackson, Miss., 1976.

Klein, Viola. *The Feminine Character: History of an Ideology* [1946]. Urbana, Ill., 1972.

Kotalik, Jiri, et al. *Mucha, 1860–1939: Peintures, Illustrations, Affiches, Arts Décoratifs*. Paris: Editions des Musées Nationaux, 1980.

Kruissink, Rits. *Montmartre: Van Tempel tot Tingel Tangel*. The Hague, 1960.

Kurtz, Charles M., ed. *Official Illustrations from the Art Gallery of the World's Columbian Exposition*. Philadelphia, 1893.

Lacambre, Geneviève, and Rohan-Chabot, Jacqueline de. *Le Musée du Luxembourg en 1874*. Paris: Editions des Musées Nationaux, 1974.

Langer, William L. *Political and Social Upheaval, 1832–1852*. New York, 1969.

Legrand, Francine-Claire. *Symbolism in Belgium*. Trans. Alistair Kennedy. Brussels, 1972.

Lethève, Jacques. *La Vie Quotidienne des Artistes Français au XIXᵉ Siècle*. Paris, 1968.

Lietzmann, Hilda, et al. *Bibliographie zur Kunstgeschichte des 19. Jahrhunderts: Publikationen der Jahre 1940–1966*. Munich, 1968.

Linton, E. Lynn. "The Partisans of the Wild Women." *The Nineteenth Century,* vol. 31, no. 181 (March 1892), 455–64.

Lister, Raymond. *Victorian Narrative Paintings*. New York, 1966.

Lloyd, Trevor. *Suffragettes International: The World-Wide Campaign for Women's Rights*. London, 1971.

Logan, Olive. *A Propos of Women and Theatres*. New York, 1869.

Lombroso-Ferrero, Gina. *The Soul of Woman; Reflections on Life*. New York, 1923.

Longo, Lawrence D. "The Rise and Fall of Battey's Operation: A Fashion in Surgery." *Bulletin of the History of Medicine,* 53, no. 2, (Summer 1979), 244–67.

Louÿs, Pierre. *Mimes des Courtisanes/Dialogues of the Courtesans*. Illus. Edgar Degas; dual language edn. Trans. Guy Daniels. Paris, 1973.

Low, Will H. *A Painter's Progress*. New York, 1910.

———, *A Chronicle of Friendships, 1873–1900*. New York, 1908.

Ludovici, Anthony M. *Man: An Indictment.* London, 1927.

Ludwig, Horst. *Malerei der Gründerzeit—Bayerische Staatsgemälde Sammlungen, Neue Pinakothek, München; Katalog.* Munich, 1977.

Ludwig, Horst, et al. *Münchner Maler im 19. Jahrhundert.* 4 vols. Munich, 1981–83.

Maas, Jeremy. *Victorian Painters.* New York, 1969.

McClelland, Donald R. *Where Shadows Live: Alice Pike Barney and Her Friends* (pamphlet). Washington, D.C.: National Collection of Fine Arts, 1978.

McColl, D. S. *Nineteenth-Century Art.* Glasgow, 1902.

McCormack, W. J. *Sheridan le Fanu and Victorian Ireland.* Oxford, 1980.

MacDonald, Gus. *Camera: Victorian Eyewitness.* London, 1979.

McNally, Raymond T., and Florescu, Radu. *In Search of Dracula.* New York, 1972.

MacPherson, C. B. *The Political Theory of Possessive Individualism, Hobbes to Locke.* Oxford, 1962.

Mann, Arthur. *Yankee Reformers in the Urban Age: Social Reform in Boston, 1880–1900.* Cambridge, Mass., 1954.

Marholm, Laura [Laura (Mohr) Hansson]. *Zur Psychologie Der Frau* [1897]. 2nd enl. edn. Berlin, 1903.

Marks, John George. *Life and Letters of Frederick Walker, A.R.A.* London, 1896.

Marrett, Cora Bagley, "On the Evolution of Women's Medical Societies." *Bulletin of the History of Medicine,* 53, no. 3 (Fall 1979), 434–48.

Masur, Gerhard. *Prophets of Yesterday: Studies in European Culture, 1890–1914.* New York, 1961.

Mathieu, Pierre-Louis. *Gustave Moreau.* Boston, 1976.

Mauclair, Camille. *The Great French Painters and the Evolution of French Painting from 1830 to the Present Day.* Trans. P. G. Konody. London, 1903.

Maurenbrecher, Hulda. *Das Allzuweibliche; Ein Buch von neuer Erziehung und Lebensgestaltung.* Munich, 1912.

May, Henry F. *The End of American Innocence (1912–1917).* New York, 1959.

Mencken, H. L. *In Defense of Women.* New York, 1918.

——. *Prejudices.* Ist and IInd series. London, 1921.

Menegazzi, Luigi. *Il Manifesto Italiano, 1882–1925.* Milan, n.d.

Meynell, Wilfrid, ed. *The Modern School of Art.* London, n.d. [ca. 1884].

Michaud, Guy. *Message Poétique du Symbolisme.* Paris, 1947.

Millett, Kate. *Sexual Politics.* New York: Avon, Equinox Books, 1971.

Milner, John. *Symbolists and Decadents.* London, 1971.

Mitchell, Yvonne. *Collette: A Taste for Life.* New York, 1975.

Moers, Ellen. *Literary Women: The Great Writers.* Garden City, N.Y. Anchor Books, 1977.

Montgomery, Walter. *American Art and American Art Collections.* 2 vols. Boston, 1889.

Monteverdi, Mario, et al. *La Pittura Italiana dell'Ottocento.* 3 vols. Milan, 1975.

Moody, Helen Watterson. *The Unquiet Sex.* First paper: "The Woman Collegian," *Scribner's,* 22 (1897), 150–56. Second paper: "Women's Clubs," 22 (1897), 486–91. Third paper: "The Case of Maria," 23 (1898), 234–42.

Moore, George. *Confessions of a Young Man.* New York, n.d.

——. *Modern Painting.* Enl. edn. London, 1898.

Morand, Paul: MacAvoy, Edouard; and Desbruères, Michel. *Romaine Brooks.* Special issue of *Bizarre,* no 46 (March 1968).

Morton, Frederic. *A Nervous Splendour: Vienna 1888–1889.* New York, 1979.

Mosse, George L. *Nazi Culture: Intellectual, Cultural, and Social Life in the Third Reich.* New York, 1966.

Mosse, W. E. *Liberal Europe: The Age of Bourgeois Realism, 1848–1875.* London, 1974.

Mucha, Jiri. *Alphonse Mucha, the Master of Art Nouveau.* Prague, 1966.

"Musiciennes du Silence." *Le Petit Journal des Grandes Expositions,* no. 119. Réunion des Musées Nationaux, Paris, 1982.

Muther, Richard. *The History of Modern Painting.* 4 vols. Rev. edn. Continued by the author to the end of the XIX Century. London, 1907.

Nebehay, Christian M. *Ver Sacrum, 1898–1903.* Trans. Geoffrey Watkins. New York, 1977.

Neumann, Erich. *Armor and Psyche: The Psychic Development of the Feminine—A Commentary on the Tale by Apuleius.* Trans. Ralph Mannheim. Princeton, N.J. 1956.

Nicoll, John. *The Pre-Raphaelites.* London, 1970.

Nietzsche, Friedrich. *Beyond Good and Evil: Prelude to a Philosophy of the Future.* Trans. Walter Kaufmann. New York: Vintage, 1966.

——— *The Birth of Tragedy and The Genealogy of Morals.* Trans. Francis Golffing. Garden City, N.Y.: Anchor Books, 1956.

———. *Thus Spoke Zarathustra.* Trans. Walter Kaufmann. New York, 1966.

Nochlin, Linda. "The Imaginary Orient." *Art in America,* 71, no. 5 (May 1983), 118–31, 187–89.

———. *Realism.* Baltimore, Md., 1971.

Nordau, Max. *The Conventional Lies of Our Civilization.* Chicago, 1886.

———. *On Art and Artists.* Trans. W. F. Harvey. Philadelphia, n.d.

Normandy, Georges. *Le Nu au Salon, Année 1912.* Paris, 1912.

Nouveau, Germain. "Sphinx." *Mercure de France,* 3, no. 21 (Sept. 1891), 132–33.

Nova, Didier, and Sternberg, Jacques. *Les Chefs d'Oeuvre du Kitsch.* Paris, 1971.

Olander, William R. "Fernand Khnopff's *Art or the Caresses:* The Artist as Androgyne." *Marsyas: Studies in the History of Art,* 18 (1975–76), 45–59.

Ormond, Leonee, and Ormond, Richard. *Lord Leighton.* New Haven, Conn., 1975.

Pach, Walter. *Ananias, or the False Artist.* New York, 1928.

Paladilhe, Jean, and Pierre, José. *Gustave Moreau.* Paris, 1971.

Pankhurst, Richard. *Sylvia Pankhurst, Artist and Crusader.* London, 1979.

Panofsky, Dora, and Panofsky, Erwin. *Pandora's Box: The Changing Aspects of a Mythical Symbol.* 2nd edn. New York, 1965.

Parker, Rozsika, and Pollock, Griselda. *Old Mistresses: Women, Art, and Ideology.* New York, 1981.

Parrington, Vernon L. *Main Currents in American Thought.* Vol. 3: *The Beginnings of Critical Realism in America,* 1860–1920. New York, 1930.

Pearsall, Ronald. *Night's Black Angels: The Many Faces of Victorian Cruelty.* New York, 1975.

———. *Tell Me, Pretty Maiden: The Victorian and Edwardian Nude.* Exeter, 1981.

———. *The Worm in the Bud: The World of Victorian Sexuality.* New York, 1969.

Pellicer, A. Cirici. *El Arte Modernista Catalan.* Barcelona, 1951.

Pennell, Elizabeth Robins, and Pennell, Joseph. *The Life of James McNeill Whistler.* 6th rev. edn. Philadelphia, 1925.

Peters, Robert. *The Crowns of Apollo: Swinburne's Principles of Literature and Art.* Detroit, Mich., 1965.

Petersen, Karen, and Wilson, J. J. *Women Artists: Recognition and Reappraisal from the Early Middle Ages to the Twentieth Century.* New York, 1976.

Petteys, Chris. *International Dictionary of Women Artists Born Before 1900.* Boston, 1985.

Phillpotts, Beatrice. *Mermaids.* New York, 1980.

Pica, Vittorio. *L'Arte Mondiale a Venezia nel 1899.* Bergamo, 1899.

———. *L'Arte Mondiale alla IV Esposizione di Venezia.* Bergamo, 1901.

Pichon, Yann le. *The World of Henry Rousseau.* New York, 1982.

Picon, Gaetan. *The Birth of Modern Painting.* New York, 1978.

Pierrot, Jean. *The Decadent Imagination, 1880–1900.* Trans. Derek Coltman. Chicago, Ill., 1981.

Platte, Hans. *Deutsche Impressionisten.* Gutersloh, 1971.

Pohl, Claudia, et al. *A. Böcklin, 1827–1901.* 2 vols. Darmstadt: Mathildenhöhe, 1977.

Pollock, Griselda. *Mary Cassatt.* London, 1980.

Prause, Marianne. *Bibliographie zur Kunstgeschichte des 19. Jahrhunderts; Publikationen der Jahre 1967–1979.* Munich, 1984.

Praz, Mario. *The Romantic Agony* [1933]. 2nd edn. Trans. Angus Davidson. New York: Meridian Books, 1956.

Priestley, J. B. *The Edwardians.* New York, 1970.

Rachilde [Marguerite Eymery Vallette]. "Questions Brûlantes." *La Revue Blanche,* 11 (1896), 193–200.

Raitt, A. W. *Prosper Mérimée.* New York, 1970.

——. *Villiers de l'Isle-Adam et le Mouvement Symboliste.* Paris, 1965.

Redon, Odilon. *The Graphic Works of Odilon Redon.* Introd. Aflred Werner. New York: Dover, 1969.

Reff, Theodore. "The Influence of Flaubert's Queen of Sheba on Later Nineteenth-Century Literature." *Romanic Review,* 65, no. 4 (Nov. 1974), 249–65.

Reich, Wilhelm. *The Mass Psychology of Fascism.* Trans. Vincent R. Carfagno. New York, 1970.

Remak, Joachim. *The Origins of World War I (1871–1914).* New York, 1967.

Renan, Ary. *Gustave Moreau (1826–1898).* Paris, 1900.

Restany, Pierre. "Le Merveilleux et Méconnu Gustave Moreau." *Planète,* no. 16 (May-June 1964), 81–94.

Reventlow, Franziska Gräfin zu. *Tagebücher, 1895–1910.* Munich, 1971.

Rewald, John. *The History of Impressionism.* 4th rev. edn. New York, 1973.

Reynolds, Graham. *Victorian Painting.* New York, 1967.

Roberts-Jones, Philippe, et al. *Belgian Art, 1880–1914.* New York: Brooklyn Museum, 1980.

——. *Beyond Time and Place: Non-Realist Painting in the Nineteenth Century.* Oxford, 1978.

Romein, Jan. *The Watershed of Two Eras: Europe in 1900.* Trans. Arnold Pomerans. Middletown, Conn., 1978.

Rose, Marilyn Gaddis. "The Daughters of Herodias in *Hérodiade, Salomé,* and *A Full Moon in March.*" *Comparative Drama,* 1, no. 3 (Fall 1967), 172–81.

Rosen, Ruth, and Davidson, Sue, eds. *The Maimie Papers.* Old Westbury, N.Y., 1977.

Rosenblum, Robert. *Modern Painting and the Northern Romantic Tradition: Friedrich to Rothko.* New York, 1975.

Rudorff, Raymond. *The Belle Epoque: Paris in the Nineties.* New York, 1973.

Rugoff, Milton, *Prudery and Passion: Sexuality in Victorian America.* New York, 1971.

Rzewuski, Stanislas. "Les Enfants: Fragment d'Une Autobiographie Féminine." *La Revue Blanche,* 5 (1893), 70–78.

Sachs, Samuel; Clifford, Timothy; and Botwinick, Michael. *Victorian High Renaissance.* Minneapolis, Minn. Institute of Arts, 1978.

Saint-Gaudens, Homer. *The American Artist and His Times.* New York, 1941.

Salmon, André. *L'Air de la Butte.* Paris, 1945.

Salon Artiste, Le. *Album de Dessins Originaux d'Après les Oeuvres Exposées.* 2 vols. Paris, 1885–86.

Scharlach, Bernice. "The Abduction of a Maiden: The Theft of 'Elaine.' " *San Francisco Sunday Examiner and Chronicle Magazine,* August 17, 1980, pp. 34–35.

Scheffler, Karl. *Die Frau und die Kunst.* Berlin, 1908.

Schiller, Francis. *A Moebius Strip: Fin-de-Siècle Neuropsychiatry and Paul Moebius.* Berkeley, Calif. 1982.

Schneir, Miriam, ed. *Feminism: The Essential Historical Writings.* New York: Vintage Books, 1972.

Schnitzler, Arthur. *My Youth in Vienna.* Trans. Catherine Hutter. New York, 1970.

Schorske, Carl E. *Fin-de-Siècle Vienna: Politics and Culture.* New York, 1980.

Schreiner, Olive. *Dreams.* Boston, 1891.

————. *Woman and Labour.* Ed. Adèlemarie van der Sprey and Adriaan van der Sprey. Johannesburg, South Africa, 1975.

Schubring, Paul. " 'Das Zelt': Max Klinger's neuer Radierzyklus." *Die Kunst,* 32 (1916–17), 95–104.

Schuré, Edouard. "L'Oeuvre de Gustave Moreau." *La Revue de Paris,* 7, pt. 6 (Nov.-Dec. 1900), 587–622.

Schurr, Gérald, *1820–1920; Les Petits Maîtres de la Peinture.* 5 vols. Paris, 1975–84.

[Scrutton, Hugh]. *The Taste of Yesterday.* Liverpool: Walker Art Gallery, 1970.

Secrest, Meryle. *Between Me and Life: A Biography of Romaine Brooks.* Garden City, N.Y., 1974.

Seldes, Gilbert. *The Stammering Century.* New York, 1928.

Severi, Rita. "Oscar Wilde e il Mito di Salome." *Rivista di Letteratura Moderne e Comparate,* 34, no. 1 (March 1981), 59–71.

Seward, A. C., ed. *Darwin and Modern Science.* Cambridge, 1909.

Shattuck, Roger. *The Banquet Years: The Origins of the Avant-Garde in France, 1885 to World War I.* Garden City, N.Y.: Anchor Books, 1961.

————, et al. *Henri Rousseau.* New York: Museum of Modern Art, 1985.

Sheldon, F. "Various Aspects of the Woman Question." *The Atlantic Monthly,* 17 (Oct. 1866), 425–34.

Sheldon, George William. *Ideals of Life in France, or How the Great Painters Portray Woman in French Art.* New York, 1890.

————. *Recent Ideals of American Art.* New York, 1888–90.

Showalter, Elaine. *A Literature of Their Own: British Women Novelists from Brontë to Lessing.* Princeton, N.J., 1977.

Simmel, Georg. "Zur Philosophie der Geschlechter." In *Philosophische Kultur.* Lepizig, 1911.

Simmons, Edward *From Seven to Seventy.* New York, 1922.

Skinner, Cornelia Otis. *Elegant Wits and Grand Horizontals.* Cambridge, Mass., 1962.

————. *Madame Sarah.* Cambridge, Mass., 1966.

Smith, Page. *Daughters of the Promised Land: Women in American History.* Boston, 1970.

Smith, Paul Jordan. *The Soul of Woman; An Interpretation of the Philosophy of Feminism.* San Francisco, 1916.

Smith, T. R., ed. *The Woman Question.* New York, 1918.

Smith-Rosenberg, Carroll. *Disorderly Conduct: Visions of Gender in Victorian America.* New York, 1985.

————. "The Female World of Love and Ritual." *Signs,* 1, no. 1 (Autumn 1975), 1–29.

————. "The Hysterical Woman: Sex Roles and Role Conflict in 19th-Century America." *Social Research,* 39, no. 4 (Winter 1972), 652–78.

Sonstroem, David. *Rossetti and the Fair Lady.* Middletown, Conn., 1970.

Soria, Regina, et al. *Perceptions and Evocations: The Art of Elihu Vedder.* Washington, D.C.: National Collection of Fine Arts. 1979.

Spalding, Frances. *Magnificent Dreams: Burne-Jones and the Late Victorians.* Oxford, 1978.

————. *Roger Fry, Art and Life.* London, 1980.

Sparrow, W. Shaw. *Women Painters of the World from the Time of Catherina Vigri, 1413–1463, to Rosa Bonheur and the Present Day.* London, 1905.

Spencer, Herbert, *The Study of Sociology* [1873]. Introd. by Talcott Parsons. Ann Arbor, Mich.: University of Michigan Press, 1961.

Spencer, Robin. *The Aesthetic Movement: Theory and Practice.* London, 1972.

Stern, Karl. *The Flight From Woman.* London, 1966.

Stevens, Maryanne, ed. *The Orientalists: Delacroix to Matisse.* London: Royal Academy of Arts, 1984.

Stevenson, Robert Louis. *Virginibus Puerisque and Other Papers* [1881]. New York, 1906.

Stewart, Walter A. *Psychoanalysis: The First Ten Years, 1888–1898.* London, 1969.

Strachey, Lytton. *The Really Interesting Question and Other Papers.* Ed. Paul Levy. London, 1972.

Strahan, Edward [Earl Shinn]. *The Chefs-d'Oeuvre d'Art of the International Exhibition, 1878.* Philadelphia, 1878.

——. *Modern French Art.* New York, 1881.

Strindberg, Auguste. "L'Exposition d'Edvard Munch." *La Revue Blanche,* 10 (1896) 525–26.

Strouse, Jean. *Alice James: A Biography.* Boston, 1980.

Sumner, William Graham, and Keller, Albert Galloway. *The Science of Society.* 4 vols. New Haven, Conn., 1927.

Swanson, Vern G. *Alma-Tadema: The Painter of the Victorian Vision of the Ancient World.* New York, 1977.

Swinburne, Algernon Charles. *A Pilgrimage of Pleasure; Essays and Studies.* Boston, 1913.

Tannenbaum, Edward R. *1900: The Generation Before the Great War.* Garden City, N.Y., 1976.

Tawney, R. H. *Religion and the Rise of Capitalism* [1926]. New York: Mentor Books, 1947.

Taylor, Hilary, *James McNeill Whistler.* New York, 1978.

Thornwell, Emily. *The Lady's Guide to Perfect Gentility.* New York, 1856.

Tolstoy, Leo. *On Life, and Essays on Religion.* Trans. Aylmer Maude. Tolstoy Centenary Edition, vol. 12. London, 1934.

Trask, John E. D., and Laurvik, J. Nilsen. *Catalogue de Luxe of the Panama-Pacific International Exposition.* 2 vols. San Francisco, 1915.

Trudgill, Eric. *Madonnas and Magdalens: The Origins and Development of Victorian Sexual Attitudes.* London, 1976.

Uitti, Karl D. *La Passion Littéraire de Remy de Gourmont.* Paris, 1962.

Venturi, Lionello. *Cézanne.* New York, 1978.

Vicinus, Martha, ed. *Suffer and Be Still: Women in the Victorian Age.* Bloomington, Ind., 1972.

Villeneuve, Roland. *Le Diable: Erotologie de Satan.* Paris, 1963.

——. *Le Musée de la Bestialité.* Paris, 1973.

Vogt, Paul. *Was Sie Liebten: Salonmalerei im XIX. Jahrhundert.* Cologne, 1969.

Volta, Ornella. *Le Vampire.* Paris, 1962.

Vorst, Marie van. *Modern French Masters.* Paris, 1904.

Voss, Heinrich, *Franz von Stuck, 1863–1928: Werkkatalog der Gemälde, mit einer Einführung in seinen Symbolismus.* Munich, 1973.

Wald, Carol, and Papachristou, Judith. *Myth America: Picturing Women, 1865–1945.* New York, 1975.

Waldberg, Patrick. *Eros Modern Style.* Paris, 1964.

Waldron, George B. "Five Hundred Years of the Anglo-Saxon." *McClure's Magazine,* 12, no. 2 (Dec. 1898), 185–88.

Walker, Robert H. *Life in the Age of Enterprise.* New York: Capricorn Books, 1971.

Walters, Ronald G. *Primers for Prudery: Sexual Advice to Victorian America.* Englewood Cliffs, N.J., 1974.

Walton, William. *The Art and Architecture of the World's Columbian Exposition.* 2 vols. Philadelphia, 1893.

Wattenmaker, Richard J. *Puvis de Chavannes and the Modern Tradition.* Rev. edn. Art Gallery of Ontario, 1975.

Webb, Peter. *The Erotic Arts.* Boston, 1975.

Weber, Max. *The Protestant Ethic and the Spirit of Capitalism.* Trans. Talcott Parsons. New York, 1958.

[Weill, Alain]. *Masters of the Poster, 1896–1900.* Trans. Bernard Jacobson. New York, 1977.

Weimann, Jeanne Madeline. *The Fair Women: The Story of the Woman's Building, World's Columbian Exposition, Chicago, 1893.* Chicago, 1981.

Weininger, Otto. *Ueber die letzten Dingen.* Vienna, 1904.

Welter, Barbara. *Dimity Convictions: The American Woman in the Nineteenth Century.* Athens, Ohio, 1976.

West, Rebecca. *1900.* New York, 1982.

Whitford and Hughes Gallery. *Dreamers and Academics.* Sales catalogue. London, 1981.

———. *Fine Paintings, 1880–1930.* Sales Catalogue. London, 1983.

———. *Moments et Folies de la Femme Fatale.* Sales catalogue. London, 1985.

———. *Peintres de l' Ame.* Sales catalogue. London, 1984.

———. *Visions Rich and Strange, 1880–1920.* Sales catalogue. London, 1982.

Wickes, George. *The Amazon of Letters: The Life and Loves of Natalie Barney.* New York, 1976.

Wilcox, Ella Wheeler, *Men, Women and Emotions.* Chicago, 1893.

Willey, Basil. *More Nineteenth-Century Studies: A Group of Honest Doubters.* New York, 1956.

Williamson, Audrey. *Artists and Writers in Revolt; The Pre-Raphaelites.* London, 1976.

Wilson, Edmund. *Axel's Castle: A Study in the Imaginative Literature of 1870–1930.* New York, 1931.

Wilson, Richard; Pilgrim, Dianne; and Murray, Richard N. *The American Renaissance, 1876–1917.* New York: Brooklyn Museum, 1979.

Wood, Christopher. *Olympian Dreamers: Victorian Classical Painters, 1860–1914.* London, 1983.

———. *The Pre-Raphaelites.* New York, 1981.

———. *Victorian Panorama: Paintings of Victorian Life.* London, 1976.

Young, G. M. *Victorian England: Portrait of an Age.* London, 1953.

Zagona, Helen Grace. *The Legend of Salomé and the Principle of Art for Art's Sake.* Paris, 1960.

Zeldin, Theodore. *France 1848–1945: Ambition and Love.* Oxford, 1979.

Zeman, Z. A. B. *Twilight of the Habsburgs: The Collapse of the Austro-Hungarian Empire.* London, 1971.

Ziff, Larzer. *The American 1890s: Life and Times of a Lost Generation.* Lincoln, Nebr. 1966.

Zinn, Howard. *A People's History of the United States.* New York, 1980.

Zola, Emile. *The Drunkard [L'Assommoir].* Trans. Arthur Symons. London, 1958.

———. *Mon Salon; Manet; Ecrits Sur L'Art.* Ed. Antoinette Ehrard. Paris: Garnier-Flammarion, 1970.

Zürcher, Hanspeter. *Stilles Wasser: Narziss und Ophelia in der Dichtung und Malerei um 1900.* Bonn, 1975.

Sources and Credits

Frontispiece: Private collection.

PREFACE. *Die Kunst,* 5 (1901–2), 442.

CHAPTER I. 1. *Great Pictures in Private Galleries* (London, 1905), vol. 2. 2. *Hals,* Masters in Art Series (Boston, 1900). 3. *Hogarth,* Masters in Art Series (Boston, 1902). 4. *The Art Journal,* 66 (1904), facing p. 283. 5. Samuel Isham, *The History of American Painting,* with supplementary chapters by R. Cortissoz (New York, 1927), p. 477. 6. Maurice Hamel, *Salons de 1904* (Paris, 1904), facing p. 49. 7. Ripley Hitchcock, H. C. Ives et al., *The Art of the World at the World's Columbian Exposition* (New York, 1894), facing p. 119. 8. *Royal Academy Picltures* (London, 1902). 9. *Die Kunst,* 1 (1899–1900), 218. 10. *Die Kunst für Alle,* 13 (1898), facing p. 316.

CHAPTER II. 1. Henri Frantz, *Le Salon de 1900* (Paris, 1900), facing p. 15. 2. *Royal Academy Pictures* (London, 1891). 3. *The International Studio,* 50 (1913), 243. 4. *Die Kunst,* 23 (1910–11), 295. 5. *Roll,* Peintres d'aujourdhui, no. 1 (Paris, 1910). 6. Marcel Montandon, *Segantini* (Bielefeld, 1904), p. 82. 7. *Die Kunst für Alle,* 12 (1897), 201. 8. *Die Kunst,* 3 (1900–1901), 447. 9. Ernest Maindron, *Les Affiches Illustrées* (Paris, 1896). 10. O. von Schleinitz, *William Holman Hunt* (Bielefeld, 1907). 11. *The Art Journal,* 51 (1889), facing p. 129. 12. Rev. Ebenezer Cobham Brewer, *Character Sketches of Romance, Fiction and the Drama,* rev. Am. edn., 1896, ed. Marion Harland. vol. 2, facing p. 364. 13. *The Art Journal,* 27 (1865), facing p. 332. 14. *The Magazine of Art* (1893), 197. 15. *Je Sais Tout,* vol. 3, pt. 2 (1907), 673. 16. *Catalogue Illustré, Société Nationale des Beaux-Arts, Salon de 1913.* 17. Rev. Ebenezer Cobham Brewer, *Character Sketches,* vol. 1, facing p. 48. 18. *Die Kunst für Alle,* 13 (1898), 337. 19. Hitchcock, *The Art of the World at the World's Columbian Exposition,* p. 141. 20. W. Graham Robertson, *Time Was* (London, 1931), facing p. 108. 21. *The American Magazine of Art,* 7 (1916), 367. 22. *Famous Art Reproduced* (New York, 1900), p. 172. 23. *the International Studio,* 24 (1904), 13. 24. *Catalogue Illustré, Société Nationale des Beaux-Arts, Salon de 1913.* 25. *The Magazine of Art* (1900), 391. 26. Edward Strahan (Earl Shinn), *The Art Treasures of America* (Phildelphia, 1879), vol. 3, facing p. 64. 27. *The Magazine of Art,* 3 (1880), 5. 28. Photo: National Collection of Fine Arts,. Smithsonian Institution. 29. *Jugend,* no. 17 (1889), 266. 30. Marie van Vorst, *Modern French Masters* (Paris, 1904), p. 123. 13. André-Charles Coppier, *Les Eaux fortes de Besnard* (Paris, 1920), p. 14. 32. Hans Rosenhagen, *A. von Keller* (Bielefeld, 1912), p. 55. 33. *The Magazine of Art,* 8 (1885), 398. 34. *Royal Academy Pictures* (London, 1895). 35. *Catalogue Illustré, Société National des Beaux-Arts, Salon de 1899.* 36. *The International Studio,* 2 (1897), 89. 37. *Royal Academy Pictures* (London, 1895). 38. *L'Illustration,* no. 2464 (May 17, 1890), 451.

CHAPTER III. 1. *The Magazine of Art* (1900), 55. 2. *Catalogue, Whistler Memorial Exhibition,* New Gallery (London, 1905). 3. Isidore Spielmann, *Souvenir, Fine Arts Section, Franco-British Exhibition,* 1908. 4. *The Magazine of Art* (1894), facing p. 220. 5. Malcolm Bell, *Sir Edward Burne-Jones* (London, 1905), facing p. 8. 6. *Great Pictures in Private Galleries,* vol. 1 (London, 1905). 7. *Moore,* Masters in Art Series (Boston, 1908). 8. *The Magazine of Art* (1887), facing p. 360. 9. *Royal Academy Pictures* (London, 1891). 10. *Les Letters et les Arts,* vol. 1 (1888). 11. *The*

International Studio, 45 (1911), vii.   12. *The Magazine of Art* (1896), 368.   13. *Die Kunst*, 27 (1212–13), 147.   14. *The International Studio*, 49 (1913), lxix.   15. *The International Studio*, 7 (1899), 38.   16. *Catalogue of the Fine Arts Department, Tennessee Centennial Exposition*, 1897.

CHAPTER IV.   1. *The Art Journal*, 45 (1883), facing p. 332.   2. *The International Studio*, 31 (1907), 177.   3. *Die Kunst*, 25 (1911–12), 395.   4. *Die Kunst*, 5 (1901–2), 209.   5. Vittorio Pica, *La Galleria d'Arte Moderna di Venetia* (Bergamo, 1909), facing p. 132.   6. *The International Studio*, 28 (1906), xci.   7. Photo: National Museum of American Art, Smithsonian Institution.   8. Maurice Hamel, *The Salons of 1902* (Paris, 1902), facing p. 8.   9. *The International Studio*, 51 (1914), 320.   10. Victor Nadal, *Le Nu au Salon* (Paris, 1902).   11. *The International Studio*, 34 (1908), 41.   12. Strahan, *The Art Treasures of America*, vol. 1.   13. Private collection.   14. *The International Studio*, 43 (1911), 225.   15. Marcel Montandon, *Segantini* (Bielefeld, 1904).   16. Lucien Pissarro, *Rossetti* (London, n.d.).   17. *The International Studio*, 44 (1911), 182.   18. *The International Studio*, 28 (1906), cv.   19. *die Kunst*, 9 (1904), 17.   20. Private collection.   21. Paul Mantz, *Le Salon de 1889* (Paris, 1889), facing p. 47.   22. *The International Studio*, 20 (1903), 45.   23. Armand Sylvestre, *Le Nu au Salon des Champs Elysées* (Paris, 1896).   24. William Walton, *The Chefs-d'Oeuvre of the Exposition Universelle, 1900* (Philadelphia, 1901), vol. 5, facing p. 56.   25. Strahan, *The Art Treasures of America*, vol. 1, facing p. 66.   26. Sylvestre, *Le Nu au Salon des Champs Elysées* (Paris, 1895).   27. Nadal, *Le Nu au Salon* (Paris, 1903).   28. *the International Studio*, 11 (1900), 88.   29. Alfred Kuhn, *Lovis Corinth* (Berlin, 1925).   30. Nadal, *Le Nu au Salon* (Paris, 1905), vol. 2.   31. Brewer, *Character Sketches*, vol. 1, facing p. 48.   32. *Die Kunst*, 29 (1913–14), 381.   33. Walton, *The Chefs-d'Oeuvre of the Exposition Universelle, 1900*, vol. 1, facing p. 54.   34. Antonin Proust, *Le Salon de 1891* (Paris 1891), facing p. 10.   35. *Royal Academy Pictures* (London, 1895).   36. *Royal Academy Pictures* (London, 1890).   37. *The Magazine of Art* (1898), 60.   38. Walton, *The Chefs-d'Oeuvre of the Exposition Universelle, 1900*, vol. 6, facing p. 54.   39. *Catalogue Illustré, Exposition Nationale des Beaux-Arts, Salon de 1893* (Paris, 1893), p. 206.

CHAPTER V.   1. *Art and Progress*, 5 (1914), 158.   2. *The International Studio.* 70 (1920), 29.   3. *Royal Academy Pictures* (London, 1900).   4. *Royal Academy Pictures* (London, 1912).   5. *Jugend* (1896), no. 50, 815.   6. Paul Jamot, *Degas* (Paris, 1924).   7. *Die Kunst*, 11 (1904–5), facing p. 97.   8. *Book of American Figure Painters* (Philadelphia, 1886).   9. *Die Kunst*, 11 (1904–5), 341.   10. *Burne-Jones*, Masters in Art Series (Boston, 1901).   11. Illustration for Roviralta, *Boyres Baixes* (Barcelona, 1902).   12. *Catalogue Illustré, Société Nationale des Beaux-Arts, Salon de 1912* (Paris, 1912).   13. Louis Vauxcelles, *The Salons of 1907* (Paris, 1907). English text, facing p. 55.   14. *Die Kunst*, 1 (1899–1900), 307.   15. *Die Kunst*, 11 (1904–5), 441.   16. *Les Arts*, no. 65 (May 1907), 5.   17. *Die Kunst*, 13 (1904–5), p. 475.   18. *Die Kunst*, 29 (1913–14), 515.   19. *The International Studio*, 8 (1899), 252.   20. *Gazette des Beaux-Arts*, vol. 51, pt. 2 (1909), facing p. 334.   21. Photo: National Museum of American Art, Smithsonian Institution.   22. Gleeson White, ed., *Master Painters of Great Britain* (New York, 1909), p. 365.   23. *Famous Pictures Reproduced from Renowned Paintings*, 2nd edn. (Chicago, 1917).   24. Photo: Barry Friedman Gallery, New York.   25. Jarno Jessen, *Rossetti* (Bielefeld, 1905).   26. Armand Sylvestre, *Le Nu au Salon—Champ de Mars* (Paris, 1895).   27. *Les Arts*, no. 77 (May 1908), 14.   28. *L'Illustration*, no. 3244 (April 29, 1905), 266.   29. J. M. dos Reis Jr., *Historia da Pintura do Brasil* (Sao Paulo, 1944), p. 294.   30. William Walton et al., *Art and Architecture of the World's Columbian Exposition* (Philadelphia, 1893), vol. 2, facing p. 20.   31. Collection Vance Kondon, San Diego.   32. *Deutsche Kunst und Dekoration*, 33 (1913–14), 433.   33. *La Plume*, 17 (1905), 481.   34. *The International Studio*, 10, no. 38 (April 1900), 104.

CHAPTER VI.   1. *The International Studio*, 22 (1904), ccxxxiv.   2. *Art and Progress*, 2 (1911), 364.   3. *Royal Academy Pictures* (London, 1899).   4. Cosmo Monkhouse, *British Contemporary Artists* (New York, 1899), p. 113.   5. *Catalogue of the Ninth Annual Exhibition, Carnegie Institute* (Pittsburgh, 1904).   6. Jarno Jessen, *Rossetti* (Bielefeld, 1905).   7. *The Magazine of Art* (1896), 468.   8. Private collection.   9. Etienne Moreau-Nélaton, *Manet raconté par lui-même* (Paris, 1926), vol. 2.   10. *The Art world*, 1, no. 5 (Feb. 1917), 349.   11. Photo: Sotheby's New York.   12. *Die Kunst*, 13 (1905–6), 406.   13. *Die Kunst*, 25 (1911–12), 178.   14. *The Century*, 83 (1911–12), 301.   15. *Les Arts*, no. 65 (May 1907), 9.   16. *Royal Academy Pictures* (London, 1906).   17. *Les Arts*, no. 90 (June 1909), 27.   18. Georges Normandy, *Le Nu au Salon* (Paris, 1912).   19. *The Century*, 81 (1910–11), 255.   20. *Royal Academy Pictures* (London, 1906).   21. *Die Kunst*, 23 (1910–11) 297.   22. *Die Kunst für Alle*, 7 (1892).   23. *Catalogue Illustré, Exposition des Artistes Français, Salon de 1897* (Paris, 1897), p. 225.   24. *The Magazine of Art* (1894), 225.   25. *Die Kunst*, 24 (1908–9), p. 184.   26. *Famous Pictures Reproduced from Renowned Paintings*, 2nd edn. (Chicago, 1917), p. 259.   27. *The Magazine of Art* (1899), 368.   28. *The International Studio*, 49 (1913), 17.   29. *Royal Academy Pictures* (London, 1895).   30. *Die Kunst*, 2 (1899–1900), 93.   31. *Die Kunst*, 24 (1908–9), 185.

CHAPTER VII.   1. *Die Kunst*, 25 (1911–12), 422.   2. *Die Kunst*, 23 (1910–11), 223.   3. *Die Kunst*, 25 (1911–12), 104.   4. *Die Kunst*, 27 (1212–13), 511.   5. *Die Kunst*, 25 (1911–12), 577.   6. Paul Mantz, *Salon de 1889* (Paris, 1889), facing p. 2.   7. *Royal Academy Pictures* (London, 1896).   8. *The Studio*, 5 (1895), 23.   9. *Salon de 1884* (Paris, 1884), facing p. 76.   10. *Die Kunst*, 13 (1905–6), 139.   11. Hamel, *The Salons of 1902*, facing p. 9.

CHAPTER VIII.   1. Walton, *The Chefs-d'Oeuvre of the Exposition Universelle, 1900*, vol. 6, facing p. 67.   2. *Die Kunst für Alle*, 11 (1895–96), 342.   3. *Die Kunst*, 25 (1911–12), 275.   4. Private collection.   5. *Die Kunst für Alle*, 11

(1895–96), facing p. 184. 6. Isham, *The History of American Painting,* p. 573. 7. *The International Studio,* 54 (1915), 198. 8. *Die Kunst,* 27 (1912–13), 457. 9. *Die Kunst,* 15 (1906–7), 465. 10. Feldman Collection, Del Ray, Florida. 11. *Royal Academy Pictures* (London, 1891). 12. *Die Kunst,* 11 (1904–5), 359. 13. *Royal Academy Pictures* (London, 1894). 14. *Die Kunst,* 3 (1900–1901), 519. 15. Alfred Kuhn, *Lovis Corinth* (Berlin, 1925), p. 159. 16. *Royal Academy Pictures* (London, 1897). 17. Julius Meier-Graefe, *Cézanne und sein Kreis* (Munich, 1920), facing p. 120. 18. Sylvestre, *Le Nu au Salon des Champs Elysées* (Paris, 1890). 19. *The Studio,* 3 (1894), 162. 20. Walton, *The Chefs-d'Oeuvre of the Exposition Universelle, 1900,* vol. 2, facing p. 20. 21. Nadal, *Le Nu au Salon* (Paris, 1906). 22. *Die Kunst,* 11 (1904–5), 394. 23. *The International Studio,* 21 (1904), 135. 24. Walton, *The Chefs-d'Oeuvre of the Exposition Universelle, 1900,* vol. 3, facing p. 36. 25. *Catalogue Illustré, Société Nationale des Beaux-Arts, Salon de 1911* (Paris, 1911). 26. *Die Kunst,* 7 (1902–3), 439. 27. Private collection. 28. *The International Studio,* 55 (1915), L. 29. Ludwig Pietsch, *Contemporary German Art* (New York, 1888), vol. 1, facing p. 20. 30. *The Art Journal,* 58 (April 1896), 106. 31. *The International Studio,* 37 (1909), facing p. iii. 32. Charles Dana Gibson, *Americans* (London, 1900). 33. Monkhouse, *British Contemporary Artists,* p. 71.

CHAPTER IX. 1. Strahan, *The Art Treasures of America,* vol. 1, facing p. 52. 2. Meier-Graefe, *Cézanne und sein Kreis,* facing p. 160. 3. *The International Studio,* 53 (1914), 71. 4. Friedrich Fuchs, *Venus: Die apotheose des Weibes* (Berlin, 1905), vol. 2, chap. 4, p. 15. 5. *Die Kunst,* 15 (1906–7), 85. 6. *Die Kunst,* 7 (1902–3), 99. 7. Heinrich Alfred Schmidt, *Arnold Böcklin* (Munich, 1919). 8. *Die Kunst,* 23 (1910–11), 21. 9. *Die Kunst,* 23 (1910–11), 8. 10. *Die Kunst,* 13 (1905–6), 387. 11. Private collection. 12. Nadal, *Le Nu au Salon* (Paris, 1905), vol. 1. 13. Private collection. 14. Julius Meier-Gaefe, *Modern Art* (New York, 1908), vol. 1, facing p. 220. 15. *Catalogue Illustré, Société Nationale des Beaux-Arts, Salon de 1899,* 17. 16. Nadal, *Le Nu au Salon* (Paris, 1903). 17. Jamot, *Degas.* 18. *The Magazine of Art* (1891 [Dec. 1890]), 41. 19. *Royal Academy Pictures* (London, 1912). 20. *Die Kunst,* 11 (1904–5), 441. 21. Georges Ollendorf, *Le Salon de 1887* (Paris, 1887), facing p. 82. 22. *The International Studio,* 22 (1904), 97. 23. *The International Studio,* 43 (1911), 203. 24. *Die Kunst für Alle,* 10 (1894–5), 339. 25. *The Century,* 82 (1911), 889. 26. Mantz, *Salon de 1889,* facing p. 20. 27. Adolf Rosenberg, *Lenbach* (Bielefeld, 1905). 28. *Die Kunst für Alle,* 10 (1894–95), 22. 29. *Jugend* (1898), no. 2, p. 23. 30. Engraving for Gabriele D'Annunzio's *Il Piacere,* 1889. 31. *Die Kunst,* 25 (1911–12), 87. 32. *Die Kunst,* 7 (1902–3), 419. 33. *Les Arts,* no 77 (May 1908), 17. 34. *Catalogo, XI Esposizione Internazionale d'Arte della Città di Venezia,* 1914. 35. *Die Kunst,* 5 (1901–2), 374. 36. *Jugend* (1914), no. 18, 547. 37. Nadal, *Le Nu au Salon* (Paris, 1905), vol. 2. 38. *Famous Art Reproduced* (New York, 1900), p. 120. 39. Sylvestre, *Le Nu au Salon—Champ de Mars* (Paris, 1889). 40. *Scribner's Monthly,* 12 (1892), p. 744. 41. Frontispiece for Joséphin Péladan, *Istar* (Paris, 1888). 42. Barry Friedman Gallery, New York. 43. *Die Kunst,* 21 (1909–10), 387. 44. *The Studio,* 4 (1895), 125. 45. *Die Kunst,* 13 (1905–6), 15. 46. *Die Kunst,* 9 (1903–4), 37. 47. Nadal, *Le Nu au Salon* (Paris, 1905), vol. 2. 48. Sylvestre, *Le Nu au Salon* (Paris, 1888). 49. *Die Kunst,* 5 (1901–2), 221. 50. Pietsch, *Contemporary German Art,* vol. 1. 51. Jean Dubay and Pierre MacOrlan, *Félicien Rops* (Paris, 1928). 52. Sylvestre, *Le Nu au Salon—Champ de Mars* (Paris, 1889). 53. *Die Kunst,* 33 (1915–16), facing p. 285. 54. *Royal Academy Pictures* (London, 1893). 55. *Famous Pictures Reproduced* (Chicago, 1917). 56. Photo: National Museum of American Art, Smithsonian Institution. 57. *The International Studio,* 67 (1919), xxvi. 58. Gustave Kahn, *Félicien Rops et Son Oeuvre* (Paris, n.d.). 59. Nadal, *Le Nu au Salon* (Paris, 1905), vol. 1. 60. Sylvestre, *Le Nu au Salon—Champ de Mars* (Paris, 1895). 61. *Die Kunst,* 9 (1903–4), 21. 62. *Die Kunst für Alle,* 13 (1897–98), 89. 63. *Die Kunst,* 25 (1911–12), 59. 64. *Die Kunst,* 7 (1902–3), 483. 65. *Die Kunst,* 29 (1913–14), 173. 66. *Ver Sacrum,* I, no. 12 (1898), 12.

CHAPTER X. 1. *Die Kunst,* 17 (1907–8), 569. 2. *The Magazine of Art* (1898), 496. 3. *Die Kunst,* 9 (1903–4), 460. 4. Private collection. 5. *Deutsche Kunst und Dekoration, 33 (1913–14), 312.* 6. *The Art of 1897,* supplement to *The Studio* (London, 1897). 7. Dubay and MacOrlan, *Félicien Rops,* p. 181. 8. Christian Brinton, *Impressions of the Art of the Panama-Pacific Exposition* (New York, 1916), p. 193.

CHAPTER XI. 1. *L'Illustration,* no. 3244 (April 29, 1905), 258. 2. *The International Studio,* 55 (1915), xxxii. 3. Walton, *The Chefs-d'Oeuvre of the Exposition Universelle, 1900,* vol. 5, facing p. 62. 4. *Die Kunst für Alle,* 9 (1894), facing p. 316. 5. *The Magazine of Art* (1896), 166. 6. Giulio de Frenzi (Luigi Ferzoni), *Ignacio Zuolaga* (Rome, 1912), p. 27. 7. *Die Kunst,* 21 (1909–10), facing p. 433. 8. *Deutsche Kunst und Dekoration,* 37 (1915–16), 135. 9. Sylvestre, *Le Nu au Salon des Champs Elysées* (Paris, 1896). 10. *Die Kunst,* 19 (1908–9), facing p. 441. 11. *Die Kunst,* 11 (1904–5), 305. 12. *Die Kunst,* 5 (1901–2), 437. 13. *Die Kunst,* 17 (1907–8), 421. 14. *Die Kunst,* 35 (1916–17), 77. 15. *Les Lettres et les Arts,* 1, no. 1 (1886), facing p. 144. 16. *Deutsche Kunst und Dekoration,* 37 (1915–16), 202. 17. Private collection. 18. *Die Kunst,* 29 (1913–14), facing p. 169. 19. *Die Kunst,* 29 (1913–14), 177. 20. Georges Olmer and Saint-Juirs, *Salon de 1886 (Paris, 1886), facing p. 34.* 21. *Die Kunst,* 7 (1902–3), 258. 22. *Die Kunst für Alle,* 12 (1896–97), facing p. 312. 23. *Die Kunst,* 25 (1911–12), 289. 24. *Die Kunst,* 25 (1911–12), facing p. 173. 25. Friedrich Fuchs, *Venus: Die Apotheose des Weibes* (Berlin, 1905), vol. 2, chap. 8, p. 13. 26. *Die Kunst,* 24 (1908–9), 6. 27. Antonin Proust, *Le Salon de 1898* (Paris, 1898), facing p. 40. 28. Private collection. 29. *The Studio,* 4 (1894), 29. 30. *Jugend* (1897), no. 18, 285. 31. *Die Kunst,* 17 (1907–8), facing p. 224.

Index

Page numbers in italics refer to illustrations.